CRC Series in Chromatography

Editors-in-Chief

Gunter Zweig, Ph.D. and Joseph Sherma, Ph.D.

General Data and Principles
Gunter Zweig, Ph.D. and
Joseph Sherma, Ph.D.

Lipids
Helmut K. Mangold, Dr. rer. nat.

Hydrocarbons
Walter L. Zielinski, Jr., Ph.D.

Carbohydrates
Shirley C. Churms, Ph.D.

Inorganics
M. Qureshi, Ph.D.

Drugs
Ram Gupta, Ph.D.

Phenols and Organic Acids
Toshihiko Hanai, Ph.D.

Terpenoids
Carmine J. Coscia, Ph.D.

Amino Acids and Amines
S. Blackburn, Ph.D.

Steroids
Joseph C. Touchstone, Ph.D.

Polymers
Charles G. Smith,
Norman E. Skelly, Ph.D.,
Carl D. Chow, and Richard A. Solomon

**Pesticides and Related
Organic Chemicals**
Joseph Sherma, Ph.D. and
Joanne Follweiler, Ph.D.

Plant Pigments
Hans-Peter Köst, Ph.D.

CRC
Handbook
of
Chromatography

Steroids

Author

Joseph C. Touchstone, Ph.D.

Professor
Department of Obstetrics
and Gynecology
Hospital of the University
of Pennsylvania
Philadelphia, Pennsylvania

Editors-in-Chief

Gunter Zweig, Ph.D.
President
Zweig Associates
Washington, D.C.

Joseph Sherma, Ph.D.
Professor of Chemistry
Lafayette College
Easton, Pennsylvania

CRC Press, Inc.
Boca Raton, Florida

Library of Congress Cataloging-in-Publication Data

Touchstone, Joseph C.
  Steroids.

  (Handbook of chromatography)
  Includes bibliographies and index.
  1. Steroid hormones—Analysis—Handbooks, manuals,
etc. 2. Steroids—Analysis—Handbooks, manuals, etc.
3. Chromatographic analysis—Handbooks, manuals, etc.
I. Zweig, Gunter. II. Sherma, Joseph. III. Title.
IV. Series: CRC handbook of chromatography., [DNLM:
1. Chromatography—handbooks. 2. Steroids—analysis—
handbooks. QU 39 T722s]
QP572.S7T68 1986   574.19′243   85-17066
ISBN 0-8493-3014-9

This book represents information obtained from authentic and highly regarded sources. Reprinted material is quoted with permission, and sources are indicated. A wide variety of references are listed. Every reasonable effort has been made to give reliable data and information, but the author and the publisher cannot assume responsibility for the validity of all materials or for the consequences of their use.

Direct all inquiries to CRC Press, Inc., 2000 Corporate Blvd., N.W., Boca Raton, Florida, 33431.

©1986 by CRC Press, Inc.
International Standard Book Number 0-8493-3014-9

Library of Congress Card Number 85-17066
Printed in the United States

SERIES PREFACE

The Handbook of Chromatography — Steroids has been assembled by Professor Joseph C. Touchstone of the School of Medicine, University of Pennsylvania, Department of Obstetrics and Gynecology. This handbook of chromatography is another volume in a continuing series which originated in 1972 as a two-volume set and is still available as the *Handbook of Chromatography (General).* It became obvious when it was contemplated to write a second edition, that the field of chromatography for all classes of organic and inorganic compounds had grown so rapidly that it would be desirable to produce handbooks devoted to a single class or closely related classes of compounds. Each of these handbooks would be prepared by a volume editor, assisted in some cases by an editorial board of experts in the respective field. As a result of this decision, the following handbooks have been published during the ensuing years: *Drugs* (two volumes), *Carbohydrates, Polymers, Phenols and Organic Acids, Pesticides, Amines and Amino Acids, Terpenoids, and Lipids* (two volumes). Other volumes which are in preparation will cover the following subjects: pigments, hydrocarbons (GLC), peptides, nucleic acids and derivatives, and inorganics.

We invite our readers to suggest other titles and potential editors suitable as a *Handbook of Chromatography.*

Gunter Zweig, Ph.D.
Joseph Sherma, Ph.D.
Editors-in-Chief

PREFACE

The subject of steroids encompasses a large number of variations of the parent compound cyclopentanoperhydrophenanthrene. This volume is divided into sections relating to preparation, separation, and detection. Since success in any mode of separation, even in chromatography depends on analyte/junk ratio, some emphasis has been placed on preparation of the sample. The volume is limited to the separation aspects. Perhaps a second volume will be compiled related to the qualitative and quantitative aspects including nuclear magnetic resonance and mass spectrometry. It was deemed advisable to omit these topics since eight classes of steroids covering the material would be too large for one volume.

Each section has information on name and structure for that class, as well as some discussion related to specific methodology for preparation of the particular sample for chromatographic analyses. This handbook shoud enable those interested in steroid methodology to obtain a head start. LC, GC, and TLC separation techniques are described for each of the classes of steroids.

Joseph C. Touchstone
Author

THE EDITORS-IN-CHIEF

Gunter Zweig, Ph.D., received his undergraduate training at the University of Maryland, College Park, where he was awarded the Ph.D. in biochemistry in 1952. Two years following his graduation, Dr. Zweig was affiliated with the late R. J. Block, pioneer in paper chromatography of amino acids. Zweig, Block, and Le Strange wrote one of the first books on paper chromatography which was published in 1952 by Academic Press and went into three editions, the last one authored by Gunter Zweig and Dr. Joe Sherma, the co-Editor-in-Chief of this series. *Paper Chromatography* (1952) was also translated into Russian.

From 1953 till 1957, Dr. Zweig was research biochemist at the C. F. Kettering Foundation, Antioch College, Yellow Springs, Ohio, where he pursued research on the path of carbon and sulfur in plants, using the then newly developed techniques of autoradiography and paper chromatography. From 1957 till 1965, Dr. Zweig served as lecturer and chemist, University of California, Davis and worked on analytical methods for pesticide residues, mainly by chromatographic techniques. In 1965, Dr. Zweig became Director of Life Sciences, Syracuse University Research Corporation, New York (research on environmental pollution), and in 1973 he became Chief, Environmental Fate Branch, Environmental Protection Agency (EPA) in Washington, D.C. From 1980—1984 Dr. Zweig was Visiting Research Chemist in the School of Public Health, University of California, Berkeley, where he was doing research on farmworker safety as related to pesticide exposure.

During his government career, Dr. Zweig continued his scientific writing and editing. Among his works are (many in collaboration with Dr. Sherma) the now 11-volume series on *Analytical Methods for Pesticides and Plant Growth Regulators* (published by Academic Press); the pesticide book series for CRC Press; co-editor of *Journal of Toxicology and Environmental Health*; co-author of basic review on paper and thin-layer chromatography for *Analytical Chemistry* from 1968 to 1980; co-author of applied chromatography review on pesticide analysis for *Analytical Chemistry*, beginning in 1981.

Among the scientific honors awarded to Dr. Zweig during his distinguished career are the Wiley Award in 1977, Rothschild Fellowship to the Weizmann Institute in 1963/64; the Bronze Medal by the EPA in 1980.

Dr. Zweig has authored or co-authored over 80 scientific papers on diverse subjects in chromatography and biochemistry, besides being the holder of three U.S. patents. In 1985, Dr. Zweig became president of Zweig Associates, Consultants in Washington, D.C.

Jospeh Sherma, Ph.D., received a B.S. in Chemistry from Upsala College, East Orange, N.J., in 1955 and a Ph.D. in Analytical Chemistry from Rutgers University in 1958. His thesis research in ion exchange chromatography was under the direction of the late William Rieman, III. Dr. Sherma joined the faculty of Lafayette College in September 1958, and is presently Charles A. Dana Professor of Chemistry in charge of two courses in analytical chemistry. At Lafayette he has continued research in chromatography and had additionally worked a total of 12 summers in the field with Harold Strain at the Argonne National Laboratory, James Fritz at Iowa State University, Gunter Zweig at Syracuse University Research Corporation, Joseph Touchstone at the Hospital of the University of Pennsylvania, Brian Bidlingmeyer at Waters Associates, and Thomas Beesley at Whatman, Inc., Clifton, N.J.

Dr. Sherma and Dr. Zweig co-authored or co-edited the original Volumes I and II of the *CRC Handbook of Chromatography*, a book on paper chromatography, seven volumes of the series *Analytical Methods for Pesticides and Plant Growth Regulators*, and the Handbooks of Chromatography of drugs, carbohydrates, polymers, and phenols and organic acids. Other books in the pesticide series and further volumes of the *CRC Handbook of Chromatography* are being edited with Dr. Zweig, and Dr. Sherma has co-authored the handbook on pesticide chromatography. A book on quantitative TLC was edited jointly with Dr. Touchstone, and a general book on TLC was co-authored with Dr. B. Fried. Dr. Sherma has been co-author of eight biennial reviews of column liquid and thin layer chromatography (1968—1982) and the 1981 and 1983 reviews of pesticide analysis for the journal *Analytical Chemistry*.

Dr. Sherma has written major invited chapters and review papers on chromatography and pesticides in *Chromatographic Reviews* (analysis of fungicides), *Advances in Chromatography* (analysis of nonpesticide pollutants), Heftmann's *Chromatography* (chromatography of pesticides), Race's *Laboratory Medicine* (chromatography in clinical analysis), *Food Analysis: Principles and Techniques* (TLC for food analysis), *Treatise on Analytical Chemistry* (paper and thin layer chromatography), *CRC Critical Reviews in Analytical Chemistry* (pesticide residue analysis), *Comprehensive Biochemistry* (flat bed techniques), *Inorganic Chromatographic Analysis* (thin layer chromatography), and *Journal of Liquid Chromatography* (advances in quantitative pesticide TLC). He is editor for residues and elements for the *JAOAC*.

Dr. Sherma spent 6 months in 1972 on sabbatical leave at the EPA Perrine Primate Laboratory, Perrine, Fla., with Dr. T. M. Shafik, and two additional summers (1975, 1976) at the USDA in Beltsville, Md., with Melvin Getz doig research on pesticide residue analysis methods development. He spent 3 months in 1979 on sabbatical leave with Dr. Touchstone developing clinical analytical methods. A total of more than 230 papers, books, book chapters, and oral presentations concerned with column, paper, and thin layer chromatography of metal ions, plant pigments, and other organic and biological compounds; the chromatographic analysis of pesticides; and the history of chromatography have been authored by Dr. Sherma, many in collaboration with various co-workers and students. His major research area at Lafayette is currently quantitative TLC (densitometry), applied mainly to clinical analysis and pesticide residue and food additive determinations.

Dr. Sherma has written an analytical quality control manual for pesticide analysis under contract with the U.S. EPA and has revised this and the EPA Pesticide Analytical Methods Manual under a 4-year contract jointly with Dr. M. Beroza of the AOAC. Dr. Sherma has also written an instrumental analysis quality assurance manual and other analytical reports for the U.S. Consumer Product Safety Commission, and is currently preparing two manuals on the analysis of food additives for the U.S. FDA, both of these projects also in collaboration with Dr. Beroza of the AOAC.

Dr. Sherma taught the first prototype short course on pesticide analysis with Henry Enos of the EPA for the Center for Professional Advancement. He was editor of the Kontes TLC quarterly newsletter for 6 years and also has taught short courses on TLC for Kontes and the Center for Professional Advancement. He is a consultant for numerous industrial companies and federal agencies on chemical analysis and chromatography and regularly referees papers for analytical journals and research proposals for government agencies. At Lafayette, Dr. Sherma, in addition to analytical chemistry, teaches general chemistry and a course in thin layer chromatography.

Dr. Sherma has received tow awards for superior teaching at Lafayette College and the 1979 Distinguished Alumnus Award from Upsala College for outstanding achievements as an educator, researcher, author, and editor. He is a member of the ACS, Sigma Xi, Phi Lambda Upsilon, SAS, and AIC.

THE AUTHOR

Joseph C. Touchstone, Ph.D., is Professor of Research Obstetrics and Gynecology and Professor of Research Surgery at the School of Medicine, at the University of Pennsylvania, where he has worked since 1952.

Dr. Touchstone is a graduate of S.F. Austin University (Texas), B.S., Purdue University (Indiana), M.S., and St. Louis University (Missouri), Ph.D. He is a member of a number of professional societies, has founded two organizations for interaction in chromatography, and is on the editorial board of the *Journal of Liquid Chromatography*. He has written eight books in the field of chromatography and 250 papers in various journals. Current interests are the separation of various components of biological fluids, particularly steroids and phospholipids.

Dr. Touchstone has lectured widely at national and international meetings. He has organized symposia and chaired various meetings.

TABLE OF CONTENTS

Chapter 1

INTRODUCTION

The use of chromatography in the analysis of steroids has increased immeasurably in recent years, especially so since high performance liquid chromatography (HPLC) has provided a relatively nondestructive means of separation. The analysis has been made more efficient by the use of automation and instrumentation as microprocessors are becoming more common.

Steroids have an important number of biological functions. There are innumerable classes and types of steroids, both natural and synthetic, and the number increases considerably when the different metabolites are considered. In most biomedical applications the ratio of analyte to the extraneous material present is so poor that prior extraction is necessary. In some cases the analysis can be carried out with unextracted starting material. In use recently are a number of "aids to extraction" or sample preparation devices, usually small disposable columns that provide sufficient alteration of the ratio so that chromatography can be carried out. Many workers in the field fail to realize this and wonder why overloaded chromatographic systems do not provide the desired separation. With this in mind, each chapter of this volume provides methodology to prepare samples for chromatographic separation using the procedures listed in that section. Some of the extraction methods can present specific separations of classes and further enhance the results of the final analysis.

Probably the greatest drawback in analytical methods for steroids is the fact that there are so many examples of analogy or isomerism or compounds of different structure, but of similar chromatographic characteristics, thus making separation all the more difficult. Some compounds are hydrophilic while others are lipophilic. Others are very unstable, while there are those which are stable. These factors contribute to the inherent difficulty in separating the various steroids.

No individual chromatographic method can separate all the steroids in a class. Thus, there are an increasing number of reports in the literature of separations of steroids describing the analysis of only a few steroids within a class. Since most investigators will be searching only in the individual class of his/her specialization, the particular organization of the handbook described above has been chosen. The users of this book should realize that reliance on only one of the chromatographic techniques available will somewhat restrict the degrees of freedom. A well-equipped analytical laboratory will have equipment for thin layer (TLC), gas (GC), and HPLC readily available. For both qualitative and quantitative analysis, TLC has a number of advantages because a number of samples and reference compounds can be tested simultaneously. Neither GC nor HPLC have this facility. This volume gives examples of separations in all three disciplines and provides the user with immediate information as to the capabilities available for a specific problem. Unlike other reviews, the directions for each step are presented. Not all available methods are presented, but those given are representative of the example under consideration.

Throughout the volume standard abbreviations are used. Rather than have a glossary at the end of the book, each section has its specialized notations. All solutions unless otherwise noted are on a volume/volume basis.

Chapter 2

SAMPLE HANDLING AND PREPARATION

Handling of the Samples

Since many readers of this handbook may be involved in the assay of biological samples (mainly samples in clinical laboratories), something should be said of the problems of sample handling. Those in the quality control laboratories or pharmaceutical companies do not experience these problems, but should be aware of them if they are involved in the analysis of metabolites. Data on the stability of hormones in blood and urine under different storage contitions are scarce. Statements such as "the specimens were transported to the laboratory in dry ice and stored at $-20°C$ until analyzed", have been published in papers describing a new assay method, might be adopted as a part of a procedure often without any justification or subsequent validation. Craig[1] has shown that urinary 17-oxysteroids, 17-hydroxy-corticosteroids, 11-oxy-17-hydroxycorticosteroids, 11-deoxy-17-hydroxycorticosteroids, pregnanediol, total estrogens, and estrogens from the urine of pregnant women do not deteriorate even in an unpreserved specimen kept at room temperature for at least 1 week. The main advantage of adding a preservative and storing in a refrigerator for these assays would appear to be in making an analysis less distasteful for the operator. When measured by a fluorimetric technique urinary cortisol levels rose considerably in a specimen kept at room temperature, but when maintained at 4 or 0°C no significant change was noted after 1 week. Care must be taken in interpreting these results since storage temperature is the only parameter which was considered. Although the initial sample was made by pooling several urine specimens, it may not be truly representative. For example, the occurrence of certain drugs or their metabolites which were not present in the urine pool employed in a study may influence the stability of the steroids studied. Much more work is needed in this area before it will be possible to specify precise storage conditions for blood and urine samples. The following general rules can be applied to the transport and storage of specimens prior to analysis:

1. The specimens should be protected from fluctuations in temperature during transport by placing them in a well-insulated container.
2. If there is any doubt about the stability of the hormone, the specimen should be kept frozen. If this is done, a sufficient number of aliquots should be taken to allow for the initial assay and any further determinations that are necessary. This procedure ensures that a freshly thawed specimen is available for each analysis. Under no circumstances should specimens be refrozen after thawing.
3. Bacterial growth in urine should be prevented by the addition of a suitable bacteriostatic agent. For general use, 1% boric acid, which is available in tablet form is suitable. Organic solvents such as chloroform and toluene have also been used.
4. The validity of unsubstantiated statements about procedures for collection and storage that have appeared in original communications should be established. Following these instructions too closely may lead to unnecessary work.

Sample Preparation

Throughout this volume the problems of sample preparation are discussed for each compound class; however, in the case of clinical and biological samples, there are a number of general methods for preparation of samples. These are described herein. Quantitative estimation of steroid hormones in a biological sample, be it plasma or urine, is usually not possible without a preliminary separation of interfering components. Usually the first step of hormone determinations involves liquid-liquid partition or separating steroids from water-soluble components by shaking with an organic solvent. To achieve a nearly quantitative isolation, this procedure has to be repeated several times.

There are a number of sample preparation columns described in the recent literature. These involve applying the sample in its original solution to a small column and then eluting with a suitable solvent. Among these are ClinElut, Sepak, and Extrelute. These columns are very useful and avoid emulsions usually encountered in the extraction of biological samples. Since these techniques are more recent, the user is often faced with development of the procedure for his specific use.

Because these methods are relatively fast, they are quite useful in preparing samples for radioimmunoassay. The Extrelut and ClinElut columns are packed with kieselguhr of granular structure with high pore volume or specially prepared cellulose of high absorbency power.

Simple partition for preparation of samples is not sufficient for biological samples when gas chromatography-mass spectrometry (GC-MS) or precise quantitation are to be carried out. The most severe drawback is the requirement for an extensive purification of the biological samples, especially when the steroids are present in low concentrations. Radioimmunoassay is, therefore, the only practical method when one or a few steroids are to be analyzed in a number of samples. The recoveries of entire groups must be high and nonselective. The samples have to be sufficiently pure so that overloading and loss of resolution on the GC columns are avoided. This is particularly important when capillary columns are used, which is necessary in analyses of complex mixtures. The aim of the work in an analytical laboratory is to develop simpler and more rapid isolation procedures in order to increase the practical usefulness of GC-MS or other analytical methods in studies of metabolic profiles of steroids in biological fluids and tissues. The methods in this section are general. Specific methods are given in the section of each class of steroids.

General Procedures for Purification of Steroids

Purification procedures are based on a separation of steroids and contaminating materials into neutral, acidic, and basic compounds of different polarity. All compounds are considered soluble in organic solvents or mixtures of water and alcohols. Thus, desalting is always the first step, i.e., an extraction by solvents, Amberlite® XAD-2 or Lipidex® 1000. The resulting extract is then purified by filtration through column beds which either retain the steroids or let them pass with the solvent front.

The purification of small amounts of steroids requires the use of highly inert chromatographic media. Adsorption chromatography is, therefore, less suitable because of the risk of irreversible adsorption of chemical transformations. While classical partition chromatography is a mild technique, it is impractical to use systems based on two immiscible solvent phases which require maintenance of very constant equilibrium conditions. An intermediate system, liquid-gel chromatography, in which a stable two-phase system can be generated when a solvent mixture is added to a substituted Sephadex® gel, is suitable.[2] In addition to providing a wide choice of straight- or reversed-phase partition systems, substituted Sephadex® and Lipidex® gels can be used for attachment of specific functional groups.[3] This made possible the synthesis of inert ion exchangers,[4] which permits rapid ion exchange in organic solvents in which steroids and lipids are readily soluble.

By synthesis of appropriate gel derivatives with high capacity the filter columns can be made small, which reduces the time required for each purification step. In general, the filters consist of 200 to 600 mg gel for purification of steroids in 20 mℓ urine or 10 mℓ plasma.

The first step of purification of the extracted steroids is to remove organic bases by filtration through a lipophilic cation exchanger in H⁺ form in aqueous methanol.[5] Sulfoethyl Sephadex® LH-20 has been synthesized for this purpose, but its capacity was too low to permit the use of a small column. Small columns of sulfopropyl Sephadex® (Pharmacia) can be used in 72% methanol to serve the same purpose. Removal of organic bases is necessary for optimal performance of the next step of purification: filtration through an anion exchanger. However, in some applications when the anion exchanger is a strong base, filtration through a cation exchanger may not be necessary, and the time required to purify the steroids is reduced. Whether or not this step is required must be tested in the individual application of the method.

The second step of purification is filtration through a hydrophobic strong anion exchanger, triethylaminohydroxypropyl Lipidex 5000 (TEAP-Lipidex) in OH⁻ form. A solvent is used that generates a reversed-phase system. A mixture of methanol, water, and chloroform (80:30:15, by volume) was used,[6] but 72% methanol has been found to be equally effective and more compatible with the preceding filtration through the cation exchanger. The reversed-phase system serves to remove nonpolar lipids from the extract.[7] The efficiency may be tested with labeled progesterone and cholesterol which should separate by a wide margin, progesterone being eluted near the solvent front. If the sample size is not too large, the effluent from the cation exchange column may be passed directly through the anion exchanger without prior concentration.

When neutral steroids have been eluted with the solvent front, nonpolar lipids are eluted from the column with methanol-chloroform (80:15, v/v) Acidic steroids elute with a series of acids and buffers. Phenolic steroids are recovered by saturation of the solvent (methanol-chloroform) with CO_2. This is a mild and convenient mode of elution.[6] The unconjugated phenolic steroids obtained by this are highly purified and can be directly analyzed by glass capillary columns GC-MS after conversion to trimethylsilyl (TMS) ethers (see chapter on derivatives for GC).

Other acidic steroids and steroid conjugates are eluted with solvents similar to those described by Setchell et al.,[5] who used a weak anion exchanger to separate the conjugated steroids in urine. A hydrophobic strong anion exchanger, TEAP-Lipidex, can be used for purification of steroids from many types of biological materials. The different groups of conjugated steroids are cleaved by suitable enzymatic and solvolytic procedures, and the liberated steroids are purified again by filtration through TEAP-Lipidex. Efforts to optimize the various steps have resulted in a procedure by which different groups of steroids can be separated and purified. When only unconjugated steroids are to be analyzed, they can be isolated more rapidly.

Derivatization is the final step prior to GC-MS analysis, depending on the steroid. O-Methyloxime trimethylsilyl ether (MO-TMS) derivatives prepared according to Thenot and Horning[7] are the most suitable derivatives for general use in analyses of metabolic profiles. The reaction mixture is conveniently purified and reagents are removed by filtration through Lipidex 5000 in a nonpolar solvent.[6] The filtration is rapid and constitutes an important part of any purification scheme for neutral steroids prior to derivtization. Derivatives other than MO-TMS derivatives may have advantages in analyses of specific steroids and give higher sensitivity and specificity in quantitations by selected ion detection. It is likely that filtration through Lipidex can be used also in these cases to increase purity of the steroid to be analyzed. Some of these methods are described in Chapters 3 to 5.

Selected Procedures for Isolation of Groups of Steroids

Most studies do not require analysis of total profiles of steroids. It is convenient to have methods for isolation of groups of special interest. As described above, the general method permits separate isolation of unconjugated, phenolic, and different classes of conjugated steroids. Thus, it includes selective isolation of estrogens with a free phenolic hydroxyl group. Using the hydrophobic strong anion exchanger TEAP-Lipidex for isolation of the steroids, it is possible to analyze picogram amounts of estrogens in rat uterus nuclei quantitatively by GC-MS.

Another group of important steroids is the neutral 3-ketosteroids. A method for selective isolation of these compounds based on the observation that their oximes are positively charged in methanol and are retained by a lipophilic strong cation exchanger[8] is described. Oximes of other ketosteroids are retained to a much lesser degree, permitting a rapid isolation of the 3-oximes by filtration through a small column of cation exchanger. The original method utilized sulfoethyl Sephadex® LH-20.[5] The low capacity of this ion exchanger is a drawback. An ion exchanger, sulfohydroxypropyl Sephadex® LH-20 is now available which has ten times higher capacity and the polarity of which can be adjusted by introduction of alkyl chains. The method for isolation of 3-ketosteroids thus, involves four filtration steps: (1) filtration of the biological sample through the cation exchanger in H^+ form, (2) formation of oximes and filtration of the reaction mixture through an anion exchanger to remove reagent and purify the sample, (3) filtration of the oxime through the cation exchanger in H^+ form, and (4) derivatization and filtration through Lipidex 5000. Following the synthesis of TEAP-Lipidex, this anion exchanger is used instead of the weak lipophilic anion exchanger. Furthermore, the derivatization procedure is based on the observation that oxime functions may be exchanged.[9] Originally, the oximes were converted into TMS ether derivatives, the gas chromatographic properties of which are not comparable to those of MO-TMS derivatives. However, the oximes are readily converted into MO-TMS derivatives,[9] and the analysis can then be performed as described for the general method.

Synthetic steroids possessing an ethynyl group constitute another group of biologically important steroids. Methods have been studied which would be compatible with our general scheme for isolation of steroids. The ability of a cation exchanger in Ag^+ form to bind ethyl steroids has been utilized.[10] Published procedures have numerous practical disadvantages and a detailed study of silver ion chromatography resulted in an isolation method which fits directly onto the general method described in the previous section. The fractions of neutral and phenolic steroids through a column of TEAP-Lipidex are passed through a small column (8 × 20 mm) of SP-Sephadex® in Ag^+ form. This column is packed in 72% methanol on top of a 30-mm layer of SP-Sephadex® in H^+ form and has a high capacity to bind ethynyl steroids. The latter steroids are displaced from the column by elution with hexene in methanol, and the ion exchanger in H^+ form serves as a filter to remove silver ions which may form insoluble salts with the steroids. The steroids are then derivatized as described in the general chapter. The yield of added ethynyl steroids added to urine or plasma is about 85% throughout the procedure.

Since it may not be necessary to subfractionate the ethynyl steroids into neutral, phenolic, and conjugated, a simpler isolation procedure has also been developed. Ethynyl steroids of low and medium polarity are extracted from urine or an enzymatic hydrolysate by filtration through Lipidex 1000. The methanol elute is then passed through a column of SP-Sephadex® in H^+ form and ethynyl steroids are then extracted by the silver column described above. Following conversion into MO-TMS derivatives and filtration through Lipidex 5000, the steroids are analyzed by GC-MS. Norethisterone was added to determine the recovery, which is 85 to 90% throughout the entire procedure.

Extraction of steroids using Extrelute (method of Wehner and Handke[11]) — The steroid extractions were carried out with Extrelute ready-to-use columns (Merck, Dahrmstadt). These columns were filled with kieselguhr of granular structure and high pore volume. After cleaning and refilling they can be reused. For extraction of the samples, 1 mℓ plasma was diluted to 20 mℓ with water and then transferred to the column. After distribution of the aqueous phase on the column (about 15 min were required), it was eluted with an organic solvent. Within 10 to 15 min the steroids were extracted from the stationary phase. The aqueous phase was retained on the column. For 5 hormones of different polarity (progesterone, corticosterone, estrone, estradiol, and estriol) the elution solvent diethyl ether was used. The data (averages of 5 individual determinations) for each hormone show recoveries to be 96% higher when 40 to 50 mℓ of ether are used to elute the added hormones from the column. These results are reproducible. Formation of emulsions is avoided. With estrogens, the diluted plasma, after being placed on the column, are first extracted with benzene-petroleum ether (1:1) (80 mℓ) to wash out or retain the estrone and estradiol. There is less than 1% estriol. Subsequent elution with 50 mℓ ether gave recoveries of 92% estriol in the 10—50 mℓ fraction.

Extraction with ClinElut for aldosterone for radioassay (method of Nickel[12]) — Combine 1 mℓ serum with radiolabeled trace aldosterone plus 6 mℓ distilled water; vortex; let stand for 10 min. Then add 5 mℓ dichloromethane to ClinElut column #CE1010-S. To this add the dilute serum followed by three separate 10-mℓ aliquots of dichloromethane, waiting 2 min after each addition. Collect the combined elutants and evaporate to dryness. Reconstitute with appropriate solvent according to the analytical procedure. Recoveries from plasma and urine are 95%.

Extraction of steroid conjugates with Amberlite® XAD-2 (method of Bradlow[13]) — A 7 × 35 cm column with a coarse fritted disc layered with 1—2 cm sand, was half-filled with water. 1 kg Amberlite® XAD-2, slurried in water, was poured into the column. The resin was washed by flowing distilled water from the bottom of the column at a rate sufficient to suspend the resin. The overflow was removed by a tube attached to a water aspirator. As soon as all the fines had been removed from the top (15—20 min) and the water had drained from the column, it was ready for use. Urine (2 ℓ) was percolated through the resin under gravity flow (45—60 min), followed by 4 ℓ water, which was allowed to drain completely. The conjugates are converted to their trimethylammonium salts by washing with 0.5 vol pH 7.2. 0.5 M trimethylamine sulfate was followed by a 0.5 vol water. Residual water was displaced with 200—300 mℓ petroleum ether. The conjugates were eluted by portion with 5 ℓ methanol. The methanol was concentrated *in vacuo* in a flash evaporator. The combined residue usually contained 10% or less of the urine solids. The material was soluble in water, alcohol, and tetrahydrofuran, among other solvents. The column was regenerated by washing with water as described above, and used repeatedly with no significant loss of capacity. (The column should be kept wet when not in use because the resin deteriorates on drying.) Using a single 1 kg column, 4 ℓ urine can be processed in 1 day with only occasional attention. Quantitive recoveries for steroid conjugates are possible with this method. In another example, 1 kg XAD-2 in a 7 × 35 cm column was used for extraction of 2 ℓ urine. Smaller volumes can be processed by scaling down the proportions.

Separation of steroid conjugates from urine using ion exchangers (method of Setchell et al.[5]) — Amberlite® XAD-2 and Amberlyst A-15 were supplied by Rohm & Haas. Amberlite® XAD-2,200 g was washed with 2 ℓ each 2 M sodium hydroxide, water, 2 M HCl, water, acetone, ethanol and water, refluxed for 4 hr in water, and finally washed with distilled water prior to use. Fines were removed by decantation or by backwashing of columns. Amberlyst A-15 was prepared in the H$^+$ form by washing with 2 M HCl in 72% ethanol, followed by 72% ethanol. Sephadex® LH-20 can be obtained from Pharmacia Fine Chemicals. The material was sieved and the fractions 100—140 or 140—170 mesh used. Lipidex 5000 was obtained from Packard Beck Ltd. (Holland).

Preparation of Ion Exchange Gels

Sulfoethyl Sephadex® LH-20 (SE-LH-20) — Sephadex® LH-20 (50 g) was suspended in 0.17 M sodium isopropoxide (1 ℓ and sodium hydroxide (25 g) in a powdered form was added. The mixture was stirred for 24 hr under N$_2$ (avoiding the use of a magnetic stirrer nich destroys the Sephadex® beads). Sodium-2-bromoethane sulfonate (8.5 g) was then added 5 times at 8 hr intervals with continued stirring under N$_2$. The product was filtered and washed with 1 ℓ each of 96% ethanol, 72% ethanol, 1 M sodium hydroxide in 72%

ethanol, 72% ethanol (until neutral), 0.5 *M* HCl in 72% ethanol, and 72% ethanol (until neutral). About 1 g of this material was washed with ethanol and dried to constant weight for titration. The gel was suspended in 20 mℓ 0.25 *M* sodium chloride in 72% ethanol and was titrated with 0.1 *M* sodium hydroxide in 72% ethanol. The capacity was usually between 0.15—0.30 mEq/g.

Diethylaminohydroxypropyl Sephadex® LH-20 (DEAP-LH-20) — Sephadex® LH-20 was used as a starting material to prepare chlorodroxypropyl Sephadex® LH-20.[15] The product had a chlorohydroxypropyl content of 23.2% (w/w) and was reacted with diethylamine.[16] The resulting ion exchanger was washed with 50% aqueous ethanol, 0.2 *M* acetic acid in 72% ethanol, and 72% ethanol until neutral. The gel was stored in the acetate form in 72% ethanol. For determination of ion exchanging capacity, 1 g gel in OH⁻ form was suspended in 20 mℓ 0.25 *M* sodium chloride in 72% ethanol and titrated with 0.1 *M* HCl in 72% ethanol, after the addition of a known excess of sodium hydroxide solution. The gel used in this study had a capacity of 0.9 mEq/g dry gel, and could bind about 0.5 m/mol cholic acid per gram.

Diethylaminohydroxypropyl-hydroxyalkyl Sephadex® LH-20 (DEAPHA-LH-20) — Sephadex® LH-20 was reacted with Nedox 114 and the hydroxyalkyl content of the product was 16.6% (w/w). This was used as the starting material for the preparation of a chlorohydroxypropyl derivative (23.6% [w/w] substitution) which was then reacted with diethylamine to yield the DEAPHA-LH-20 gel with a capacity of 0.6 mEq/g.

Preparation of Ion Exchange Columns

Glass columns (200 × 4 mm) with a 50-mℓ reservoir and ground glass joint at the top were fitted with a Teflon® end piece covered with Teflon® gauze, 70-μm mesh size, to hold the column bed. The lipophilic Sephadex® gels, 0.6 g, were allowed to swell in 72% ethanol for about 30 min, and the column bed was packed with a pressure of 0.3 kg/cm², resulting in a bed height of 5 to 10 cm. During chromatography the pressure was adjusted after each solvent change to maintain a flow rate of approximately 25 mℓ/hr.

Analytical Procedure of Steroids in Urine

The analysis is as follows: Urine (25 mℓ) was passed through a column of Amberlite® XAD-2 resin (bed size, 25 × 1 cm).[18] After washing the column with water (50 mℓ), the steroids were recovered by elution with ethanol (180 mℓ). Distilled water (70 mℓ) was added to the ethanol extract and the sample passed through a column of SE-LH-20 cation exchanger (4 g; bed size 20 × 1 cm), using a pressure of 0.5 kg/cm² to produce a flow of 30 to 50 mℓ/hr. The effuent was taken to dryness, the residue redissolved in a small volume of 72% ethanol, and transferred to a column containing DEAP-LH-20 (0.6 g) in the acetate form in 72% ethanol. After application of the sample, pressure was applied to the column (0.3 kg/cm²) to produce a flow rate of approximately 25 mℓ/hr, and the neutral steroids were eluted from the column with 25 mℓ 72% ethanol. The solvent was then changed to 0.25 *M* formic acid in 72% ethanol (25 mℓ) and the glucuronide conjugates were recovered. Steroid monosulfates were eluted with 25 mℓ 0.3 *M* acetic acid-potassium acetate, pH 6.3 in 72% ethanol. Finally, steroid disulfates were recovered by elution with 25 mℓ 0.5 *M* potassium acetate solution in 72% ethanol adjusted to pH 10 with potassium hydroxide.

Neutral fraction — The neutral fraction was taken to dryness, redissolved in 10 mℓ sodium acetate buffer (pH 4.5, 0.5 *M*), incubated with 0.3 mℓ sodium acetate buffer (pH 4.5, 0.5 *M*), 0.3 mℓ sodium acetate buffer (pH 4.5, 0.5 *M*), and 0.3 mℓ *Helix pomatia* digestive juice for 16 hr at 37°C. Following incubation the steroids were extracted on a column of Amberlite® XAD-2 as described above. The extract was dissolved in 72% ethanol (25 mℓ) and passed through a DEAP-LH-20 column (0.6 g) to recover the neutral compounds liberated by enzyme preparation. The solvent was evaporated, the sample dissolved in ethanol (2 mℓ), and stored in small vials until required for analysis.

Glucuronide fraction — This was taken to dryness and redissolved in 10 mℓ sodium acetate buffer (pH 4.5, 0.5 M). Glucuronidase (2 mℓ) was added and the sample incubated at 37°C for 48 hr. Following hydrolysis, the steroids were extracted on a column of Amberlite® XAD-2, and the liberated steroids isolated by filtration through DEAP-LH-20 in an identical manner to that of the neutral steroid fraction.

Mono- and disulfate fractions — These fractions were taken to dryness, and distilled water added — 15 mℓ to the monosulfate fraction and 25 mℓ to the disulfate fraction to give a 0.5 M concentration of potassium acetate. The pH was adjusted to 4.5 by addition of HCl, *Helix pomatia* digestive juice (0.3 mℓ) was added, and the sample was incubated at 37°C for 16 hr. Following hydrolysis, the steroids were extracted on Amberlite® XAD-2 as described below. The ethanol extract was taken to dryness, 30 mℓ acidified ethyl acetate was added, and the sample was incubated at 39°C for 16 hr. Following this solvolysis, the ethyl acetate was washed to neutrality with sodium bicarbonate (8%) followed by water. The combined washings were passed through a column of Amberlite® XAD-2 to recover any steroids extracted during the washing procedure and the XAD-2 extract was added to the ethyl acetate phase. The combined sample was then passed through a column of DEAPH-LH-20 as before, to recover the neutral steroids liberated during hydrolysis and solvolysis. The sample was stored in ethanol (2 mℓ) until required for analysis. Recoveries for metabolites of cortisol and prenolone were 90% or more using these procedures. The samples can, if necessary, be derivatized by GC-MS or other analytical procedure.

The extraction found in Table 1 can be used to prepare samples for the study of urine profiles. In the case of GC, derivatization is usually recommended. The flow scheme is from Leunissen and Thijssen.[14] Note that an alkali wash will remove the estrogrens if 1 N NaOH is used instead of the 8% NaHCO$_3$.

Separation of free and conjugated steroids (method of Axelson et al.[15]) — The extraction method based on Sep-Pak C$_{18}$ cartridges can be used as a faster alternative to the use of XAD-2. The urine is filtered through a cartridge which is washed with 5 mℓ each of methanol and water prior to use, followed by 5 mℓ water. Steroids are then eluted with 8 mℓ methanol. Cation exchange — 3 mℓ water is added to the eluate from the extraction step, and the solution is filtered (flow rate 0.7 mℓ/min) through SP-Sephadex® in H$^+$ form (40 × 4 mm, packed in 72% methanol) followed by a rinse with 3 mℓ 72% methanol.

Anion exchange — the effluent from SP-Sephadex® is passed through TEAP-LH-20 in OH$^+$ form (40 × 4 mm, packed in 72% methanol, flow rate 0.5 mℓ/min). The effluent is collected with the 4-mℓ rinse of 72% methanol. Neutral (uncharged) steroids are present in this fraction. Unconjugated phenolic steroids are then eluted with 8 mℓ 72% methanol saturated with CO$_2$, the flow rate being maintained by a pressure of CO$_2$. Monoglucuronides are then eluted with 10 mℓ 0.4 M formic acid, monosulfates with 10 mℓ 0.3 M acetic acid-potassium acetate at pH 6.5, and disulfates with 15 mℓ potassium acetate-potassium hydroxide, 0.5 M in acetate disulfates (pH 10.0). All solutions are made with 72% methanol.

Glucuronide fraction — the fraction is taken to dryness *in vacuo* and dissolved in about 5 mℓ of the purified enzyme solution. Following incubation for 1 hr at 62°C the reaction mixture is extracted with Sep-Pak C$_{18}$ or Amberlite® XAD-2 as above. The triethylamine sulfate wash is omitted in the XAD-2 extraction. The methanol eluate is directly passed through a column of TEAP-LH-20 in OH$^-$ form (40 × 4 mm, packed in methanol). The effluent and a rinse with 4 mℓ methanol contain the liberated neutral steroids. Phenolic steroids are eluted with 8 mℓ methanol saturated with CO$_2$ as above. The fractions are kept in methanol at 4°C until analyzed.

Sulfate fractions — the fractions are concentrated *in vacuo* to 2 mℓ (monosulfates) and 3 mℓ (disulfates). Water (5 mℓ) is added and the pH is adjusted to 4.5 with concentrated HCl (20 μℓ to the mono- and 200 μℓ to the disulfate fractions). Enzymatic hydrolysis can be performed at this stage. The fractions can be extracted with Sep-Pak C$_{18}$ or XAD-2 as

Table 1
EXTRACTION OF STEROIDS WITH XAD-2

XAD-2 column: 10 × 1 cm, 8 g XAD-2 (300—1000 μm)

10 mℓ Urine
20 mℓ Distilled water (wash)
40 mℓ Methanol (elution)
Evaporate under N_2 at 60°C

HYDROLYSIS AND SOLVOLYSIS

10 mℓ Acetate buffer (pH 5.0), 100 μℓ *Helix pomatia* juice, 18 hr at 37°C,
pH 1.0 with conc. HCl, 1.5 g NaCl, 25 mℓ ethyl acetate

Aqueous phase   Organic phase
discard   18 hr at 45°C

ALKALI WASH

2 × 5 mℓ 8% $NaHCO_3$ and 2 × 5 mℓ distilled water (1 *N* NaOH to remove estrogens)

Aqueous phase   Organic phase
XAD-2 extraction
Methanol

Evaporate under N_2 at 60°C

DERIVATIZATION (MO-TMS ESTERS)

Lipidex-5000 column (70 × 4 mm,   Dissolve in 1 mℓ methylene chloride;
0.25 g dry gel); solvent: hexane-   wash with 1 mℓ 0.1 *N* H_2SO_4 and
pyridine-HMDS-dimethoxypropane   2 × 1 mℓ distilled water, then
(97:1:2:10, v/v/v/v) N_2 pressure,   dry over anhydrous $NaSO_4$; inject
0.5 atm; flow rate, 3 mℓ/min; collect into GC
3.5 mℓ, evaporate under N_2 at 60°C,
redissolve in 1 mℓ hexane; then
chromatography

described for urine; an additional wash of XAD-2 with 6 mℓ water is added to remove
triethylamine sulfate. Water (1 mℓ) is added to the methanol eluate and the solution is filtered
through SP-Sephadex®. The combined effluent is concentrated *in vacuo* at 25°C to 100 μℓ.
Distilled tetrahydrofuran, 5 mℓ, acidified with 5 μℓ 4 *M* aqueous sulfuric acid is added.
Following incubation for 1 hr at 50°C, 1 mℓ methanol is added and the solution is passed
through TEAP-LH-20 in OH⁻ form. The 80 × 4 mm column is packed in methanol and
washed with 3 mℓ tetrahydrofuran-methanol (5:1, v/v) prior to application of the sample.
The column is washed with 6 mℓ of the latter solvent. The neutral steroids are recovered
in the combined effluent. Phenolic steroids are eluted with the same solvent saturated with
CO_2 under CO_2 pressure. Solvents are removed *in vacuo* and the fractions are stored in
methanol at 4°C until analyzed.

REFERENCES

1. **Craig, A.,** Ph.D. thesis, Brunel University, 1974.
2. **Ellingbol, J., Nyström, E., and Sjovall, J.,** *J. Lipid Res.,* 11, 266, 1970.
3. **Ellingbol, J., Alme, B., and Sjovall, J.,** *Acta Chem. Scand.,* 24, 463, 1970.
4. **Alme, B. and Nystrom, E.,** *J. Chromatogr.,* 59, 45, 1971.
5. **Setchell, K. D. R., Alme, B., Axelson, M., and Sjovall, J.,** *J. Steroid Biochem.,* 7, 615, 1976.
6. **Axelson, M. and Sjovall, J.,** *J. Steroid Biochem.,* 8, 683, 1977.
7. **Thenot, J. P. and Horning, E. C.,** *Anal. Lett.,* 5, 21, 1972.
8. **Axelson, M. and Sjovall, J.,** *J. Chromatogr.,* 126, 705, 1976.
9. **Axelson, M.,** *Anal. Biochem.,* 86, 133, 1978.
10. **Ercoli, A., Vitali, R., and Gardi, R.,** *Steroids,* 3, 479, 1964.
11. **Wehner, R. and Handke, A.,** *Clin. Chim. Acta,* 93, 429, 1979.
12. **Nickel, K. L.,** *Clin. Chem.,* 23, 885, 1977.
13. **Bradlow, L.,** *Steroids,* 30, 581, 1977.
14. **Leunissen, W. J. J. and Thijssen, J. H. H.,** *J. Chromatogr.,* 146, 365, 1978.
15. **Axelson, M., Sahlberg, B. L., and Sjovall, J.,** *J. Chromatogr.,* 224, 355, 1981.

Chapter 3

DERIVATIZATION FOR GAS CHROMATOGRAPHY

That gas chromatography (GC) could ever be used for analysis of steroids is surprising. These compounds are generally not volatile and are usually thermolabile. The development of suitable columns and supports was a very significant problem. However, for the most part, steroids must be derivatized before GC can be carried out. The methods described here will give the most commonly used derivatives that have been employed to stabilize the compounds as well as increase the volability. Nevertheless, there are examples of successful GC in which no derivative was prepared.

Derivatives can be prepared that allow sensitive detection. These derivatives (usually halogenated) were developed for use with the electron capture detector. There are a number of reasons for derivatizing steroids prior to GC. In high pressure liquid (HPLC) or thin layer chromatography (TLC) the tendency is to avoid derivatization since these procedures can be time consuming and are not always quantitative. In GC, the derivative may be more stable or volatile, less absorbed, easier to separate, and in many cases, easier to detect. Derivatization is often required for detection in HPLC and in most cases, derivatization is necessary when GC of steroids is attempted because of their poor thermal stability.

In certain cases the derivative can be prepared on the column in the gas chromatograph, but this is rare, and thus, the discussions of methodologies given in this section will be limited to derivatives prepared prior to injection in the gas chromatograph. The steroid may be derivatized by attaching a chromophore of high molar extinction coefficient at 254 nm for use with these detectors. Other derivatives may be used to take advantage of the variable wavelength on fluorescence detectors now available. For highest sensitivity the preparation for a fluorescent derivative is desirable. Reagents for derivatization should have several characteristics.

1.   Only one derivative must be formed of the parent compound, otherwise, quantitation is poor.
2.   Reaction must be complete. Reproducibility suffers if the reaction is different each time it is performed.
3.   The by-products should not interfere with the separation or mask peaks in the chromatograms.
4.   The derivatives must not interfere with differences in separation of the parent compounds; however, in some cases derivatization can result in better preparation and increase the specificity of detection.

In addition to the above characteristics derivatization can be carried out in two modes as described below.

Precolumn derivatization has certain advantages:

1.   Better control of reaction conditions since there are no restrictions on the mobile phase used for separation. Reaction kinetics are not hindered, whereas, in post-column conditions, the slower reaction can hinder operation.
2.   Derivatization can serve as a cleanup step, and more flexible extraction may eliminate interfering substances. Furthermore, excess reagent is removed.
3.   Chromatographic properties can sometimes be improved.

The kinetics of post-column reactions have to be high ($<$ 30 sec). Furthermore, the reagent must have no detectability in the range of the derivative. Ideally the derivative, when coupled with the parent compound, should form a chromophore with absorption and fluorescence of entirely different characteristics.

In spite of this there are many steroids which need to be derivatized for their detection at the nano- or picogram level. For UV detection the steroid may be derivatized by attaching a chromophore of high molar extinction coefficient at 254 nm for use with these detectors. Other derivatives may be used to take advantage of the variable wavelength on fluorescence detectors now available. For highest sensitivity the preparation for a fluorescent derivative is durable. Reagents for derivatization should have several characteristics:

1.   Only one derivative must be formed of the parent compound, otherwise, quantitation is poor.
2.   Reaction must be complete; reproducibility suffers if the reaction is different each time it is performed.
3.   The components should not interfere with the separation of mask peaks in the chromatograms.
4.   The derivatives must not interfere with differences in separation of the parent compounds; however, in some cases derivatization can result in better separation and increase the specificity of detection.

In addition to the above characteristics derivatization can be carried out in two modes as described below.

Precolumn derivatization has certain advantages:

1.   Better control of reaction conditions since there are no restrictions on the mobile phase used for separation. Reaction kinetics are not hindered, whereas, in post-column conditions the slower reaction can hinder operation.
2.   Derivatization can serve as a cleanup step, and more specific extraction may eliminate interfering substances. Furthermore, excess reagent is removed.
3.   Chromatographic properties can sometimes be improved.

Post-Column Derivatization

Post-column derivatization is becoming more practical and useful due to the advent of accessories for the HPLCs instrument. Advantages of this mode include:

1.   Artifacts less likely to interfere since the separation has already taken place.
2.   The possibility of using different detection modes on the same example is apparent. For example, detection can be carried out initially in the UV, which can derivatize and detect at another wavelength or use fluorimetry.

The disadvantages are that the mobile phase can strongly affect the reaction for derivatization. This means that every time a reagent is selected a set of conditions must be evaluated for effectiveness. Since the flow rate of HPLC is relatively high, the kinetics of post-column reactions have to be high (< 30 sec) as well. Furthermore, the reagent must have no detectability in the range of the derivative. Ideally, the derivative, when coupled with the parent compound, should form a chromophore with absorption and fluorescence of entirely different characteristics.

No single reagent or group of reagents can give optimum results in all cases; however, there are a number of general purpose derivatives that can be used. These are described herein. Some of then can be used either as pre- or post-column derivatives.

Preparation of Azobenzene Sulfonates[1]

For derivatization of the estrogens via their 4-axobenzenesulfonyl (ABS) derivatives heat

0.1 to 10.0 μg estrone, estradiol, or estriol and 1.0 mℓ 0.1% ABS chloride in Me$_2$CO with 0.85 mℓ Me$_2$CO and 0.05 mℓ 0.02 N NaOH for 30 min at 50°C, dilute with 20 mℓ Et$_2$O after cooling, extract once with 5 mℓ 0.01 N NaOH and three times with 5 mℓ of H$_2$O, filter the organic phase through Na$_2$SO$_4$, and evaporate to dryness under N. Thin-layer chromatograph the estrogen on Silica Gel G with CHCl$_3$-C$_6$H$_6$-EtOH (18:2:1), CHCl$_3$-dioxane (94:6), or cyclohexane-EtOHAc (3:1) by scanning detection at 313 nm. The recovery will be 94.2 to 96.9%, the sensitivity almost 50 ng estrogen, and the precision ±5.3% for 1.0 μg (6 runs). Allow the reaction mixture to stand at room temperature for 24 hr. Evaporate the reagent under a stream of dry nitrogen at room temperature. The residue is then dissolved in 0.1 mℓ of carbon tetrachloride (or other suitable solvent) and 2 μℓ injected into the gas chromatograph. The introduction of 10 μℓ directly into the mass spectrometer using the direct inlet system will serve to obtain mass spectra.

Silylation of Hydroxy Function

There are a number of silylation derivatives which can be made with steroids, making them more volatile and more amenable to GC. The various hydroxyl sites of steroids may be converted to trimethylsilyl derivatives depending on their relative steric hinderance. They fall into three groups unhindered, moderately hindered, and highly hindered. Three separate types of reagents can be used according to these classifications. General procedures follow.

Unhindered — To 5 to 10 mg sample in a screw-capped vial add 0.5 mℓ N,O-bis-(trimethylsilyl)-acetamide (BSA). Add an appropriate amount of solvent (acetonitrile) to attain complete solution. In the case of slow-reacting steroids, catalysis can be accomplished by adding a drop or two of trimethylchlorosilane.

Moderately hindered (and unhindered) — (11-β-ol-Steroids will react, the 17α-OH will not react). Place 0.1 to 5 mg of sample in small screw-capped vial (Reacti-vial is suitable). Add 0.2 to 0.4 mℓ of N,O-bis(trimethylsilyl)-acetamide plus trimethylchlorosilane (5:1). If necessary (as in the case of corticoids), add 0.1 to 0.2 mℓ of pyridine. Shake well to dissolve. In some cases the reaction will be complete in a few minutes (unhindered) but complete reaction is attained in 6 to 24 hr at room temperature or 3 to 6 hr at 60°C.

Esterification with 4-Nitrobenzoyl Chloride[4]

The esterification of digitalis glycosides with nitrobenzoyl chloride provides a derivative with a more favorable detection wavelength (260 nm) than the underivatized steroid. This derivative is prepared as follows: a 100-mg amount of 4-NBCl is dissolved per milliliter of pyridine with gentle warming. This is prepared fresh daily. The glycosides are dissolved in pyridine. The reaction is carried out in stoppered 10-mℓ centrifuge tubes. To 50 μℓ of a glycoside solution containing not more than 0.5 mg of the glycoside, 150 μℓ reagent solution is added. The mixture is well shaken and allowed to react for 10 min at room temperature. After this period, the reaction is shown to be quantitative by TLC. The centrifuge tubes are embedded in a beaker filled with sand at a temperature of 50°C. This sand bath is placed in a desiccator and the pyridine is removed by a water suction vacuum within 10 min. The centrifuge tubes are flushed with a stream of air or nitrogen and 2 mℓ 5% sodium carbonate solution, which also contains 5 mg 4-dimethylaminopyridine is added. The excess of reagent is hydrolyzed after 5 min of shaking or treatment in an ultrasonic bath. A blank treated simultaneously should yield a clear solution (the derivatives are poorly soluble in water and give turbid solutions). The derivatives are then extracted with 2 mℓ chloroform and treated with 2 mℓ 5% sodium hydrogen solution and twice with 3 mℓ 0.05 N HCl containing 5% sodium chloride. This procedure gives quantitative isolation of the derivatives and complete exclusion of excess of reagent and pyridine. Dinitrobenzoyl chloride can also be used as the reagent here.

Table 1
GROUPS SILYLATED BY DIFFERENT REAGENTS AFTER 24 HR AT ROOM TEMPERATURE[1]

Silylating agent	OH						C=O			Side chain				
	3	11β	16β	17β	20	21	3	17	20	CH₃ CH₂ COH	CH₃ CHOH COH	CH₂OH C=O CH	CH₃ C=O COH	CH₂OH C=O COH
1. HMDS	±	−	−	−	−	+	−	−	−	−	−	−	−	−
2. TMCS	±	−	−	−	+	±	−	−	−	−	−	−	−	−
3. HMDS + TMCS	+	−	+	+	+	+	−	−	−	−	−	−	−	−
4. BSA	+	+	+	+	+	+	−	−	−	−	−	−	−	−
5. BSA + TMCS	+	−	+	+	+	+	−	−	−	−	−	−	−	−
6. TMSDEA	+	+	+	+	+	+	−	−	−	−	−	−	−	−
7. TMSDEA + TMCS	+	+	+	+	+	+	−	−	−	−	−	−	−	−
8. BSTFA	+	+	+	+	+	+	−	−	−	−	−	−	−	−
9. BSTFA + TMCS	+	−	+	+	+	+	−	−	−	−	−	−	−	−
10. MSTFA	+	+	+	+	+	+	±	±	±	−	−	±	−	±
11. MSTFA + TMCS	+	+	+	+	+	+	−	+	−	−	−	+	+	+
12. MSTFA + KOAc	+	−	+	+	+	+	−	−	−	−	−	+	+	+
13. TMSI	+	−	+	+	+	+	−	−	−	+	+	−	−	−
14. TMSI + MSTFA	+	−	+	+	+	+	−	−	−	+	+	−	−	−

Note: + = Positive reactions; − = negative reations; ± = semi-quantitative reactions.

Reagents for Table 1: 1. Hexamethyldisilazane; 2. trimethylchlorosilane; 3. 1 + 2; 4. N,O-bis(trimethylsilyl) acetamide; 5. 4 + 2; 6. trimethylsilyldiethylamine; 7. 6 + 2; 8. N,O-bis(trimethylsilyl) trifluoracetamide; 9. 8 + 2; 10. N-methyl-N-trimethylsilyltrifluoroacetamide; 11. 10 + 2; 12. 10 + potassium acetate; 13. trimethylsilylimidazole; and 14. 13 + 10.

Dimethylsilyl (DMS) and chloromethyldimethylsilyl (CDMS) derivatives — These were prepared as above using tetramethyldisilazane (TMDS) or di(chloromethyl)-tetramethyldisilazane (CMTMDS) and either dimethylchlorosilane (DMCS) or chloromethyl-dimethylchlorosilane (CMDMCS) in place of the BSTFA and TMCS, respectively.

tert-**Butyldimethylsilyl (TBDMS) derivatives** — A 10-$\mu\ell$ portion of a mixture of *tert*-butyldimethylchlorosilane and imidazole in dimethylformamide (Applied Science Labs, State College, Pa.) was added to the sample (10 μg) and the mixture was heated at 60°C for 10 min. Pyridine, acetonitrile, or dimethylformamide (4 parts), the appropriate chlorosilane (2 parts), and either diethylamine (1 part) or imidazole (1 part) were mixed, cooled, and centrifuged to remove the precipitate; 10 $\mu\ell$ of these reagents were added to the samples (10 μg) and the mixtures were heated at 60°C for 10 min.

Derivatives for Electron Capture Detection in GC

The electron capture detector is highly sensitive for steroids derivatized with halogens. Steroids in a sample can be labeled for electron capture detection by attaching a polyhalogen group. This section gives a number of methods that provide for this procedure. This type of derivatization usually imparts volatility and stability as well as electron capture detectability, although the derivatives can also be detected by flame ionization and argon ionization detectors. These reagents overlap those of the previous group of derivatization methods.

The most common derivatives for this purpose are the perfluoroacyl derivatives of alcohols and amines. Of these the heptafluorobutyrate imparts the greatest sensitivity, although in special cases other derivatives may be better; for example, the pentafluoropropionyl ester may be more volatile and be more suitable for analysis.

A general method for preparation of fluorinated derivatives — To a 1-mℓ Reacti-vial add 2 to 3 mg sample; 150 $\mu\ell$ of 1:4 solution of reagent of choice in tetrahydrofuran. Seal and heat at 100 to 120°C for 30 min. Inject directly into the gas chromatograph. This procedure can be used with BTFA, MBTFA, BHFBA, or MBHFBA and is successful with most hydroxylated steriods.

Derivatization using heptafluorobutyric anhydride — In a 1-mℓ Reacti-vial add 5 ng phenolic steroid, 0.5 mℓ dry benzene, 100 $\mu\ell$ 0.1 *M* solution of trimethylamine in benzene, and 10 $\mu\ell$ HFBA. Allow to stand 10 min at room temperature. Wash benzene with 0.5 mℓ phosphate buffer, pH 6.0. Inject required aliquot. This method will derivatize most hydroxylic steroids and estrogens.

A second method requires the following: In a 1-mℓ Reacti-vial add 100 μg estrogen sulfate (sample or extract) or other sulfate, 200 $\mu\ell$ benzene (redistilled daily), and 200 μg heptafluorobutyric anhydride. Heat in block at 70°C for 0.5 hr. Evaporate to dryness under a nitrogen stream in a warm water bath. Add a predetermined amount of benzene for injection of aliquots into the gas chromatograph. This will directly derivatize steroidal sulfates. It will also transform free steroids containing hydroxyl groups such as 16-hydroxy epiandrosterone or androsterone to the halogenated ester.

Use of Flophemesyl Derivatives for Hydroxy Steroids

The fluoroalkyldimethyl ethers have excellent volatility upon GC, but respond poorly to electron capture detection in many cases. The pentafluorophenyldimethylsilyl ether (flophemesyl ether) has been found to be 20 times more sensitive to electron capture detection than the chloromethyl dimethylsilyl ether, but only slightly less volatile.[4] The introduction of one flophemesyl group permits detection in the nanogram range while the introduction of two groups extends the detection to the picogram level.

Flophemesylimidazole (R = $C_3H_4N_2$) was prepared in two stages. A solution of pentafluorophenyllithium[3] in diethyl ether at −70°C was added slowly to dichloromethyldimethylchlorosilane (Pierce and Warriner, Chester, U.K.). The mixture was allowed to reach room temperature and filtered. Removal of solvent and distillation of the residue gave dichloromethylpentafluorophenyldimethylsilane (I,R = CHCl$_2$, yield 45%). This product was dissolved in dry dimethylformamide, and cooled to − 10°C under N. Dropwise addition of a solution of imidazolyllithium in dimethylformamide, followed by filtration and distillation, gave flophemesylimidazole (yield 30% by GC).

Flophemesyldiethylamine (I,R = N(C$_2$H$_5$)$_2$) was prepared by addition of a solution of flophemesyl chloride in hexane to a 2 M excess of dry, redistilled diethylamine, dissolved in hexane, and cooled to − 70°C under N. The mixture was allowed to reach room temperature, filtered, and the filtrate distilled under vacuum. A lower yield (32%) was obtained when the bromosilane was used in place of the chlorosilane (67%). The colorless product darkened on standing to a pale orange with precipitation of some polymeric material, but this did not impair its properties as a silylating reagent.

For forming the sterol ethers, flophemesylamine (20 μℓ) was added to the sterol (1 to 5 mg) in pyridine (20 μℓ, dried over barium oxide) at room temperature in a Reacti-vial and allowed to stand for 15 min before GC using flame ionization detection (FID). For ECD the solvent was removed under vacuum and the residue redissolved in purified benzene or hexane.

A general procedure for formation of these derivatives is given below. It is advantageous to protect the ketone function with the formation of the methoxione before reaction with this class of derivatives, as it is with other silyl reagents.

Sterol (1 to 5 mg) was dissolved in dry pyridine (20 μℓ, dried over barium oxide) at room temperature in a Reacti-vial; 20 μℓ chlorosilane in 200 μℓ pure dry hexane was added per 50 μℓ diethylamine. The mixture was allowed to stand for 20 min and the solids settled by centrifugation. The supernatant liquid was added to 1 to 5 mg of cholesterol in screw-capped Reacti-vials, which were heated at 60°C for 3 hr. The solvent was removed by a stream of dry nitrogen and the silyl ether dissolved in ethyl acetate for flame ionization detection (FID) or hexane for ECD following GC. Complete conversion was achieved in all cases.

Dimethylthiophosphinic Esters[1]

Dimethylthiophosphinic chloride reacts with hydroxy steroids in the presence of triethylamine to produce the corresponding dimethylthiophosphinic esters according to:

$$\text{ROH} + \text{CIP}\underset{\text{CH}_3}{\overset{\text{S}\;\;\;\;\text{CH}_3}{\lessgtr}} \xrightarrow{\text{N(C}_2\text{H}_5)_3} \text{ROP}\underset{\text{CH}_3}{\overset{\text{S}\;\;\;\;\text{CH}_3}{\lessgtr}} + \text{HCl}$$

The hydroxy steroid (nanogram amounts) is treated with 25 μℓ dimethylthiophosphinic chloride solution and 25 μℓ triethylamine solution for 1 to 3 hr at 90°C under N. The volatile components are removed *in vacuo* (approximately 0.4 torr). The residue is dissolved in 100 μℓ methanol and 30 mg NaHCO$_3$ are added. The mixture is treated on a vortex mixer and incubated for 30 min at 50°C. The supernatant is transferred to a TLC plate and developed with chloroform-ethyl acetate (4:1 v/v). The derivative is localized with the radio-TLC scanner and eluted with methanol. The eluate is dried under a stream of nitrogen and dissolved in acetone. An aliquot is used for determination of the recovery rate in the liquid scintillation counter and 1 to 2 μℓ samples are injected into the gas chromatograph together with an appropriate internal standard.

Preparation of *t*-Butyldimethyl Silyl Ethers[2]

t-Butyldimethylsilyl imidazole: imidazole (1.7 g, 25 mmol) in ether (40 mℓ) was treated with sodium wire (0.6 g, 25 mmol). After the vigorous reaction subsides, the ether was distilled off and replaced with toluene (40 mℓ). The mixture was then refluxed for 1 hr, cooled, and treated dropwise with *t*-butyldimethylsilyl chloride (7.6 g, 50 mmol) in ether (10 mℓ), followed by stirring under reflux for 1 hr. After the mixture had cooled, it was filtered and the solvent evaporated *in vacuo* to give *t*-butyldimethylsilyl imidazole (BDMSI) (3.0 g, 56%) b.p. 86 to 90°C/0.7 mm.

t-Butyldimethylsilyl chloride can be purchased from P.C.R. (Gainesville, Fla.). *t*-BDMS derivatives: (a) Acid catalysis — the steroid (<0.5 mg) was allowed to react with 50 μℓ BDMSI at 60°C for 1 hr in the present of pyridinium hydrobromide (<1 mg). Prior to silylation some steroids were converted to the *O*-methoxime derivatives in the usual manner and silylated as above. The derivatized steroids were isolated either by extraction[1] or by filtering through Sephadex® LH-20[2] (b) Base catalysis — the steroid (<0.5 mg) was dissolved in toluene (50 μℓ) in the presence of anhydrous potassium acetate (1 mg) and then treated with BDMSI (10 μℓ). After heating the sample at 95°C for 2 hr, the mixture was cooled and the toluene evaporated. The derivative was isolated by filtering through Sephadex® LH-20[2] or by extraction of the crude extract with *n*-heptane.

Trimethylsilyl Ether-Enol-Trimethylsilyl Ether[9]

All reactions were carried out in tightly capped small test tubes. Noncatalyzed silylation was carried out on 50 μg steroid with 200 μℓ BSA or BSTFA overnight at room temperature. For the base-catalyzed reaction, 5 to 40 mg potassium (or sodium) acetate in methanolic solution was evaporated to dryness under a nitrogen stream in the reaction tube; 50 to 100 μg steroid was then added. After complete evaporation of the solvent, 200 μℓ silylating agent (BSTFA, BSA, or TSIM) was added; 2 to 4 μℓ of the mixture were either directly injected into the gas chromatograph (or the gas chromatograph-mass spectrometer) or evaporated under a nitrogen stream and taken up in hexane (for direct injection MS), carbon tetrachloride (for NMR study), or HMDS (for GC or storage). For the quantitative study of the reactions, cholesterol butyrate (50 to 100 μg) was added as an internal standard prior to the silylating agent.

Preparation of Oxime Derivatives[6] For Blockage of Ketone Groups: General Methods

Oximes, methoxime, benzyloxime — A mixture of 4 mg hydroxylamine hydrochloride, 5 mg methoxylamine hydrochloride, or 9 mg *O*-benzylhydroxylamine hydrochloride, 50 μℓ pyridine, and 0.1 mg 5α-androstan-17-one was heated for 30 min at 60°C. The solutions of MO and BO derivatives were used directly for GC and GC-MS studies.

Trimethylsilyloxime and oxime acetate — After formation of the oxime as above the pyridine was evaporated with a stream of nitrogen; 100 μℓ trimethylsilylimidazole or 100 μℓ of acetic anhydride was added to obtain the trimethylsilyloxime or oxime acetate. The reaction was allowed to proceed for 3 hr at 60°C. Solutions were used directly for injection into the gas chromatograph. Dissolve 2 mg steroid in 0.5 mℓ pyridine. Add 1 mg methoxyamine hydrochloride. Heat 3 hr at 60°C; add 2 mℓ water and extract 3 times with 5-mℓ portions of benzene; combine benzene. Wash in turn with 1 *N* HCL, 5% Na-HCO₃, and water. Dry over Na₂SO₄ and evaporate with a stream of nitrogen. Reconstitute in a suitable solvent for injection. This procedure has been used for protection of the 20-keto group in corticoids as well as 17-ketosteroids.

Steroid methoximes are prepared by treating the appropriate ketosteroid with an anhydrous solvent (usually pyridine) with excess methoxyamine hydrochloride at room temperature overnight or at 65°C for 3 hr. Solvent was removed under vacuum and the residue partitioned between ethyl acetate and a solution of sodium chloride (10%) and HCl (5%) in water. The ethyl acetate was washed with a solution of sodium chloride (10%) and sodium bicarbonate (5%) in water, dried over molecular sieves, and evaporated to give the methoximes.

Formation of Cyclic Butyryl Boronates

Dissolve 1 to 2 mg sample in 200 $\mu\ell$ dimethylfluoramide. To this add the equivalent weight of *n*-butyryl boronic acid. Keep at room temperature for as long as required to complete the reaction. Follow the progress by injecting an aliquot into the gas chromatograph. For anhydrous conditions, which are required in some cases, a few drops of dimethoxypropane can be added. This procedure is useful for preparation of derivatives of the glycolic forms of the corticosteroids.

Pentafluorobenzyl Oximes as Steroid Derivatives[10]

Preparation of steroid-*O*-(2,3,4,5,6-pentafluorobenzyl)-oximes (steroid-*O*-PFBO) — A stock solution of the reagent, PFBHA-HCl, was made in distilled pyridine to contain 50 mg/mℓ. This solution was stable for weeks when stored in the refrigerator and protected from light. Suitable dilutions were made with distilled pyridine for derivatization at the microgram and nanogram levels. These diluted solutions were also stable for several days in the refrigerator.

Derivatives in milligram quantities for flame ionization detection — From a stock solution of the steroid in benzene, an aliquot containing 1 mg was transferred to a tapered 5-mℓ glass stoppered centrifuge tube. The solvent was removed under a stream of N_2 while the tube was immersed in water at about 40 to 45°C; 0.2 mℓ of the reagent were added. The tube was swirled to assure contact with the steroid, stoppered, and heated in a suitable bath at 65°C for 0.5 hr. Pyridine was then removed by nitrogen jet with the tube heated at 40 to 45°C (bath). Cyclohexane (1 mℓ) was added and the tube shaken on a vortex. Water (1 mℓ) was then added, the tube again shaken, and the phases allowed to separate. The organic phase was separated and dried over anhydrous sodium sulfates (aliquots can then be injected into the GC). When 1 to 10 μg steroid is used, 0 to 2 mℓ aliquot containing 250 μg reagent is used. After removal of the used pyridine, the derivative is dissolved in 0.1 mℓ cyclohexane and washed with 0.1 mℓ of water. For 12.5 to 50 ng of steroid, 25 μg reagent in 10 $\mu\ell$ pyridine is used. The residues from the reaction are dissolved in 0.5 mℓ cyclohexane and washed with 0.5 mℓ water. If it is desired to derivatize 0.1 to 5 ng steroid, 5 μg reagent in 10 mℓ pyridine is used. Too much reagent masks the resulting chromatograms because of the broad bands.

Preparation of Butyloximes and Pentyloximes[7]

These derivatives can be prepared in pyridine solution using 0.5 mg steroid as described for the other oximes and the prerequisite alkyl hydroxylamine.

Methoxime-Trimethylsilyl Ethers[8]

Reactions were carried out in micro vials (0.3 mℓ with Teflon® cap liners). Methoximes were prepared by adding 50 $\mu\ell$ of a solution of methoxylamine hydrochloride (Applied Science Laboratories, Inc.) in pyridine containing 100 mg/mℓ reagent to the steroid (0.1 mg). After reaction as indicated at 60 or 100°C, the pyridine was evaporated with a nitrogen stream. The silyl donor TSIM (or BSA in some instances, Pierce Chemical Co.) was added (50 $\mu\ell$), and the solution was heated as indicated. The solutions were used directly in GC or GC-MS studies.

The most satisfactory general condition involved MO formation at 60°C for 15 min, and persilylation with TSIM at 100°C for 2 hr, without removal of excess methoxyamine hydrochloride.

Dissolve 2 mg sample in the reagent (2% methoxylamine-HCl in pyridine) and allow to stand overnight at room temperature. Dilute with 2 mℓ saturated aqueous sodium chloride. Extract with 3-mℓ portions of ethyl acetate; combine ethyl acetate. Wash in turn with salt-saturated 0.1 *N* HCl, saturated NaCl solution, salt-saturated 5% $NaHCO_3$, and saturated salt solution. Dry over magnesium sulfate. Concentrate with nitrogen stream to 0.5 mℓ.

Heptafluorobutyrylimidazole and trifluoroacetylimidazole have been used when acylating with anhydrides such as heptafluorobutyric, pentafluoropropionic, or trifluoroacetic anhydrides. The method below gives good yields.

Dissolve 50 μg (250 μg for FID) of sample in 0.5 mℓ derivatization grade benzene in a 5 mℓ Activial. Add 0.1 mℓ 0.05 M TMA in benzene followed by 10 μℓ desired anhydride. Cap the vial and heat to 50°C for 15 min. Cool and add 1 mℓ 5% aqueous ammonia solution. Shake for 5 min, separate the benzene layer, and inject directly into the gas chromatograph. Unlike other base catalysts, TMA does not cause interfering peaks at high EC sensitivities.

N-Methyl-bis-trifluoroacetamide has been used to acylate hydroxyl groups as follows: to 5 to 10 mg sample in a 3 or 5 mℓ TR-vial add 0.5 mℓ MBTFA and 0.5 mℓ pyridine. Cap and heat to 65°C for 1 hr with occasional shaking. Cool and inject directly in the gas chromatograph.

REFERENCES

1. **Gleispach, H.**, *J. Chromatogr.*, 91, 407, 1974.
2. **Jacob, K. and Vogt, W.**, *J. Chromatogr.*, 150, 339, 1978.
3. **Blair, I. A. and Phillipou, G.**, *J. Chromatogr.*, 16, 201, 1978.
4. **Morgan, E. D. and Poole, C. F.**, *J. Chromatogr.*, 104, 354, 1975.
5. **Morgan, E. D. and Poole, C. F.**, *J. Chromatogr.*, 89, 225, 1974.
6. **Phillipou, G.**, *J. Chromatogr.*, 129, 384, 1976.
7. **Thenot, J. P. and Horning, E. C.**, *Anal Lett.*, 4, 683, 1972.
8. **Thenot, J. P. and Horning, E. C.**, *Anal. Lett.*, 5, 21, 1972.
9. **Baillie, T. A., Brooke, C. J. W., and Horning, E. C.**, *Anal. Lett.*, 5, 351, 1972.
10. **Pfaffenberger, C. D., Malinak, L. R., and Horning, E. C.**, in *Advances in Chromatography 1978*, Zlatkis, A., Ed., Gordon & Breach Science Publ., New York, 1978, 318.
11. **Chambaz, E. M., Defaye, G., and Madane, C.**, *Anal. Chem.*, 45, 1090, 1973.
12. **Kosby, K. T., Kuiser, D. G., and Van Der Silk, A. L.**, *J. Chromatogr. Sci.*, 13, 97, 1975.

Chapter 4

DERIVATIZATION FOR HIGH PERFORMANCE LIQUID CHROMATOGRAPHY (HPLC)

Unlike gas chromatography preparation of derivatives in liquid chromatography is rare. Derivatization, however, can solve many problems. The reasons for this are (1) to improve stability during handling, (2) allow chromatography to take place, (3) improve separation, and (4) increase detectability. The most important of these is the use of derivatization to increase detectability. The great flexibility of HPLC and also TLC lies in the choice of mechanisms for separation (ion exchange, partition, and absorption), and in the ease of ability to modify both the stationary and mobile phases. This means that any compound can be separated without modification.

Derivatization common in GC results in increased stability, sometimes easier separation, and (sometimes more important) increased detectability. An example of this is the formation of halogenated derivatives which greatly enhance the value of the electron capture detector.

Detectability has always been a limiting factor in LC. Early detectors for LC were of fixed wavelength type and had limited sensitivity depending on the structures of the compound being analyzed. More recent variable wavelength and fluorescence detectors greatly extended the range of these detectors. With these instruments available the level of sensitivity was dictated by the ability to label or derivatize the molecule of interest. Strong UV chromophores or fluorophores have become increasingly available.

UV Absorbing Derivatives

Many steroids, particularly the corticoids, have structural characteristics such that they may be detected with the common single wavelength UV detectors. This detector is usually unaffected by external conditions. It can be sensitive at the nanogram levels depending on the compound under consideration. It is also rugged and inexpensive. In spite of this there are many steroids which need to be derivatized in order to be detected at the nano- or picogram level. For UV detection the steroid may be derivatized by attaching a chromophore of high molar absorptivity at 254 nm for use with these detectors. Other derivatives may be used to take advantage of the variable wavelength or fluorescence detectors now available. For highest sensitivity, the preparation for a fluorescent derivative is desirable. Reagents for derivatization should have several characteristics:

1. Only one derivative must be formed from the parent compound, otherwise, quantitation is poor.
2. The reaction must be complete. Reproducibility suffers if the reaction is different each time it is performed.
3. The by-products should not interfere with the separation or mask peaks in the chromatograms.
4. The derivatives must not interfere with separation of the individual compounds; however, in some cases derivatization can result in better separation and increase the specificity of detection.

In addition to the above characteristics, derivatization (pre-column and post-column) can be carried out in two modes as described below. Pre-column derivatization has certain advantages:

1. Better control of reaction conditions since there will be no restrictions on the mobile

phase used for separation. Reaction kinetics are not hindered, whereas, in post-column conditions, the slower reaction can hinder operation.

2.   Derivatization can serve as a cleanup step, and more flexible extraction may eliminate interfering substances. Furthermore, excess reagent is removed.

3.   Chromatographic properties can sometimes be improved.

The rate of the reactions has to be rapid (< 30 sec) as not to hinder mobile phase flow. Furthermore, the reagent must not separate in the range of the derivative. Ideally, the reagent, when coupled with the parent compound, should form a derivative with absorption of fluorescence of entirely different characteristics.

Post-column derivatization has been discussed in Chapter 3.

Derivatization with dansyl chloride (method of Oertel and Penzes[1]) — For the determination of fluorescence derivatives via the 5-(dimethylamino)-1-naphthalenesulfonyl (dansyl) derivatives, the following can be used: Heat 0.01 to 1.0 µg estrogen and 0.1 mℓ 0.01 dansyl chloride in Me_2CO with 0.9 mℓ Me_2CO and 0.1 mℓ 0.1 N NaOH for 30 min at 50°C, dilute with 20 mℓ C_6H_6, wash once with 2.5 mℓ 0.1 N NaOH and 3 times with 2.5 mℓ H_2O. Filter the organic phase and evaporate under N jet. The purification is performed on silica gel without binder using $CHCl_3$-C_6H_6 (9:1) or $CHCl_3$-C_6H_6-EtOH (18:2:1). Determination can be achieved by direct fluorimetry or by fluorimetry after elution with $CHCl_3$ (excitation at 362 nm, emission at 517 nm). The recovery will be 93.4 to 95.2%, the sensitivity $\sim$ 5 ng steroid, and the precision 5 to 7%.

Preparation of azobenzene sulfonates (method of Oertel and Penzes[1]) — For formation of 4-azobenzenesulfonyl (ABS) estrogen derivatives, heat 0.1 to 10.0 µg estrone, estradiol, or estriol and 1.0 mℓ 0.1% ABS chloride in acetone with 0.85 mℓ acetone and 0.05 mℓ 0.02 N NaOH for 30 min at 50°C, dilute with 20 mℓ diethyl ether after cooling, extract once with 5 mℓ 0.1 N NaOH and 3 times with 5 mℓ H_2O, filter the organic phase through Na_2SO_4, and evaporate to dryness under N_2. Perform TLC on Silica Gel G with development by chloroform-benzene-ethanol (18:2:1:), chloroform-dioxane (94:6), or cyclohexane-ethyl acetate (3:1). Detect the estrogen by scanning at 313 nm. The recovery will be 94.2 to 96.9%, the sensitivity almost 50 ng estrogen and the precision $\pm$ 5.3% for 1.0 µg (six runs). HPLC can give equivalent results.

Formation of fluorescent hydrazones of ketosteroids with isonicotinylhydrazine for post-column detection — 16 mM solution of isonicotinyl hydrazine (INH) in methanol is prepared; 80 mM aluminium chloride is prepared by dissolving $AlCl_3$·$6H_2O$ in methanol. These solutions are stable for several weeks. The column effluent can first be monitored by a UV detector at 254 nm. It is then added to a mixture of INH and aluminum chloride solution as follows. Each reagent solution, heated in a PTFE tube and immersed in a water bath maintained at 70°C is delivered separately by a pump at half the flow rate of the effluent. The reaction mixture is passed through a reaction coil in the same water bath and is monitored by a fluorescence detector with excitation at 360 nm. A secondary filter that transmits above 450 nm provides selectivity. The reaction times for the systems of Hitachi® Gel 3011 column and Zerbax-Sil® column will be about 2.0 and 1.2 min, respectively.

Esterification with 4-nitrobenzoyl chloride (method of Nachtmann et al.[4]) — The esterification of digitalis glycosides with nitrobenzoyl chloride (NBCl) provides a derivative with a more favorable detection wavelength (260 nm) than the underivatized steroid. This derivative is prepared as follows: The 4-NBCl reagent is prepared at 100 mg/mℓ in pyridine with gentle warming. This is prepared fresh daily. The glycosides are dissolved in pyridine. The reaction is carried out in stoppered 10-mℓ centrifuge tubes. To 50 µℓ glycoside solution containing not more than 0.5 mg glycoside, 150 µℓ reagent solution is added. The mixture is well shaken and allowed to react for 10 min at room temperature. After this period, the reaction is shown to be quantitative by TLC. The centrifuge tubes are embedded in a beaker filled with sand at a temperature of 50°C. This sand bath is placed in a desiccator and the pyridine is removed by a water suction vacuum within 10 min. The centrifuge tubes are flushed with a stream of air or nitrogen and 2 mℓ of 5% sodium carbonate solution, which also contains 5 mg 4-dimethylaminopyridine are added. The excess of reagent is hydrolyzed after 5 min of shaking or treatment in an ultrasonic bath. A blank treated simultaneously should yield a clear solution (the derivatives are poorly soluble in water and give turbid solutions). The derivatives are then extracted with 2 mℓ chloroform and treated with 2 mℓ 5% sodium bicarbonate solution and twice with 3 mℓ 0.05 N HCl containing 5% sodium chloride. This procedure gives quantitative isolation of the derivatives and complete exclusion of excess of reagent and pyridine. Dinitrobenzoyl chloride can also be used as the reagent.

Reaction of steroidal ketones with *p*-nitrobenzyloxamine (PNBA) hydrochloride[5] — The formation of oximes of ketonic compounds is common. This derivative is a common one in GC. The following method has been used with 17-ketosteroids. This can be scaled down depending on the amount of sample available. Add steroid: 0.5 mg and PNBA: 10 mg. Dissolve in 50 mg of pyridine. Heat at 60°C for 30 min. Remove pyridine with an N jet in a warm water bath. Take up the reaction mixture in methylene chloride (1 mℓ). Wash with dilute HCl and water before injecting an aliquot. This type of derivative permits detection at the 1 to 10 ng level.

REFERENCES

1. **Oertel, G. W. and Penzes, L.,** *Z. Anal. Chem.,* 252, 306, 1970.
2. **Umberger, E. J.,** *Anal. Chem.,* 22, 768, 1956.
3. **Horikawa, R., Tanimura, T., and Tamura, Z.,** *J. Chromatogr.,* 168, 526, 1976.
4. **Nachtmann, F., Spitz, H., and Frei, R. W.,** *J. Chromatogr.,* 122, 293, 1976.
5. **Theriot, J. P. and Horning, E. C.,** *Anal. Lett.,* 4, 683, 1971.

Chapter 5

DETECTION OF STEROIDS ON TLC

General Nondestructive Detection Methods

Widely used methods include:

- *Water*:

  Spray the chromatogram with water until the plate is translucent. The separated zones (must be water insoluble) usually lipids, appear as white opaque spots against a dark background. This is not a very sensitive method.[1]

- *Iodine*:

  Iodine has been found to be very helpful for detection without destruction. Place the chromatogram in a closed chamber containing a few crystals of iodine. The chamber can be warmed on a hot plate to vaporize the iodine. Yellow to brown zones will appear. The chromatogram can also be sprayed with a 1% solution of iodine in methanol or ethanol. The zones must be marked since the iodine will eventually evaporate.[2]

- *Fluorescent indicators*:

  The use of fluorescent indicators in thin layers is common.[3] The fluorescent bromine test indicates unsaturated or other compounds which react with bromine. The dried chromatogram is sprayed with a solution of 0.05% fluorescein in water. The plate is then exposed to bromine vapor by inverting it over vapors from an open bottle of bromine. The fluorescein is converted to eosin as a red dye except where unsaturated compounds absorb the bromine. View under the shortwave UV light in a dark room.[4]

- *Rhodamine B*:

  Rhodamine B can be sprayed for help in locating materials separated by thin layer chromatography (TLC). Spray a 0.05% alcoholic solution of Rhodamine B on the chromatogram; a 0.2% aqueous solution has also been used.[2] Cholesterol and other lipids show up as purple spots on a pink background. The color is different with different compounds: The chromatograms are viewed in a dark room under shortwave UV light.

Other Reagents

Other fluorescent reagents that have been used are aqueous solutions of sodium fluorescein,[5] a 0.005 to 0.01% solution of morin in ethanol,[6] and a 0.2% solution of $2',7'$-dichlorofluorescein in ethanol.[6] Here the detection is only viable if the separated steroid absorbs in the region of the UV light used to activate the fluorescence or if the concentration is high enough, dark zones will appear in the fluorescent layer.

Commercial Fluorescent Plates

The suppliers of commercial thin layer plates have made available "GF" layers. The fluorescent indicator is incorporated into the silica gel before spreading on the support. Zinc sulfide and zinc silicate are most often used. The separated material will be visualized only if it absorbs UV light in the region of the light used to activate the phosphor in the layer. These plates have been used for quantitation of cortisol from extracts of plasma. The plates were scanned by densitometry at a wavelength of 250 nm. The steroid which absorbs in the layer appears as a blue (purple) spot in the yellow fluorescence. Table 1 gives some indication of the sensitivity that can be attained using this method.

Most of the examples given in this handbook for separation of steroids were those which involved silica gel as sorbent. Due to the inertness of this material it is suitable for carrying

Table 1
SPRAY REAGENTS FOR TLC

Adams reagent	a. Mix ethanol (60%), perchloric acid (50%), formaldehyde, and water 2:1:0:1:0:9, v/v b. Prepare a 0.1% solution of 1.2-naphtochinon-2-sulfonic acid with mixture a c. Spray with b and heat at 70—80°C and observe colors 15—60 min	For ergosterol (I), stigmasterol (II), and cholesterol (III):[78] (I) Deep rose[79]→ deep blue[80]→ blue black[81] (II) Rose[79]→ deep rose[80]→ blue[81] (III) Light rose[79]→ rose[80]→ deep rose[81]→ blue[82]
Ammonium bisulfate reagent[89]	a. Spray with a saturated aqueous solution of ammonium bisulfate b. Heat for 45 min at 200°C	Similar to sulfuric acid reaction; cholesterol and cholesterol esters,[83] cucurbitacins[84]
Ammonium thiocyanate-ferrous sulfate reagent[90]	Prepare:[77] a. 1.35% solution of ammonium thiocyanate in acetone b. 4% aqueous solution of ferrous sulfate Spray successively with a and b	Sterol hydroperoxides[85]
Antimony pentachloride	a. Spray with a 20—30% solution of antimony pentachloride-chloroform b. Heat at 110—120°C for 3—5 min	Sterols,[86,87] sterol epoxides,[87] sterol bromides,[86] and trimethylsterols;[88] spot colors,[86,88] generally yellow to brown; butyrospermol, violet α-euphorbol and dihydrobutyrospermol (ac = acetate, red)
Bismuth trichloride	a. Spray with a 33% solution of bismuth trichloride in 96% ethanol b. Heat at 90°C for 10—15 min *Remark:* can be used on reversed-phase TLC on undecane and tetradecane-impregnated layers[91]	Sterols and sterol acetates;[91] spot colors: acetates of cholesterol, stigmasterol, and β-sitosterol, violet; acetates of ergosterol and 7-dehydrocholesterol, gray; lathosterol, vitamin D₂, and acetates of lanosterol; 5-dehydroergosterol and cholestanol (ac), negative
Brady's reagent	a. 5% Solution of dinitrophenyl-hydrazine in methanol b. Concentrated sulfuric acid c. Spray with a mixture of 8 vol a, 1 of b	Ketonic sterols[93]
Cadmium chloride	a. Spray with 50% solution of cadmium chloride in 50% ethanol b. Heat at 90°C for 15 min and observe bright fluorescence at long UV light (365 nm) *Remark:* can be used on reversed-phase TLC	Brominated sterols[91]
Chlorosulfonic acid	a. Spray with a 30% solution of chlorosulfonic acid in acetic acid b. Heat at 90°C for 10—15 min *Remark:* can also be used for silica gel and aluminum oxide-impregnated papers (heat at 100°C for only 80 sec)	C₂₇ and C₂₈ sterols,[94] trimethylsterols,[95] sterol acetates;[91] spot colors:[91] 7-cholestanol and vitamin D₂, gray; acetates of cholesterol, stigmasterol, and stigsitosterol, violet; 7-ergostenol (ac), yellow green; cholestanol (ac), white; lanosterol (ac), orange brown; acetates of ergosterol, 7-dehydrocholesterol, and 5-dihydroergosterol, gray brown
Ekkert reaction	a. Spray with a 1% solution of *p*-anisaldehyde in acetic acid-sulfuric acid (98:2)	Sterols and sterol acetates;[91] spot colors: acetates of cholesterol, stigmasterol, β-sitosterol, lanosterol,

Table 1 (continued)
SPRAY REAGENTS FOR TLC

	b. Heat at 90°C for 10 min *Remark:* can be used on reversed-phase TLC, undecane, and tetradecane-impregnated layers[91]	blue purple; lathosterol, 7-ergostenol (ac), and 5-dihydroergosterol (ac), violet; ergosterol (ac) and 7-dehydrocholesterol (ac), gray green; cholestanol (ac), negative
Hammarsten-Yamasaki reaction	I.[95] a. Spray with concentrated HCl 　b. Heat for 5 min at 80°C II.[96] a. Expose plate to HCl vapor 　b. Observe in day- and UV-light (yellow fluorescence in shortwave) *Remark:* more sensitive on silica gel layers; not specific for 3-hydroxy-Δ^5-steroids	7-Hydroxycholesterol gives a blue color; specific for 3,7-dihydroxy-Δ^5- and 3-hydroxy-$\Delta^{5,7}$ steroids 3-Hydroxy-Δ^5-steroids give a pink or violet color which changes to gray and pale yellow; fluorescence in UV light (254 and 366 nm); sensitivity: 1 μg/cm^2 for cholesterol (0.6 μg/cm^2)
Iron trichloride	I. a. 10% solution of anhydrous iron trichloride in acetic acid 　b. Dissolve 0.2 mℓ a in 50 mℓ mixture acetic acid-sulfuric acid (3:2) 　c. Spray with b and heat at 110°C for 10 min II. Spray with a 5% ethanolic solution of iron-trichloride	C$_{27}$ sterols;[97] spot colors: 5-cholestene-3β,7α-diol, 5-cholestene-3β,7α-diol, blue before and after heating; 5-cholestene-3β,4β-diol, blue (violet after heating); 3,5-cholestadiene, blue violet, before and after heating; 7-dehydrocholesterol and lathosterol, gray green; 1,4-cholestadien-3-one and 4,6-cholestadien-3-one, gray rose or rose; 4-cholestene-3,6-dione, red violet; 7-oxocholesterol, cholestanol, epicholestanol, violet; 5-cholesten-3-one, 3,5-cholestadien-7-one, 3β-hydroxycholestan-6-one, cholestane-3β,5α,6β-triol, and 3β,5α-dihydroxy-cholestan-6-one, brown or brown orange Cucurbitacins of diosphenol series:[98] brown spots
Liebermann-Burchard reaction	I. a. Spray with a 30% solution of acetic anhydride in 50% sulfuric acid 　b. Heat 10 min at 90°C; for brominated steroids, spray warm plate with a[99] *Remark:* used on reversed-phase TLC, undecane and tetradecane-impregnated layers[91]	Brominated C$_{27}$ sterols[99] bright blue spots, C$_{27}$ sterols;[91] spot colors: acetates of cholesterol, stigmasterol, and β-sitosterol, violet; ergosterol (ac) and 7-dehydrocholesterol (ac), gray green; lanosterol (ac), orange purple; 7-ergostenol (ac), yellow brown; 5-dihydroergosterol (ac), purple brown; lathosterol, purple blue; vitamin D$_2$, gray brown; red color with several insect-moulting hormones as inokosterone,[102] rubrosterone,[103] cyasterone, and sengosterone[104]
	II.[88] a. Spray successively with 10% acetic anhydride-absolute ethanol and 10% sulfuric acid-absolute ethanol 　b. Heat at 120°C for 5 min	Trimethylsterols;[88] spot colors: euphol, parkeol, β-euphorbol, lanosterolbutyrospermol, dihydrobutyrospermol (ac), violet or violet gray; cycloartenone, euphone, agnosterol, brown or brown yellow; butyrospermone and euphene, gray
	III.[100] a. Spray with mixture of acetic anhydride-sulfuric acid-acetic acid 20:1:10	C$_{27}$ sterols; spot colors:[100] 5-cholestene-3β,4β-diol, 5-cholestene-3β,7β-diol, and 5-cholestene-3β,7α-

Table 1 (continued)
SPRAY REAGENTS FOR TLC

	b. Heat at 110°C for 3 min	diol, blue, before and after heating; 5-cholesten-3-one, 4-cholestene-3,6-dione, cholestanol, epicholestanol, coprostanol, cholestane-3β,5α,6β-triol, 3β,5α-dihydroxycholestan-6-one, brown; 4,6-cholestadiene-3-one, gray brown; 3β-hydroxycholestan-6-one, yellow brown; lathosterol, gray green, before and after heating; 3,5-cholestadiene, green (after heating, blue)
	IV.[101] a. Spray with mixture of acetic anhydride-sulfuric acid (4:1)	Employed on starch-bound silica gel layers[101]
Matthews reagent[105]	a. Spray with a 0.5% vanillin solution in H_2SO_4-ethanol (4:1) b. Heat at 100°C for 5 min	Cholestanols and cholestanones;[107] (5α-cholestan-1α(1β,2α,2β,3α, 3β,4α,6α, or 7α)-ol and 5β-cholestan-3α(4β or 6β)-ol: blue violet: cholestanones: yellow to violet spots Insect-moulting hormones;[108] ecdysone and 20,26-dihydroxyecdysone: turquoise; crustecdysone: yellow green
β-Naphthol-sulfuric acid	a. Spray with a 0.2% solution of β-naphthol in 4 *N* sulfuric acid b. Heat at 90°C for 10 min	Sterols and sterol acetates;[91] spot colors: lathosterol, 7-ergostenol (ac), cholesterol (ac), blue; lanosterol (ac), 5-dihydroergosterol (ac), blue purple; vitamin D_2, gray; ergosterol (ac), and 7-dehydrocholesterol (ac), gray blue; cholestanol (ac), white; stigmasterol (ac), violet; β-sitosterol (ac), purple
p-Nitrobenzyl-pyridine[106]	I. Detection of steroid alcohols by formation of tosylates *in situ*: a. Spray with 5% solution of *p*-toluene-sulfonylchloride in a solution of anhydrous pyridine-toluene 1:1 (v/v) b. Nearly dry plates are sprayed with a 2% solution of 4-(*p*-nitrobenzyl)-pyridine in acetone, heated for 20 min at 110°C and sprayed again with c c. Aqueous sodium carbonate (1 *M*)	Deep blue or purple spots; sensitivity: ergosterol, 1 μg; cholesterol, 10 μg; 5-androstene-3β,17β-diol is negative up to 50 μg
	II. Detection of steroid tosylates: spray with b and c *Remark:* I and II can only be used on cellulose layers	Deep blue or purple spots; sensitivity of steroid tosylates tested: cholesterol 0.1 μg, ergosterol and 5-androstene-3β,17β-diol 0.2 μg
Perchloric acid	Spray with concentrated perchloric acid (70%)	Vitamin D, orange brown[109]
Primuline[110]	a. 0.1% solution of primuline b. Take 1 mℓ a to 100 mℓ with a mixture acetone-water, 4:1 c. Spray plate free of solvent with b and observe at 365 nm	Cholesterol and cholesterol esters; steroids give pale yellow or pale blue spots; sensitivity: 0.5 μg (dehydroepiandrosterone, pregnanedione), 1—2 μg (Δ^4-3-oxosteroids), 2—5

Table 1 (continued)
SPRAY REAGENTS FOR TLC

	Remark: ineffective on cellulose layers	µg (estrogens)
Potassium iodide-starch test[111]	a. Add to 10 mℓ 4% aqueous solution of potassium iodide, 4 mℓ glacial acetic acid, and a little zinc powder b. 1% aqueous solution of starch c. After filtration of a, spray the layer successively with a and b	Sterolperoxides;[85] peroxides set free iodine (blue color)
Resorcinol-sulfuric acid reaction	a. Spray with 20% resorcinol solution in 96% ethanol and heat at 80°C for 5 min b. Spray with 1 *N* sulfuric acid and heat again at 90°C for 10 min *Remark:* can be used on reversed-phase TLC, undecane-, and tetra-de-cane-impregnated layers	Sterol and sterol acetates;[91] spot colors: β-sitosterol, cholesterol (ac), stigmasterol (ac), faint blue; ergosterol (ac), faint green; 7-ergostenol, vitamin D$_2$, lanosterol (ac), brown, yellow brown, or orange brown; 5-dihydroergosterol, purple brown; 7-dehydrocholesterol (ac), purple; lathosterol, white; cholestanol (ac), negative
Sulfuric acid-ceric sulfate[112] reaction	I. a. Spray with a saturated solution of ceric (IV) sulfate in 65% concentrated sulfuric acid b. Heat at 120°C for 15 min; observe colors in day-light and fluorescence in UV light *Remark:* do not use on aluminum oxide or silver nitrate-treated layers II.[113] Use instead, a 1% ceric ammonium sulfate solution in 10% sulfuric acid *Remark:* unsatisfactory on silver nitrate-treated layers	Spot colors for cholesterol, cholestanol, campesterol, stigmasterol, β-sitosterol, 4α-methyl-5α-stigmasta-7,24(28)-dien-3β-ol, cycloartenol, lophenol, parkeol: brown gray to blackish gray, with violet or red violet fluorescence
Thymol-sulfuric acid reaction	a. Spray with 20% solution thymol in 96% ethanol and heat at 80°C for 5 min b. Spray with 1 *N* sulfuric acid solution and heat again at 90°C for 10 min *Remark:* can be used on reversed-phase TLC, undecane-, and tetra-decane-impregnated layers	Sterol and sterol acetates;[91] spot colors: acetates of cholesterol, stigmasterol, and β-sitosterol, violet; ergosterol (ac) and 7-dehydrocholesterol (ac), gray brown; cholestanol (ac), white; lanosterol (ac), orange brown; 7-ergostenol (ac), yellow green; dihydroergosterol, blue; vitamin D$_2$, purple brown
Vanillin-phosphoric acid reagents[98,114]	a. Mixture 20 parts of 84% phosphoric acid and 80 parts of ethanol b. Prepare a 2% solution of vanillin in a c. Spray with b and heat at 120°C for 10—20 min *Remark:* can be used after iron chloride reagents	Cocurbitacins[114] give a violet to amethyst color; fluorescence under UV-ochre (cucurbitacins A,B,D,E,I), blue (cucurbitacin C) or red (cucurbitacins J and K)
Winogradow reaction	I.[115] a. Spray with 90% solution trichloracetic acid in HCl b. Heat at 100°C for 15 min c. Observe colors before, after heating in UV, visible light	C$_{27}$ sterols and trimethylsterols;[115] colors: 7-dehydrocholesterol: immediate-16 greenish gray (after heating, olive gray, which turns to olive brown); in UV, pink; colors developed only after heating: cholesterol,

Table 1 (continued)
SPRAY REAGENTS FOR TLC

		violet → green gray; lanosterol, brownish gray → lilac brown; 3,5-cholestadien-7-one, orange yellow
	II.[91] a. Spray with a 90% solution of trichloracetic acid in acetic acid	Sterols;[91] spot colors: β-sitosterol, gray; 7-ergostenol, yellow-brown; 5-dihydroergosterol, α-spinasterol, and symosterol, gray brown
	b. Heat at 80°C for 15 min	
	Remark: can be used on reversed-phase TLC, undecane-, and tetradecane-impregnated layers	
Zinc chloride	I. a. Spray with 30% methanolic zinc chloride solution and heat at 130°C for 30 min	Detection of sterols on normal and silver nitrate silica gel layers
	b. Examine layer under UV-light (366 nm)	
	Remark: can be oversprayed with sulfuric acid	
	II.[95] a. Spray with 60% solution of zinc chloride in glacial acetic acid (Tschugajaff reaction)	Insect moulting hormones; inokisterone: red;[102] rubrosterone: orange[103]
	b. Heat at 90°C for 10 min	

out a wide number of identifying or locating procedures. These can be performed with the sample prior to separation before or after applying it to the layer. Alternately, the technique can be carried out after chromatography. The reactions are also applicable to alumina layers, but trial and error is necessary for cellulose layers.

In Situ Reactions

Chemical reactions can be carried out directly on the layer after applying the sample. After completion of the reaction, solvent development of the chromatogram is carried out in the usual manner. In order to follow the course of completeness of the procedure it is advisable to apply both the original and the expected products as reference material. This procedure is valuable in identification work since very little material is required. If the reaction cannot be performed in this manner it is advisable to do so in a micro-reaction vial or other suitable vessel.

Listed below are a number of *in situ* procedures which are applicable to steroids. Two or more of these procedures carried out on the same sample (preferably on two different chromatograms) can be invaluable in indicating the identity of an unknown.

Oxidation

Chromic Anhydride

A saturated solution of chromic anhydride in acetic acid is made up. A drop of this solution is applied to the origin of the chromatogram at the point where the sample had been applied. Development is then performed.[7]

Hydrogen Peroxide

A solution of 30% hydrogen peroxide can be used. A drop of this solution is applied to the point of sample application. This is then exposed to UV light for 10 min.[8]

Chromic Acid

A 10% solution of chromic acid in acetic acid is made up. A drop of this is applied to the origin of the chromatogram. This will oxidize the steroid side chain.[9]

Periodate Oxidation

A 10% solution of sodium periodate is made up freshly. This is applied to the chromatogram at the starting line. Spraying is carried out twice. The plate is then heated to 50°C before development. The 17-hydroxycorticosteroids are converted to 17-ketosteroids.[1]

Reduction Reactions

Lithium Aluminum Hydride

The solution is made up as 10% lithium aluminum hydride in ether. The plate should be dry before a drop of the solution is applied to the origin of the chromatogram.[11]

Aluminum Isopropoxide

The solution is 5 g aluminum isopropoxide in benzene. The reaction is preferable carried out on the sample before application to the layer. A drop of the solution is added to the sample in a small test tube. This is heated until condensate appears on the walls of the tube. Reacti-vials are ideal for this since the vial can be heated in a heating block.[11]

Sodium Borohydride

Two solutions are required: a 10% solution of sodium borohydride in ethanol and a 0.1% sodium hydroxide solution. These are mixed in equal proportions just before use. After reaction in a small tube for 30 min the excess reagent is neutralized with 25% acid. This procedure has also been carried out directly on the thin layer.[10] Another method is the use of a 1% solution of sodium borohydride in methanol. This has been found to reduce 7-ketocholesterol to the alcohol. It has also been successful for alkaloids and sterol hydroperoxides.[12]

Ferrous Ammonium Sulfate

The reagent is made as 5% aqueous ferrous ammonium sulfate-methanol ether (2:1:1). This is applied to the sample on the origin of the chromatogram.

Hydrolysis Reactions

Acid

Expose the layer to concentrated HCl acid in a tank heated to 100°C to generate the fumes.[13] This has been used for hydrolysis of saponin and glycosides.[14] Steroid sulfates are hydrolyzed by exposing the thin layer to the atmosphere of HCl-dioxane (9:1) for 3 hr in a tank.[15] A solution of 0.1 mℓ concentrated sulfuric acid in 9.9 mℓ dioxane or acetone can also be used.[16] For hydrolysis of trimethylsilyl ethers of steroids a solution of concentrated sulfuric acid in methanol (1:1) can be sprayed on the layer.[11]

Alkali

A solution of 0.1 *N* sodium hydroxide in ethanol can be used to saponify alkaloids before chromatography by heating the sample in a small tube with a small amount of the reagent at 100°C for 1 hr or by spraying on the chromatogram.[9]

Halogenation

Bromination

A solution of 0.1% (w/v) of bromine in chloroform is used. This is spotted on top of the sample or the layer. It is used to affect separation of cholesterol from cholestanol.[18] Bromine can also be added to the mobile phase. For example, 0.5% halogen was added to acetic acid-acetonitrite (1:1) used as mobile phase for separation of cholesterol from brassicasterol.[19]

Iodination

The thin layer after application of the sample is exposed to iodine vapors in a tank. This is followed by evaporation of the excess iodine and development of the chromatogram. It is used in study of alkaloids.[20] This reaction has been used for characterization of cholesterol.[21] It has also been used for study of estrogens.[22] For quantitative work care must be taken to study completeness of the reaction.

Esterification
Trifluoroacetic Anhydride
Apply trifluoroacetic anhydride directly to the sample on the layer. Dry the plate before and after the application of the reagent. Water in the layer or the sample prevents completeness of the reaction. The reaction is usually completed at room temperature.[23] If esterification before application to the layer is desirable, a few drops of the reagent can be added to the sample dissolved in benzene. A few drops of pyridine are added and the reaction mixture heated in a sealed tube at 10°C for 1 hr. This is also applicable for preparation of heptafluorobutyrates.[24] The excess reagent is neutralized by addition of 2 N sodium carbonate. Many steroids with free hydroxyl groups have been esterified by this procedure. Acetic anhydride may also be used.

Acetic Anhydride or Acetyl Chloride
Acetic anhydride[9] or acetyl chloride[25] is applied directly to the sample on the layer. Excess reagent can be removed by blowing hot air from a drier across the layer, which must be dry before applying the reagent and before subsequent development of the chromatogram.

Hydrazones
Girard's Reagent
The solution consists of 0.1% of trimethylaceto hydrazide ammonium chloride in 10% (V/V) glacial acetic acid in methanol. This is applied to the starting point on the layer.[25] Steroidal ketones are then applied. Allow to stand in a tank for 15 hr with acetic acid vapor. Heat the plate at 80°C for 10 min before developing the chromatogram.

Nitroso Derivatization[26]
After applying the sample, expose the layer to ammonia vapors and then to nitrogen dioxide, which is prepared in the tank from metallic copper and concentrated nitric acid; do this in a hood. (Many nitroso compounds are suspected to be carcinogenic.)

Spraying Reagents for Detection and Identification
Universal Detection Reagents
A description of various charring techniques follows. Concentrated sulfuric acid can be sprayed on the chromatogram. Some steroids will react in the cold, others will show up only after heating.[27] The addition of 5% nitric acid to the sulfuric acid will aid in the visualization of steroids difficult to char. Sulfuric acid-water (1:1) can also be sprayed followed by heating of the chromatogram.[28] Table 1 shows some results using this reagent. Sulfuric acid, 2% in ethanol-water (1:1), followed by heating is often applied.[28]

Butanol containing 15% concentrated sulfuric acid has been sprayed on chromatograms for detection of bile acids. After spraying the plates are heated at 110°C for 25 to 30 min for conjugated acids or 45 to 58 min for the free acids.[29] Sulfuric acid-acetic anhydride mixtures can be used. These are 5:95 or 1:4 mixtures which are sprayed on the chromatogram, followed by heating. A modification of this is to spray first with acetic anhydride. After drying, the chromatogram is then sprayed with concentrated sulfuric acid.[30]

Sulfuric acid saturated with potassium dichromate,[31] potassium dichromate saturated in 80% sulfuric acid,[32] 3 g sodium dichromate in 20 mℓ water, and 10 mℓ sulfuric acid[33] have all been used for charring and work well with all steroids. Drying must be done before heating of the sprayed chromatogram. Heating to 220°C may be necessary. A solution of 10% sulfuric acid in ethanol has been used to induce fluorescence in the bile acids after heating at 170°C for 5 min.[34]

Fuming sulfuric acid can be used when other reagents are not aggressive enough for detection. This must be done in a closed chamber which can be heated. Be sure to carry out this operation in a hood.[35]

Ammonium sulfate or ammonium bisulfate (20% aqueous) has been sprayed on layers. Heating at 130°C for 60 min is required, or heating at a higher temperature for varying times may be necessary.[35] Ammonium sulfate or ammonium bisulfate has been incorporated into the layers. The plate can be dipped in an alcoholic saturated solution of ammonium bisulfate; after drying the chromatograms can be spotted in the usual manner. Development must be carried out with mobile phases of such solvent strength that the sulfate is not washed from the layer.[36]

Orthophosphoric acid diluted 1:1 with either water or methanol can be sprayed on the chromatogram. After drying and heating at 120°C for 15 min the different steroids show a wide range of color and fluorescence.

Cupric sulfate has been incorporated into the layer as a charring agent. A concentration of 1 to 5% of the silica gel can be used. Heating generates sulfuric acid. The plates must be dry before the charring is carried out.[37]

Sulfuryl chloride or sulfur trioxide vapors can be fumed onto the chromatograms while the plate is being treated. The plate can also be exposed to steam to generate sulfuric acid *in situ*. This provides a means for more uniform dispersion of the reagent in the layer and also avoids problems of the uneven background resulting from difficulty in spraying reagents.[38]

A charring reagent very useful for quantitation *in situ* by densitometry has been the reagent consisting of 3% cupric acetate (x/v) in 8% phosphoric acid (w/v). After spraying or dipping the chromatogram is heated at 130°C for 30 min.[39] In some cases the heating can be as short as 10 min. Cholesterol and cholesterol ester have been determined on high performance thin layers using this technique with a detection limit of 5 ng. The advantage here is the low background as compared to sprays containing sulfuric acid which usually give a brown to dark background depending on how the layer was heated. This reagent can also be used with layer formulations containing organic binder. Reagents containing sulfuric acid usually give untenable background (in relation to densitometry) because of the high background when the polymer binder is charred.

Antimony trichloride and pentachloride as a 10 to 20% solution in carbon tetrachloride solution have been used as spraying reagents for steroids. Antimony trichloride as a saturated solution in alcohol-free chloroform can be used. After spraying the dried chromatogram, it is heated at 100 to 120°C for 5 to 10 min. Another modification is the use of saturated antimony trichloride in 20% acetic anhydride in alcohol-free chloroform. This reagent reacts not only with steroids but with many other compounds.[40] A saturated solution of antimony trichloride in alcohol-free chloroform containing 10% thionyl chloride is sprayed and the chromatogram heated at 110 to 120°C to detect steroids with a Δ^4 double bond.

Phosphomolybdic acid as a 5% solution in alcohol can be sprayed on the chromatogram, which is then heated at 120°C until the blue color appears. Exposure of the developed chromatogram to ammonia vapors bleaches the light blue background and leaves the reacted zones as dark blue areas.[41]

This reagent can also be incorporated into the layer. The plate is dipped in the solution and allowed to stand for 30 sec. It is then dried at room temperature before sample application. Development is carried out in the usual manner. Then the chromatogram is heated at 120°C until the zones show color. All classes of steroids and estrogens react.

Silver nitrate-impregnated sorbents are useful in the TLC of sterols and their acetates. The layers are dipped in a mixture of 0.25 g silver nitrate, 2 mℓ water, and 25 mℓ phenoxyethanol, which is increased to 500 mℓ with acetone. The plate is allowed to stand in the dipping solution for 30 sec. The plates should be used the same day and kept in the dark before use. After development, the zones are visible after exposure to UV light in a dark room.[42] These plates are an aid to separating sterols and their acetates according to the degree of unsaturation.[43]

Other acidic charring reagents for steroids that can provide useful information are as follows:

- *p-Toluenesulfonic acid* as a 20% solution in chloroform[45] or ethanol; after spraying the plate is heated at 120°C for 15 min for the first case and at 90°C for 2 min in the second case.[46]
- *Perchloric acid* as a 20% aqueous solution can be sprayed, followed by heating of the chromatogram at 150°C for 10 min.[47]
- *Trichloroacetic acid,* 25% in chloroform, which is heated at 120°C for 5 min after spraying the dried chromatogram.[48]
- *Concentrated HCl* has been sprayed on chromatograms. Development of colored zones follows heating at varying temperatures for different times.[49] Exposure to HCl vapor in a tank at 4°C gives pink spots with Δ^4-3-hydroxy steroids while a temperature of 23 to 40°C is necessary to show the Δ^5-3-hydroxy steroid.[50]
- The *manganous chloride reagent* made up by dissolving 50 mg manganous chloride ($MgCl_2 \cdot 4H_2O$) in 15 mℓ water and 0.5 mℓ sulfuric acid can be sprayed on chromatograms. After heating at 100 to 110°C for 10 to 15 min, different colors develop. Cholesterol gives a pink color which fades rapidly. Bile acids give different colors.[51]
- *Potassium permanganate* can be used in a spray of acidic or alkaline solutions. 0.1 Potassium permanganate and 2 *N* acetic acid are mixed (1:1) before use,[52] or 2% potassium permanganate and 4% sodium bicarbonate are mixed (1:1) before use. $KMnO_4$ can be used as a 0.25 to 2% solution in water as a general reagent.[53]

Specific Spray Reagents

There are a number of spray reagents that have a degree of specificity for detection of functional groups. Judicious use of these reagents along with chromatographic mobility can lead to provisional identification of the isolated steroid. If the prerequisite reference material is available, these reagents are listed according to the functional group required for the reaction.

Steroidal Skeleton

An aqueous solution of sodium hydroxide can be used as a specific spray reagent for Δ^4-3-ketosteroids. This can be used both on paper and thin layer chromatograms. It is important to dry the chromatogram very carefully at 80 to 90°C for 3 min in order to see the fluorescence, which is activated at 365 nm.[54] Acetic anhydride and concentrated sulfuric acid as a spray have been used to detect unsaturation in sterols (Liebermann-Buchard reagent). No heating is required. The chromatogram is viewed under UV light. Many of the reagents in the section on charring reagents will be suitable for use in this sense.

Osmium tetroxide vapors fumed on the thin layer for 5 to 10 min will provide dark spots for isolated double bonds. For conjugated double bonds 1 hr exposure in the sealed chamber is required.[55]

p-Anisaldehyde-sulfuric acid mixtures are useful for detection of different structures when sprayed on the chromatogram. Both 1 mℓ concentrated sulfuric acid, 100 mℓ acetic acid, 0.1 mℓ anisaldehyde, and 1 mℓ concentrated sulfuric acid, 50 mℓ glacial acetic acid, 0.5 mℓ anisaldehyde have been used. Heating at 110 to 120°C for 5 to 10 min is required.[56]

Table 1 gives some of the characteristic results using this reagent (reagent reacts well with the corticosteroids). View both under visible and UV light. Characteristic colors will be seen. This reagent also reacts with the bile acids.[57]

Cinnamaldehyde is dissolved in alcohol to make a 1% solution. Spray the plate with this solution and heat at 90°C for 5 min. Then spray with a solution of sulfuric acid in acetic anhydride (12% H_2SO_4 in acetic anhydride); heat further at 90°C until color development takes place.[58]

Prepare p-hydroxybenzaldehyde as a 2% solution in methanol. Mix 10 vol of this with 1 vol 50% sulfuric acid before use. Heat for a few minutes at 105°C in an oven. This reacts well with 3-ketosteroids.[58]

Resorcinaldehyde spray is made before use by mixing equal volumes of 0.5% resorcin-aldehyde in acetic acid and 5% sulfuric acid in acetic acid. After spraying, the chromatogram is heated at 110°C until color develops.[59]

Ammonium vanadate, 1.62 g, is dissolved in 125 mℓ concentrated sulfuric acid; cool and add to 125 mℓ ice water. Just before use dilute this solution 1:10 with water. After spraying, the chromatogram is heated at 110°C for 3 min.[54]

For the Zimmerman reagent to detect 17-ketosteroids make up two solutions: (1) 1% m-dinitrobenzene in ethanol and (2) 5 N potassium hydroxide. Mix two parts of 1 with one part of 2. After spraying dry in a current of hot air, this gives a violet color. It is not specific for 17-ketosteroids since it will also react with some 3-ketosteroids, 6-keto, 16-keto, and 20-ketones. This reaction can also be used for detection of corticosteroids after oxidation of the side chain with chromic acid.[55]

2,4-Dinitrophenylhydrazine, 0.4 g dissolved in 100 mℓ 2 N HCl forms a spray useful for detection of ketosteroids. The corticosteroids also react. Little heating is required. The detectability is dependent on the nature and location of the keto group. In some cases the reaction is completed by heating at 50 to 60°C for a few minutes. To decrease the background to enable optimal results by scanning densitometry for quantitation the plate can be sprayed with a dilute solution of permanganate.[53]

Isonicotinic acid hydrazide as a 1% solution in ethanol containing 1% acetic acid is a compound that requires no heating. A variation of this is a solution of 2 g t-hydrazide in 500 mℓ water containing 2.5 mℓ concentrated HCl. Reaction times at room temperature vary with the nature of the steroid.[64]

p-Phenylenediamine in a 0.9% butanolic solution containing 1.5% phthalic acid can also be used for detection of unsaturated ketosteroids. Yellow spots appear after heating at 100°C for 2 min.[65]

Naphthoquinine-perchloric acid spray solutions require 0.1% 1,2-naphthoquinone-2-sulfuric in water, 60% perchloric, 40% formaldehyde, and water. Mix these 2:1:0.1:0.9 before spraying. After spraying heat at 70 to 80°C until colors form. Heating too long causes charring. Some steroids can be detected in levels as low as 0.03 µg.[66]

For the Kedde reagent two solutions are required:[67] 2% dinitrobenzene in methanol and 2 N potassium hydroxide. Mix 1:1 just before use; no heating is required. Glycosides appear as purple spots. Steroidal glycosides react well.

The diazo salts such as diazotized sulfanilic acid and the fast violets can be used to detect estrogens on thin layers as well as paper chromatograms. These reagents are pH dependent and have low sensitivities. Examples are 0.5% solution of Fast Blue BB in water and 0.1 N sodium hydroxide. Spray first with the Fast Blue BB followed by the sodium hydroxide.[68] Using the same procedure Fast Red B can be sprayed on the chromatograms.[69] A solution of 25 mg of methylene blue dissolved in 100 mℓ 0.05 N sulfuric acid is useful as a spray for detection of steroidal sulfate. Mix this solution with an equal volume of acetone just before spraying.[70]

For the tetrazoliumchloride solutions which react with many steroids, the following are recommended: (1) dissolve 1 g 2,5-diphenyl-3-(4-styrylphenyl) tetrazoliumchloride in 100 mℓ methanol, (2) prepare 3% sodium hydroxide. Just before use, mix 1 vol (1) with 10 vols (2) and then spray. This gives a purple spot on a yellow background.[71]

A solution of 0.5% tetrazolium blue in 2.5 *N* sodium hydroxide is used for detection of reducing steroids by spraying on the chromatograms. It reacts well with the side chain of α-ketolic steroids.[72] A modification of this is the use of 4% 2,3,5-triphenyltetrazolium chloride and 1 *N* sodium hydroxide mixed (1:1) before use. After spraying of thin layers or dipping of paper chromatograms, warming at 110°C for a few minutes brings out the purple color.[73]

For the ferric chloride-potassium ferricyanide spray useful for detection of phenolic steroids, mix 0.1 *M* solution of each in 1:1 ratio just before use. This sprayed on thin layers (or paper chromatograms dipped in the solution) gives a blue color on a light blue background. Dipping a paper chromatogram in dilute water will remove the background.

Potassium iodoplatinate is made up as 2 mℓ 10% platinum chloride and 25 mℓ 8% potassium iodide diluted to 100 mℓ with water. Spraying on the chromatograms results in the appearance of brown spots.[74]

A solution of 0.05% pyrene in petroleum ether spray on silica gel layers is useful for detection of bile acids. This is useful in the sense that the pyrene can be washed out of the chromatogram by developing it with ether-petroleum ether (1:1).[75]

A spray of 5 g uranyl nitrate in 95 mℓ 10% (v/v) sulfuric acid can be used. After heating at 110°C for a few minutes, a variety of colors are observed for different steroids.[76]

Table 1 alphabetically lists a number of reagents which have been used to spray on thin layer chromatograms for the detection of steroids. The colors or results to be expected are also listed. By use of a combination of these reagents some indication of the nature of a separated material can be obtained.

REFERENCES

1. **Tate, M. E. and Bishop, C. T.,** *Can. J. Chem.,* 40, 1043, 1962.
2. **Anker, L. and Sonanini, D.,** *Pharm. Acta Helv.,* 37, 360, 1962.
3. **Miller, J. M. and Kirchner, J. G.,** *Anal. Chem.,* 26, 2002, 1954.
4. **Scora, R. W.,** *J. Chromatogr.,* 13, 251, 1964.
5. **Machata, G.,** *Mikrochim. Acta,* 9, 1960.
6. **Mangold, H. K.,** *J. Am. Oil Chem. Soc.,* 38, 708, 1961.
7. **Miller, J. M. and Mirschner, J. G.,** *Anal. Chem.,* 25, 1107, 1953.
8. **Kofoed, J. C., Fabrierkiewicz, C., and Lucas, G. H. W.,** *J. Chromatogr.,* 23, 410, 1966.
9. **Kaess, M. and Mathis, C.,** *4th International Symposium on Electrophoresis and Chromatography,* Ann Arbor Science, Ann Arbor, Mich., 1968, 525.
10. **Hamman, B. L. and Martin, M. M.,** *Steroids,* 10, 169, 1967.
11. **Kirschner, J. G.,** in *Thin Layer Chromatography,* John Wiley & Sons, Kirschner, J. G., Ed., New York, 1978, 177.
12. **Smith, L. L. and Price, J. C.,** *J. Chromatogr.,* 26, 509, 1967.
13. **Baggiolini, M. and Dewalt, B.,** *J. Chromatogr.,* 30, 259, 1967.
14. **Karlnig, T. and Wegschaider, O.,** *Planta Med.,* 21, 144, 1972.
15. **Payne, A. H. and Mason, M.,** *Anal. Biochem.,* 21, 463, 1968.
16. **Hurwitz, A. R.,** *Anal. Biochem.,* 46, 338, 1972.
17. **Curtins, H. C. and Mueller, M.,** *J. Chromatogr.,* 33, 222, 1968.
18. **Cargiel, D. J.,** *Analyst,* 87, 865, 1962.
19. **Copins Peereboom, J. W. and Beekes, H. W.,** *J. Chromatogr.,* 9, 316, 1962.

20. **Wilk, M. and Brill, U.,** *Arch. Pharmacol.*, 301, 282, 1968.
21. **Wilk, M. Taupp, W.,** *Z. Naturforsch. Teil B.* 24, 16, 1960.
22. **Brown, M. and Turner, A. B.,** *J. Chromatogr.*, 26, 518, 1967.
23. **Bennett, R. D. and Heftmann, E.,** *J. Chromatogr.*, 9, 353, 1962.
24. **Touchstone, J. C. and Dobbins, M. F.,** *J. Steroid Biochem.*, 6, 1389, 1975.
25. **Lisboa, B. P.,** *J. Chromatogr.*, 24, 475, 1966.
26. **Lisboa, B. P. and Diczfalusy, E.,** *Acta Endocrinol.*, 40, 60, 1962.
27. **Kirschner, J. G., Miller, J. M., and Kelly, G. J.,** *Anal. Chem.*, 23, 420, 1951.
28. **Privett, O. S. and Blank, M. C.,** *J. Lipid Res.*, 2, 37, 1961.
29. **Anthony, W. L. and Beher, W. T.,** *J. Chromatogr.*, 13, 567, 1964.
30. **Smith, L. L. and Foell, T.,** *J. Chromatogr.*, 9, 339, 1962.
31. **Ehrhardt, E. and Cramer, F.,** *J. Chromatogr.*, 7, 405, 1962.
32. **Privett, O. S. and Blank, M. L.,** *J. Am. Oil Chem. Soc.*, 39, 520, 1962.
33. **Ertel, H. and Horna, L.,** *J. Chromatogr.*, 7, 268, 1972.
34. **Touchstone, J. G., Dobbins, M. F., Levin, S. S., and Okamoto, B. K.,** *Fed. Proc. Fed. Am. Soc. Exp. Biol.*, 37, 1447, 1978.
35. **Martin, T. T. and Allen, M. C.,** *J. Am. Oil Chem. Soc.*, 48, 752, 1971.
36. **Touchstone, J. C. and Murawec, T.,** in *Quantitative Thin Layer Chromatography,* Touchstone, J. C., Ed., John Wiley & Sons, New York, 1973, 145.
37. **Korolczuk, J. and Kwasniewska, I.,** *J. Chromatogr.*, 88, 428, 1974.
38. **Jones, D., Bowyer, D. E., Gresham, G. A., and Howard, A. N.,** *J. Chromatogr.*, 24, 228, 1966.
39. **Touchstone, J. C., Dobbins, M. F., Hirsch, C. Z., Baldino, A. R., and Kritchevsky, D.,** *Clin. Chem.*, 24, 1496, 1978.
40. **Gaenshirt, H. and Malzacher, A.,** *Arch. Pharmacol.*, 293, 925, 1960.
41. **Wortmann, W. and Touchstone, J. C.,** in *Quantitative Thin Layer Chromatography,* Touchstone, J. C., Ed., John Wiley & Sons, New York, 1973, 24.
42. **Dobbins, M. F. and Touchstone, J. C.,** in *Quantitative Thin Layer Chromatography,* Touchstone, J. C., Ed., John Wiley & Sons, New York, 1973, 293.
43. **Vroman, H. E. and Cohen, C. F.,** *J. Lipid Res.*, 8, 150, 1967.
44. **Lisboa, B. P.,** *J. Chromatogr.*, 19, 333, 1965.
45. **Roux, D. G.,** *Nature (London),* 180, 973, 1957.
46. **Sonanini, O. and Anker, L.,** *Pharm. Acta Helv.*, 42, 54, 1967.
47. **Metz, H.,** *Naturwissenschaften,* 48, 569, 1961.
48. **Aldrich, B. J., Frith, M. L., and Wright, S. E.,** *J. Pharm. Pharmacol.*, 8, 1042, 1956.
49. **Lisboa, B. P.,** *Meth. Enzymol.*, 15, 3, 1969.
50. **Jeffery, J.,** *J. Chromatogr.*, 67, 188, 1972.
51. **Goswami, S. K. and Frey, L. F.,** *J. Chromatogr.*, 53, 389, 1970.
52. **Machata, G.,** *Mikrochim. Acta,* 79, 1960.
53. **Lisboa, B. P.,** *J. Chromatogr.*, 16, 136, 1964.
54. **Bush, I. E.,** *Biochem. J.,* 50, 370, 1952.
55. **Lisboa, B. P.,** *J. Chromatogr.*, 13, 391, 1963.
56. **Johannesen, B. and Sandel, A.,** *Medd. Nor. Farm. Selsk.*, 23, 105, 1961.
57. **Kritchevsky, D., Martak, D. S., and Rothbart, G. H.,** *Anal. Biochem.*, 5, 388, 1963.
58. **Stevens, P. J.,** *J. Chromatogr.*, 14, 269, 1964.
59. **Gower, D. B.,** *J. Chromatogr.*, 14, 424, 1964.
60. **Malaiyandi, M., Barrette, J. P., and Lanquette, M.,** *J. Chromatogr.*, 101, 155, 1974.
61. **Lisboa, B. P.,** *J. Accomatogr.*, 13, 391, 1964.
62. **Kochakian, C. D. and Stidworthy, G.,** *J. Biol. Chem.*, 199, 607, 1952.
63. **Lisboa, B. P.,** *Acta Endocrinol.*, 43, 47, 1963.
64. **Lisboa, B. P.,** *J. Chromatogr.*, 16, 136, 1964.
65. **Richter, E.,** *J. Chromatogr.*, 18, 164, 1965.
66. **Goerlich, B.,** *Planta Med.*, 9, 442, 1961.
67. **Hoerhammer, L., Wagner, H., and Bittner, G.,** *Pharm. Ztg. Ver. Apoth. Ztg.*, 108, 259, 1963.
68. **Pastuska, G.,** *Z. Anal. Chem.*, 179, 355, 1961.
69. **Crepy, O., Judas, O., and Lacher, B.,** *J. Chromatogr.*, 16, 340, 1964.
70. **Stevens, P. J.,** *J. Chromatogr.*, 14, 269, 1964.
71. **Adamec, O., Matis, J., and Galvanek, M.,** *Steroids,* 10, 495, 1963.
72. **Vaedtke, J. and Gajewska, A.,** *Steroids,* 9, 345, 1962.
73. **Nuernberg, E.,** *Arch. Pharmacol.*, 292, 610, 1959.
74. **Eastwood, M. A. and Hamilton, D.,** *Biochem. J.,* 105, 376, 1967.
75. **Gower, D. B.,** *J. Chromatogr.*, 14, 424, 1964.

76. **Paik, N. H. and Kim, B. K.,** *Yakkak Hoejl,* 13, 84, 1969; *Chem. Abstr.,* 73, 1812K, 1970.
77. **Davidek, J. and Blattna, J.,** *J. Chromatogr.,* 7, 204, 1962.
78. **Tschesche, R. and Snatzke, G.,** *Ann. Chem.,* 636, 105, 1960.
79. **Bush, I.,** *Chromatography of Steroids,* Pergamon Press, New York, 1961.
80. **Singh, E. J. and Gershbein, L. L.,** *J. Chem. Ed.,* 43, 29, 1966.
81. **Monder, C.,** *Biochem. J.,* 90, 522, 1964.
82. **Claude, J. R. and Beaumont, J. L.,** *J. Chromatogr.,* 21, 189, 1966.
83. **Copius-Peereboom, J. W. and Beekes, H. W.,** *J. Chromatogr.,* 17, 99, 1965.
84. **Morris, L. J.,** *J. Lipid Res.,* 4, 357, 1963.
85. **Wolfman, L. and Sachs, B. A.,** *J. Lipid Res.,* 5, 127, 1964.
86. **Copius-Peereboom, J. W. and Beekes, H. W.,** *J. Chromatogr.,* 9, 316, 1962.
87. **Samuel, P., Urivetzky, M., and Kaley, G.,** *J. Chromatogr.,* 14, 508, 1964.
88. **Avigan, J., Goodman, D. S., and Steinberg, D.,** *J. Lipid Res.,* 4, 100, 1963.
89. **Tschesche, R., Lampert, F., and Snatzke, G.,** *J. Chromatogr.,* 5, 217, 1961.
90. **Danielsson, H. and Einarsson, K.,** *J. Biol. Chem.,* 241, 1449, 1966.
91. **Honegger, C. G.,** *Helv. Chim. Acta,* 46, 1730, 1963.
92. **Furst, W.,** *Arch. Pharmacol.,* 298, 795, 1965.
93. **Haaki, E. and Mikkari, T.,** *Acta Chem. Scand.,* 17, 536, 1963.
94. **Furst, W.,** *Arch. Pharmacol.,* 300, 144, 1967.
95. **Eng, L. F., Lee, Y. L., Hayman, R. B., and Gerstl, B.,** *J. Lipid Res.,* 5, 128, 1964.
96. **Ikan, R. and Cudzinovski, M.,** *J. Chromatogr.,* 18, 422, 1965.
97. **Cerny, V., Joska, J., and Labler, L.,** *Coll. Czech. Chem. Commun.,* 26, 1658, 1961.
98. **Horvath, C.,** *J. Chromatogr.,* 22, 52, 1966.
99. **Shepherd, I. S., Ross, L. F., and Morton, I. D.,** *Chem. Ind. (London),* p. 1706, 1966.
100. **Cerny, V., Joska, J., and Labler, L.,** *Coll. Czech. Chem. Commun.,* 26, 1658, 1961.
101. **Cargill, D. I.,** *Analyst,* 87, 865, 1962.
102. **Eneroth, P.,** *Lipid Chromatograph Analysis,* Vol. 2, Marinetti, G. V., Ed., Marcel Dekker, New York, 1969, 149.
103. **Kuksis, A. and Beveridge, J. M. R.,** *Can. J. Biochem. Physiol.,* 38, 95, 1960.
104. **Barbier, M., Jager, H., Tobias, H., and Wyss, E.,** *Helv. Chim. Acta,* 42, 2440, 1959.
105. **Nambara, T., Imai, R., and Sakurai, S.,** *Yakugaku Zasshi,* 84, 680, 1964.
106. **Usekum, E., Jitajem, N., Weiss, E., and Reichstein, T.,** *Helv. Chim. Acta,* 48, 1093, 1965.
107. **Acker, L. and Greve, H.,** *Fette Seifen Anstrichm.,* 65, 1009, 1963.
108. **Anderson, C. A.,** *Aust. J. Chem.,* 17, 949, 1964.
109. **Neher, R. and Wettstein, A.,** *Helv. Chim. Acta,* 43, 1628, 1960.
110. **Dean, P. D. G. and Whitehouse, M. W.,** *Biochem. J.,* 98, 410, 1966.
111. **Lees, T. M., Lynch, M. J., and Mosher, F. R.,** *J. Chromatogr.,* 18, 595, 1965.
112. **Michalec, C., Sulc, M., and Mestan, J.,** *Nature (London),* 193, 63, 1962.
113. **Pinter, K. G., Hamilton, J. G., and Muldrey, J. E.,** *J. Lipid Res.,* 5, 273, 1964.
114. **Adamec, O., Matis, J., and Galvanek, M.,** *Lancet,* 81, 1962.
115. **Lisboa, B. P.,** Principe et applictaion de la chromatographie sur couche mince aux steroides, in *Chromatographie Symposium III,* Societe Belge des Sciences Pharmaceutiques, Brussels, 1964. 33.

Chapter 6

ANDROGENS

SECTION I. PREPARATIONS

Androgens (C_{19}) (see Table 1) are steroids usually designated as testicular products. They are, however, also produced in the adrenals and ovaries and may be the most abundant of the neutral steroids. Because of their fewer functional groups and resulting lower polarity than the corticoids, these compounds are somewhat more difficult to separate. It is not the purpose of this handbook to advocate one chromatographic mode over the others. Rather, it is more appropriate to describe methodology for each chromatographic discipline since there are many cases where one may be more suitable for the needs of the reader than the others. A complete description of each of the available methodology gives the answers to particular problems showing the reasons for using this book. In many cases, it is not necessary to use a combination of techniques to solve the problem at hand.

In many instances one must look for a large number of steroids in a few samples, thus the need for awareness of the available methods. In the case of the androgens (being relatively stable), gas chromatography (GC) has been quite successful without derivatization. Since many androgen metabolites are not UV-absorbing, it may be preferable to perform TLC or GC rather than high performance liquid chromatography (HPLC).

Samples from metabolic studies sometimes include radioactive substances from the labeled precursors as well as the endogenous steroids. GC in tandem with mass spectrometry (MS) for identification has become a common method. Detection limits in GC using flame ionization are not dependent on structure as is the use of the electron capture detector. TLC, as seen in Chapter 4, provides unlimited possibilities for detection as well as quantitation because of many reactions that are available both to detect as well as to change chromatographic characteristics.

Since any chromatographic method is limited somewhat by the ability to prepare suitable samples, Section I of each chapter in the remainder of the volume will be concerned with sample preparations for specific examples. The general procedures in Chapter 2 may also be useful. Sample preparation is the crucial part of the analysis and should, therefore, be performed with maximum care. All reagents used should be analytical grade, reagent grade, or better. A critical suitability test for every reagent is recommended because of possible interference by impurities. Emphasis in any metabolism study concerning steroids must be directed to hydrolysis of the conjugates.

Extraction Methods

Tissues and Bile

A general method for preparation of extracts of tissue follows that of Verbeke,[1] as diagrammed in Figure 1.

Extraction of hormones from tissue — 50 mg of minced tissue are homogenized in 50 mℓ sodium acetate buffer (0.04 M, pH 5.2) using an Ultraturrax® (3 × 20 sec). Fat tissues were also homogenized in 80 mℓ sodium acetate buffer (0.04 M, pH 5.2). After addition of 0.5 mℓ glucuronidase-sulfatase, hormone conjugates were hydrolyzed overnight at 37°C. The incubation mixture was then homogenized in 180 mℓ methanol and was spun at 10,000 g for 10 min using stainless steel containers. The supernatant was decanted and the fat removed by extracting the supernatant twice with 50 mℓ n-hexane. The hexane layers were discarded. The methanol phase was extracted successively with 150, 90, and 90 mℓ dichloromethane. The dichloromethane phases were combined in a conical flask and evaporated

Table 1
COMMON ANDROGENS AND THEIR TRIVIAL NAMES

Androstanedione	5α-Androstane-3,17-dione
Etiocholanedione	5β-Androstane-3,17-dione
Androstenedione	4-Androstene-3,17-dione
Epiethiocholanolone	3β-Hydroxy-5β-androstan-17-one
Dehydroepiandrosterone	3β-Hydroxy-5-androsten-17-one
Epiandrosterone	3β-Hydroxy-5β-androstan-17-one
Androsterone	3β-Hydroxy-5α-androstan-17-one
Androstadienedione	1,4-Androstadiene-3,17-dione
Testosterone	17β-Hydroxy-4-androsten-3-one
Etiocholanolone	3α-Hydroxy-5β-androstan-17-one
Epitestosterone	17α-Hydroxy-4-androsten-3-one
1-Dehydrotestosterone	17β-Hydroxy-1,4-androstadien-3-one
6-Dehydrotestosterone	17β-Hydroxy-4,6-androstadien-3-one
5-Dihydrotesterone	17β-Hydroxy-5α-androstan-3-one
Ethisterone	17α-Ethynyl-17β-hydroxy-4-androsten-3-one
Fluoroxymesterone	9-Fluoro-$11\beta,17\beta$-dihydroxy-17α-methyl-4-androsten-3-one
Methandriol	17α-methyl-5-androstene-$3\beta,17\beta$-diol
17α-Methyltestosterone	17α-methyl-17β-hydroxy-4-androsten-3-one
4,9,11-Methyltestosterone	17α-methyl-17β-hydroxy-4,9,(11)androstadien-3-one
19-Norethisterone	19-nor-17α-ethyl-17β-hydroxy-4-androsten-3-one

Meat, fat, kidney, or liver (50 g)
↓
Enzymic hydrolysis of homogenate
↓
Methanol
↓_____ → **Hexane (fat)**
Dichloromethane
↓
Polystyrene column
↓   ↓
Basic celite + Al₂O₃ column
↓   ↓
Estrogens   **Androgens + Gestagens**
↓   ↓
2-dimensional   **2-dimensional**
HPTLC   **HPTLC**
+   +
Fluorescence   **Fluorescence**

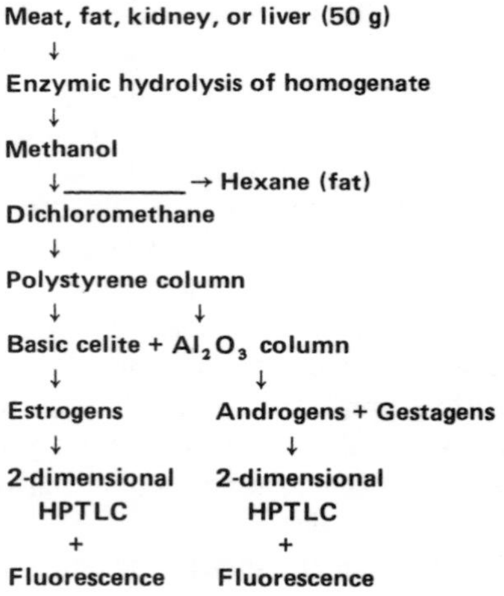

FIGURE 1. Schematic diagram for anabolic steroid detection in animal tissue.

to dryness on a rotary evaporator at 40°C. The residue was then dissolved in 20 mℓ water and percolated through an Amberlite® XAD-2 column (80 × 20 mm). The evaporation flask was rinsed with 2 × 20 mℓ distilled water. The flow rate was 5 mℓ/min for this column. The steroid was eluted by passing 40 mℓ ethanol through the column. The ethanol eluate was evaporated to dryness at 40°C and reconstituted for chromatography.

Extraction of steroids from feces (method of Erikksson[2]) — Samples were extracted twice with chloroform-methanol (1:1 v/v) and once with 80% (v/v) ethanol in water, 0.2 M with respect to ammonium carbonate. The extracts were combined and an aliquot was removed for radioactivity analysis. After evaporation of the solvents, the residue was dissolved in 70% (v/v) ethanol and passed through an Amberlyst-15 column in the sodium form. The eluate was evaporated and the dry residue was dissolved in chloroform-methanol (1:1 v/v), 0.01 *M* with respect to NaCl, and was separated in an 8 g Sephadex® LH-20 column prepared in and eluted with the same solvents. Three fractions were collected: 15 to 35 mℓ effluent (unconjugated and glucuronide fractions), 46 to 80 mℓ (monosulfate fraction), and a fraction obtained by elution of the column with 100 mℓ methanol (diconjugate fraction).

The fraction between 15 and 35 mℓ was evaporated to dryness. The residue was then partitioned between 100 mℓ each of ethyl acetate and 8.4% (w/v) sodium carbonate in water. The ethyl acetate phase was taken to dryness (free steroid fraction). The sodium bicarbonate phase was acidified with 1 *M* HCl and was extracted 3 times with equal volumes of ethyl acetate. The combined ethyl acetate phases were washed with water until neutral and were then evaporated to dryness (glucuronide fraction).

Bile samples, usually 20 to 30 mℓ, were added dropwise to 300 mℓ chloroform-methanol (1:1 v/v), kept in an ultrasonic bath. After 2 hr the extract was centrifuged, the supernatant was collected, and the sediment was extracted once more with 100 mℓ of the same solvent. An aliquot of the combined supernatants was taken for radioactivity analysis and the remaining part was used for chromatography as described for feces.

Extraction of androgens from feces and bile samples (method of Martin et al.[3]) — Gall bladder bile samples, 2.5 mℓ, were treated overnight with 10 vol acetone-ethanol (1:1 v/v) at 39°C. The dry residue obtained was suspended in methanol-water (7:3 v/v) and the suspension left at −18°C overnight. The sample was then centrifuged at 1000 *g* for 30 min at −20°C, and the resultant supernatant evaporated to dryness.

To monitor the efficiency of the extraction procedures, small amounts (50-100,000 cpm) of (^{3}H)-testosterone, (^{3}H)-testosterone glucuronide, and (^{3}H)-testosterone sulfate were added to the bile samples before processing, and to the fecal extracts before Sephadex® LH-20 chromatography. The recovery of added radioactivity was determined prior to silicic acid (or Lipidex®) chromatography by counting an aliquot of each fraction in a liquid scintillation counter with external standardization correction.

Liquid chromatography — Bile extracts and 5% of each fecal extract were then applied to 20 g columns of Sephadex® LH-20 in 3 and 2 mℓ portions of chloroform-methanol (1:1 v/v, 0.01 *M* NaCl), which was also used as an eluting solvent. Fractions of unconjugated + glucuronide, monosulfate, and disulfate steroids were collected. The unconjugated steroids and steroid glucuronides were then separated by solvent partition between ethyl acetate and 8.4% sodium bicarbonate. The steroid glucuronides were hydrolyzed using ketodase and the mono- and disulfates solvolyzed.

Batch separation of steroids — All steroid extracts were then subjected to chromatography on 200-mg columns of silicic acid (Adsorbosil) or on columns (100 × 3 mm) of Lipidex® 5000 and fractions containing the 17-oxosteroids and androstene- and androstanediols collected. The Lipidex® 5000 was prepared for use per manufacturer's directions. The extracts were applied in petroleum ether (bp 40 to 60)-chloroform (95:5 v/v) and the columns eluted stepwise with 4 mℓ of the same solvent, 4 mℓ petroleum ether (90:10 v/v) (elutes oxosteroids), and 4 mℓ of the same solvent, and 4 mℓ petroleum chloroform (60:40 v/v) (elutes androstane-androstenediols). These fractions were then evaporated to dryness, the appropriate amount of internal standard(s) for mass fragmentography added, and trimethylsilyl ethers prepared.

Mass spectrometry was carried out on an LKB 900 GC/MS fitted with an accelerating voltage alternator; the signals from 2 of the 3 available fragment ion channels were fed to separate single pen electrical recorders. The GC column contained 3% OV-210 on Gas-Chrom Q (column length, 2.7 m; column oven temperature, 220°C [isothermal] for oxosteroid determinations and 189 or 200°C [isothermal] for androstane and androstenediol determinations; the flash heater, 248°C; and the separator, 250°C). The energy of the bombarding electrons was 22.5 eV.

Testosterone metabolites in liver and extraction of metabolites (method of Shaikly[4]) — The following reaction components were contained in a total volume of 1:0 mℓ: 100 μmol potassium phosphate at pH 7.5; 3 μmol MgCl$_2$; 0.5 μmol NADPH; 1 unit glucose-6-phosphate dehydrogenase; 5 μmol glucose-6-phosphate; 125 μmol (^{14}C)-testosterone (1 μCi); and 1.25 to 5 mg liver. The reactions were begun by addition of substrate, incubated for 5 min at 37°C, and stopped by the addition of 10 mℓ methylene chloride. After vortexing for 30 sec and centrifuging to separate the phases, 5-mℓ aliquots of the CH$_2$Cl$_2$ were counted to calculate recovery (average recovery was 96%), and 9 mℓ were taken to dryness under N$_2$. The residue was dissolved in 2 $\mu\ell$ methanol, divided in half, and each 1-mℓ portion taken to dryness again. Half of the sample was dissolved in 200 $\mu\ell$ methanol and 100 $\mu\ell$ aliquot analyzed using paper chromatography; the remainder was separated by TLC. The other half of the sample was dissolved in 100 $\mu\ell$ CH$_2$Cl$_2$ and 25 $\mu\ell$ subjected to HPLC analysis.

Plasma

Extraction of plasma for determination of testosterone within (method of Luisi et al.[5]) — Usually, peripheral venous blood is withdrawn by a disposable, heparinized syringe and centrifuged. The plasma is removed and immediately processed or stored at − 18°C. An internal standard of tritiated testosterone is added to 1 mℓ male or 3 mℓ female plasma, to 3 mℓ distilled water, and to 3 vials for liquid scintillation counting. The plasma is extracted with 2 × 15 mℓ diethyl ether. The ether layers are removed and the pooled extracts dried in a test tube at 37°C under N. The extract is then reconstituted and aliquots taken for chromatography.

Extraction of plasma for determination of testosterone within (method of Erikksson[2]) — After clotting and centrifugation of blood the serum sample was extracted with 10 vol acetone-ethanol (1:1) overnight at 39°C. The extract was filtered, and the precipitate extracted once more with the same solvent. An aliquot of the pooled extracts was taken for radioactivity analysis, and the remaining portion was further treated as for fecal extracts as previously described. This is a simpler method and can sometimes save steps.

Separation from plasma of dehydroepiandrosterone and its sulfate (method of deJong and Van der Molen[6]) — Plasma was separated by centrifugation of heparinized blood and stored at − 18°C. For the estimation of unconjugated DHA 5 mℓ plasma were placed in a ground glass stoppered conical 50 mℓ centrifuge tube. A solution of 40,000 dpm (15-^{3}H) DHA in 10 $\mu\ell$ benzene-ethanol (9:1 v/v) was added and the solvent was allowed to evaporate. Following addition of 0.25 mℓ 5 M NaOH the plasma was extracted with ether (6 × 10 mℓ). The combined ether extracts were washed with water (2 × 5 mℓ) and taken to dryness at 45°C under a stream of N$_2$. The residue was then treated as outlined below. For the estimation of DHAS, 0.25 mℓ plasma was used. After addition of a solution of 3000 dpm (4-^{14}C)D HAS in 10 $\mu\ell$ ethanol, steroid sulfates were solvolyzed by adjusting this aqueous fraction to pH 1 and allowed to stand overnight at 37°C while overlayered with an equal volume of ethyl acetate. The freed steroid was extracted by repeating the above step.

Org-6001 (3α-amino-5α-androstan-2β-17-ine hydrochloride) determined in plasma (method of Sondergaard and Steiness[7]) — Into a 10 mℓ PTFE-stoppered centrifuge tube, 1 mℓ plasma and 0.5 mℓ glycine-sodium hydroxide buffer (pH 9.9) were mixed, then extracted with 2 × 7 mℓ chloroform for 15 min. The two phases were separated by centrifugation at 1500 g for 15 min. The phases were each washed with 3 mℓ 0.1 N NaOH. Samples from each phase (5 mℓ) were then combined in a conical centrifuge tube and evaporated to dryness with nitrogen at 40°C. The residue was taken up in 50 $\mu\ell$ chloroform and spotted onto a TLC plate while under a stream of nitrogen with a 10 $\mu\ell$ constriction pipette. Chloroform (20 $\mu\ell$) was then added to the tubes and spotted again. After development the spots of Org-6001 were oxidized *in situ* to produce fluorescence as follows: the plates were placed in freshly prepared ethanol-water-perchloric acid (135:135:12) for 7 sec and excess reagent removed by blotting with filter paper No. 617. The plates were then quickly placed in an oven at 90°C for 35 min on copper plates (3-mm thick) to ensure uniform heating. The plates were then cooled at room temperature for 15 min before scanning.

Extraction of testosterone from plasma for analysis by GC/MS (method of Chapman and Bailey[8]) — A 4-mℓ volume of heparinized plasma was diluted to 10 mℓ with distilled water; 0.5 mℓ 2 N NaOH were added. From a standard solution containing 1 ng/μℓ deuterium-labeled testosterone in ethanol, 100 μℓ were added to the sample. This was extracted with 30 mℓ diethyl ether 3 times and the combined extracts washed twice with 5 mℓ distilled water. This was dried over anhydrous sodium sulfate and evaporated to dryness in a rotary evaporator. Standards and extracts were reacted to form BDMS ether using procedures described in Chapter 2. The derivatized extracts were taken up in 0.2 mℓ hexane for injection for GC or GC/MS.

Urine

Analysis of urine for assay of androgens (method of Shakelton et al.[9]) — Generally, 30 mℓ urine (24-hr collections) were buffered to pH 4.6 by the addition of 3 mℓ 5 M acetate buffer (5 M acetic acid/5 M sodium acetate, 2:3), 1 mℓ *Helix pomatia* digestive juice (1,000,000 units sulfatase), and 10,000 units β-glucuronidase, followed 24 hr later by 2 mℓ ketodase (10,000 units, β-glucuronidase), the hydrolysis taking place at $-4°C$. After a total hydrolysis period of 48 hr, the freed steroids were extracted on 13 g (dry weight) columns of Amberlite® XAD-2. The columns were washed with 30 mℓ water, and the steroids were recovered by elution with 100 mℓ ethanol.

When larger quantities of minor metabolites were required, it was necessary to hydrolyze large volumes of urine, and in these cases it was advantageous to extract the steroid conjugates by Amberlite® XAD-2 prior to enzymatic hydrolysis (see Chapter 2).

Extraction of methandrinone and free and neutral steroids from urine (method of Frischkorn et al.[10]) — Urine (25 mℓ) was brought to pH 5.2 with acetic acid and any sediment removed by centrifugation. The sample was then extracted with 4×20 mℓ methylene chloride. If an emulsion formed, it was centrifuged and the lower phase collected. The combined methylene chloride extracts were washed successively with 2×10 mℓ 0.1 M sodium hydroxide, 10 mℓ 0.02 M acetic acid, and 10 mℓ distilled water, dried with anhydrous sodium sulfate, and evaporated to dryness at 40°C. The residue was taken up in 2.5 mℓ methanol, and aliquots of 10 μℓ were injected into the HPLC sample loop.

Extraction of methandienone from urine (method of Feher[11]) — To a 10-mℓ specimen of urine, 1.0 μg ethynyl-nortesterone (17-ethynyl-17-hydroxy-19-norandrost-4-en-3-one) as internal standard was added. It corrected the losses throughout the method and quantitated them. The urine was extracted with 10 mℓ water, dried over anhydrous Na_2SO_4, and evaporated. The residue was quantitatively transferred with 3×0.2 to 0.3 mℓ ether to a capillary tube with a volume of 0.3 mℓ and evaporated. The dry residue was dissolved in 10.0 μℓ *n*-dioxane, and 0.3 to 1.0 μℓ of the solution was injected into the chromatograph. Ward's[12] method provides rapid extraction and group separation of steroids in urine.

Collection of small amounts of urine from laboratory animals (method of Erikksson[2]) — Urine was collected from laboratory rats on a filter paper (Munktell paper no. 3), which was changed every 12 hr. The paper was cut into pieces, transferred to a beaker containing 96% (v/v) ethanol, ultrasonified for 2 hr, and left in the solvent for 24 hr at room temperature. An aliquot was taken for determination of radioactivity, the urine extract diluted with water until 70% (v/v) with respect to ethanol, passed through an Amberlyst-15 column in the sodium form, and further worked up as described for feces.

Extraction analysis of total steroids (method of Apter et al.[13]) — 4.4 mℓ urine in a 15-mℓ conical glass tube was heated 15 min in a boiling water bath. After cooling, 5 mℓ ether was added, the tube stoppered and shaken for 1 min. The lower aqueous phase was then removed by aspiration. The ether phase was washed once with 2.5 mℓ 1 *N* sodium hydroxide and twice with 2.5 mℓ water. After washing, the tube was centrifuged and the water was removed. The ether was evaporated under a stream of nitrogen or air in a water bath at 40 to 45°C. When necessary, a drop of absolute alcohol was added to assist in removal of water. This is suitable for chromatography. Corticoids are somewhat destroyed by this method.

The specimens were poured onto columns of Amberlite® XAD-2 resin to extract free and conjugated steroids. The columns were washed with water and the steroids recovered by elution with ethanol. The ethanol was taken to dryness and the residue dissolved in 0.5 *M* acetate buffer (pH 4.6). The steroid conjugates were hydrolyzed by two enzymes present in snail digestive juice (*Helix pomatia*).

It was found desirable to fractionate the steroid extracts prior to GC analysis to increase the sensitivity of the method by removing some of the normal endogenous steroid metabolites. Sephadex® LH-20 columns (6 g) were used with the solvent system cyclohexane-ethanol (4:1), and 10-mℓ fractions were collected and taken to dryness.

Derivatization — Trimethylsilyl ethers were prepared from the steroid fractions by dissolution in dry pyridine (200 μℓ) and treatment with 200 μℓ hexamethyldisilazane and 50 μℓ trimethylchlorosilane. The reaction was allowed to proceed overnight at room temperature and the reagents were removed under a stream of nitrogen. The residue was taken up in cyclohexane.

Preponderance of conjugated steroids (method of Fitzpatrick and Siggia[14]) — Urine solvolyzed by acidification and ethyl acetate treatment or hydrolyzed enzymatically was extracted with ethyl acetate as described and solutions were stored at −4°C until ready for analysis. Known volumes of ethyl acetate were evaporated and the residue reacted according to the same procedure as the pure steroid standards. (Care must be taken in the derivatization reaction so that the pH does not fall below 0.8, otherwise dehydroepiandrosterone may be converted to 6β-OH-3,5-cyclo-androsten-17-one or 3β-chloroandrost-5-en-17-one.) Urine can also be hydrolyzed by the method of Vestergaard,[15] which involves acid hydrolysis and simultaneous extraction of the liberated free steroid into benzene under reflux conditions. These conditions have been evaluated previously by GC and found suitable. The "Vestergaard" extract was treated analogously to the steroid standards. Urine aliquots of 200 mℓ or greater generally gave enough UV active impurity to produce a "solvent shoulder", which interfered with eluting steroid peaks.

Hydrolysis extraction and TLC of ketonic steroids (method of Hakl[16]) — 10 mℓ urine was added to 1 mℓ 2 *M* acetate buffer to adjust to pH 4.5, then heated to 100°C in conical ground glass stoppered tubes (120 × 22 mm I.D.) for 2.5 hr to hydrolyze DHEA sulfate. The mixture was cooled, then shaken for 5 min with 10 mℓ cyclohexane-diethyl ether (1:1) on a horizontal shaker at 200 strokes per minute. No emulsion was formed when cooled sufficiently before shaking. The lower phase was removed by suction and the extract shaken for 1 min with 5 mℓ 10% sodium hydrozide, then for 1 min with 10 mℓ water and centrifuged. The aqueous phase was removed and the organic phase evaporated to dryness. The residue was dissolved in 0.5 mℓ 2,4-dinitrophenylhydrazine reagent and 0.2 mℓ trichloroacetic acid. The samples were incubated in the dark at 70 to 80°C for 10 min and subsequently evaporated to dryness. The residue was dissolved in a few drops of benzene containing ethanol and applied to plates of silica gel. Chloroform-acetone (95:5) was the mobile phase for ascending for 25 min.

Analysis of conjugated steroids in urine (method of Durbeck et al.[17]) — A 50-mℓ vol urine was adjusted to pH ± 0.1 with glacial acetic acid. The unconjugated steroids were removed from the sample by extracting 4 times with 20 mℓ dichloromethane. The combined extracts were washed with 15 mℓ 0.1 *N* sodium hydroxide solution and then with 15 mℓ

0.02 N acetic acid. A trace amount of water was removed from the organic phase by filtration through a soft paper filter (the normal drying procedure with anhydrous sodium sulfate should be omitted to avoid massive contamination with phthalates or silicones). The dry residue was dissolved in 2 mℓ ethanol and the precipitated waxes separated by filtration. (After evaporation to dryness, the residue must be absolutely free from water and methanol for chromatography or derivatization.)

Importance of hydrolysis before extraction (method of Trocha[18]) — To 50-mℓ screw-top test tubes, 7 mℓ urine sample, 2 mℓ concentrated hydrochloric acid, and 7 mℓ 1,2-dichloroethane containing 30 nmol (13 μg) internal standard 8,24(5γ)-cholestanien-4,4, 14γ-trimethyl-3β-0l (65% pure) were added. The tubes were tightly capped and placed in a 75 ± 2°C water bath for 10 min. After cooling the tubes in ice water, their contents were agitated with a vortex-type mixer for 15 sec, centrifuged for 2 min at 2000 × g, and the aqueous (upper) layer aspirated. Next the dichloroethane layer was extracted with NaOH (2 mol/ℓ) by agitating the tubes for a least 15 sec on a vortex-type mixer, centrifuged, and the aqueous (upper) layer aspirated. The organic phase was washed with 5 mℓ water and this water layer was discarded. The organic extract was evaporated under an equivolume mixture of ethyl ether and ethanol and transferred to vials and re-evaporated.

Steroids in Tablets

Determination of methandrostenolone in tablets (method of Butterfield et al.[19]) — Single 5-mg tablets were place in 15 × 75 mm screw-top test tubes and crushed with a glass rod to a fine powder. An internal standard was prepared by dissolving 9.5 mg *m*-dinitrobenzene in 25 mℓ chloroform, then 20 mℓ were pipetted into a 100-mℓ volumetric flask and diluted to volume with chloroform (76 μg/mℓ). Internal standard (2.0 mℓ) was added to each tube, which was closed and placed on a rotator at 60 rpm for 20 min. The tubes were centrifuged at 3000 rpm for 5 min.

Duplicate 1 μℓ samples of the working standard solution of methandrostenolone (2.5 mℓ/mℓ) were chromatographed and peaks corresponding to this compound and the internal standard were integrated. Duplicate 1 μℓ samples of each tablet extract were chromatographed, again integrating the peaks due to internal standard and methandrostenolone in each sample.

Separation of free and conjugated steroids (method of Axelson et al.[22]) — Extraction: Urine, 2 to 5 mℓ, was diluted with 1 to 3 mℓ water and passed at a flow rate of 0.4 mℓ/min through a column of XAD-2 resin (80 × 4 mm, packed in ethanol and washed with 15 mℓ water). The column was washed with 3 mℓ water, 2 mℓ 0.5 M triethylamine sulfate, and 3 mℓ water. Steroids were eluted with 8 mℓ methanol. A flow scheme of this method is shown in Figure 2.

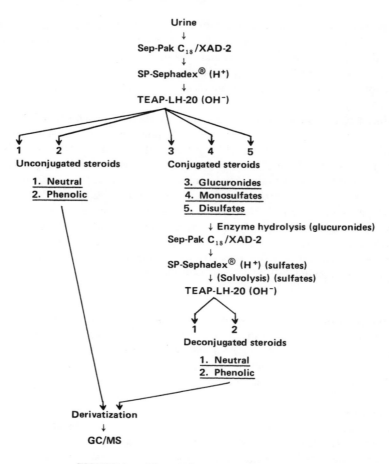

FIGURE 2.   Diagram for extraction from urine.

REFERENCES

1. **Verbeke, R.,** *J. Chromatogr.,* 177, 96, 1979.
2. **Erickson, H. and Gustafsson, J.,** *Eur. J. Biochem.,* 18, 146, 1971.
3. **Martin, F., Bhargava, A. S., and Aldercreutz, H.,** *J. Steroid Biochem.,* 8, 753, 1977.
4. **Shaikh, B., Hallmark, M. R., Issaq, H., Risser, N. H., and Kawalek, I. C.,** *J. Liquid Chromatogr.,* 2, 943, 1979.
5. **Luisi, M., Franchi, F., Menchini, G. F., Barletta, O., Fassorra, C., Ciardella, F., and Gagliardi, G.,** *Steroids Lipids Res.,* 4, 213, 1973.
6. **de Jong, F. H. and Van der Molen, H. J.,** *J. Endocrinol.,* 53, 461, 1972.
7. **Sondergaard, I. B. and Steiness, E.,** *J. Chromatogr.,* 162, 422, 1979.
8. **Chapman, J. R. and Bailey, E.,** *J. Chromatogr.,* 89, 215, 1974.
9. **Shackelton, C. H. L. and Honour, J. W.,** *Clin. Chem. Acta,* 69, 267, 1974.
10. **Frischkorn, C. G. B. and Frischkorn, H. E.,** *J. Chromatogr.,* 151, 331, 1978.
11. **Feher, T. and Sarfy, E. H.,** *J. Sports Med.,* 16, 165, 1976.
12. **Apter, C., Chagen, R., Gould, S., and Harell, A.,** *Clin. Chem. Acta,* 42, 115, 1976.
13. **Ward, R. J., Shakelton, C. H. L., and Lawson, A. M.,** *Br. J. Sports Med.,* 9, 193, 1975.
14. **Fitzpatrick, F. A. and Siggia, S.,** *Anal. Chem.,* 42, 1223, 1970.
15. **Vestergaard, P.,** *Clin Chem.,* 16, 651, 1970.
16. **Hakl, J.,** *J. Chromatogr.,* 143, 317, 1977.

17. **Durbek, H. W., Buker, I., Schenlen, B., and Telin, B.,** *J. Chromatogr.,* 167, 117, 1978.
18. **Trocha, P. and D'Amato, N. A.,** *Clin. Chem.,* 24, 193, 1978.
19. **Butterfield, A. G., Lodge, B. A., Pound, N. J., and Sears, R. W.,** *Pharm. Anal.,* 64, 441, 1975.
20. **Huettemann, R. E. and Shroff, A. P.,** *J. Chromatogr. Sci.,* 13, 357, 1975.
21. **Graham, R. E., Biehl, E. R., and Kermer, C. T.,** *J. Pharm. Sci.,* 68, 871, 1979.
22. **Axelson, M., Sahlberg, B. L., and Sjovall, J.,** *J. Chromatogr.,* 224, 355, 1981.

SECTION II. GAS CHROMATOGRAPHY

Table GC 1
SILYL ETHERS OF ANDROGENS[1]

Compound	Relative retention time (RRT)	
	1	2
1. Androst-5-en-17-one	0.31	1.39
2. Androst-5-en-17β-ol	0.23	1.39
3. Androst-5-ene-3β,17α-diol	0.41	1.00
4. 5α-Androstane-3β,17α-diol	0.41	0.48
5. 5β-Androstane-3α-17β-diol	0.36	0.38
6. Androst-5-ene-2α,17α-diol	0.34	1.16
7. 5β-Androstane-3α,17α-diol	0.27	0.33
IS Aldrin	0.12	—
IS 4-Methyl-3-deoxyestrone	—	1.28
Cholestane	1.00 (42.5 min)	1.00 (6.24 min)

Instrument: Shimadza GC-5AIFE with flame ionization
Column: 1. 2% OV-1700 Shimalite W (60—80 mesh) 6.75 m × 3 mm (steel)
2. 1.5% neopentylglycolsuccinate on Shimalite as in 1
Temperature: Oven 230°C, detector 250°C, injector 250°C
Flow: Nitrogen 54 mℓ/min

Table GC 2
SEPARATION OF IODOMETHYLDIMETHYLSILYL ETHER DERIVATIVE OF ANDROGENS[2]

Compound	RRT
A = Androsterone	0.165
D = DHEA	0.242
P = Pregnenolone	0.367
IS = 5α-androstene-3β,17β-diol-IDMSE	1.00 (21.4 min)

Instrument: Pye 104, model 74 1.5 with [63]Ni electron capture detector
Column: 1.5% Dexsil 300 on Gas Chrom Q (100—120 mesh)
Temperature: Column 250°C, detector and injector 300°C
Flow: 100 mℓ/min

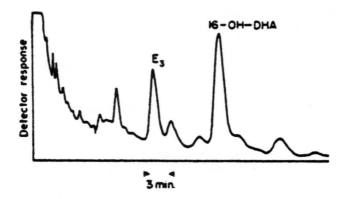

FIGURE GC 1.   Separation of heptafluorobutynates of androgens. Instrument: Glowall 310 with 12.5 μCi radium foil detector; column: 5% OV-210 + 2.5% OV-17 on Supelcoport; temperature: column 225°C, detector 240°C, injector 270°C; and flow: Argon-60 mℓ/min. (From Touchstone, J. C. and Dobbins, M. F., *J. Steroid Biochem.*, 6, 1389, 1975. With permission.)

Table GC 3
SEPARATION OF ANDROGEN DERIVATIVES[4]

Compound	RRT
Dehydroepiandrosterone 3-trimethylsilyl ether	0.47
17-(2-Chloroethyl) oxime	1.56
17-(4-Chlorobutyl) oxime	2.96
17-(5-Chloropentyl) oxime	4.06
17-(2-Bromoethyl) oxime	2.00
Cholestane	1.00 (15.1 min)
Androstane	0.10

Instrument:	Shimadzu model GC-4BM PFE with hydrogen flame detector
Column:	2% SE-30 on Gas Chrom Q (80—100 mesh)
Temperature:	Column 220°C, detector and injector 224°C
Flow:	Nitrogen 55 mℓ/min

Table GC 4
RETENTION TIMES OF ANDROGEN TRIMETHYLSILYL ETHERS[5]

Steroid	RRT
5α-Androstan-3α-ol (androstanol)	0.30
5α-Androstan-3α-ol-17-one (androsterone)	0.34
5α-Androstan-3α-ol-17-one (etiocholanolone)	0.36
5-Androsten-3β-ol-17-one (dehydroisoandrosterone)	0.40
5α-Androstan-3β-ol-17-one (epiandrosterone)	0.40
4-Androsten-3,17-dione (androstenedione)	0.41
5α-Androstan-3α-ol-11,17-dione (11-ketoandrosterone)	0.41
5β-Androstan-3α-ol-11,17-dione (11-ketoetiocholanolone)	0.42

Table GC 4 (continued)
RETENTION TIMES OF ANDROGEN
TRIMETHYLSILYL ETHERS[5]

Steroid	RRT
1,3,5(10)-Estratrien-3-ol-17-one (estrone)	0.42
5β-Pregnan-3β-ol-20-one (pregnanolone)	0.46
5α-Androstan-3α,11β-diol-17-one (11β-hydroxyandrosterone)	0.46
4-Androsten-17β-ol-3-one (testosterone)	0.47
5β-Androstan-3α,11β-diol-17-one (11β-hydroxyethiocholanolone)	0.48
1,3,5(10)-Estratrien-3,17β-diol (estradiol)	0.49
8.24,(5α)-cholestadien-4,4,14α trimethyl-3β-ol (lanosterol)	1.00

Instrument:	Perkin Elmer 900 with flame ionization
Column:	3% SE-30 on Chromosorb G (AW/DMCS) (100—120 mesh)
Temperature:	185°C for 2 min, then to 260°C at 3 min
Flow:	Nitrogen 30 mℓ/min

Table GC 5
RETENTION DATA OF TRIMETHYL
SILYL ETHER METHOXIMES OF C_{19}
STEROIDS[6]

Androgen	Retention time[a] (sec)		
	221°C	231°C	241°C
Methane	165.0	174.0	183.0
$n\text{-}C_{24}$	759.0	574.0	459.5
Androsterone	909.5	690.0	560.5
Etiocholanolone	947.0	712.0	574.5
$n\text{-}C_{25}$	964.0	706.0	549.0
Dehydroepiandrosterone	1080.5	804.5	641.5
$n\text{-}C_{26}$	1239.0	878.0	666.5
Pregnanolone	1338.0	962.5	746.5
Testosterone	1363.5	990.0	772.5
Estradiol	1460.0	1040.5	798.5

Instrument:	Modular with flame detector
Column:	0.5% SE-30 on Pipex (30 m × 0.25 mm) (capillary column)
Temperature:	221, 231, and 241°C for column
Flow:	Nitrogen at 0.005 atmos. to maintain flow

[a] Retention time relative to methane.

REFERENCES (GC)

1. **Nambara, T. and Bae, Y. H.,** *J. Chromatogr.,* 64, 239, 1972.
2. **Symes, E. K. and Thomas, B. S.,** *J. Chromatogr.,* 116, 163, 1976.

3. **Touchstone, J. C. and Dobbins, M. F.,** *J. Steroid Biochem.,* 6, 1389, 1975.
4. **Nambara, T., Iwata, T., and Kigasawa, I. C.,** *J. Chromatgr.,* 118, 127, 1976.
5. **Trocha, P. and D'Amato, N. A.,** *Clin. Chem.,* 24, 193, 1978.
6. **Mitra, G. D.,** *J. Chromatgr.,* 234, 214, 1982.

SECTION III. HIGH PERFORMANCE LIQUID CHROMATOGRAPHY

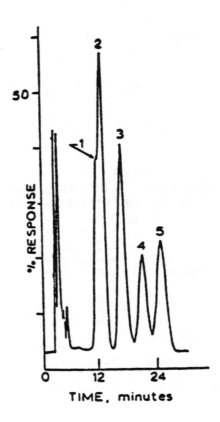

FIGURE HPLC 1.   Separation of androgens. Instrument: Varian 4100; column: Zipax 1.5% OPPN (1 m × 1.8 mm); mobile phase: isooctane; and detection: 254 nm. (1) epietiocholanolone, (2) androsterone, (3) epiandrosterone, (4) etiocholanolone, (5) dehydroepiandrosterone. (Reprinted with permission from Fitzpatrick, F. A., Siggia, S., and Dingmaryh, J., *Anal. Chem.,* 44, 2211, 1972. Copyright 1972 American Chemical Society.)

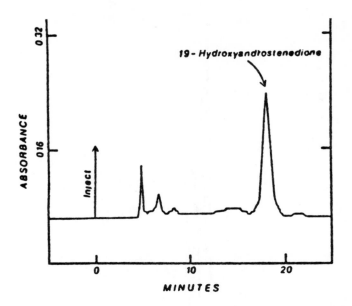

FIGURE HPLC 2. Separation of 19-hydroxyandrostenedione. Instrument: Waters Model ALC/GPC-202; column: μBondapak C_{18}, two 0.40 × 30 cm in series; mobile phase: methanol-water (50:50); and detection: 254 nm. (From Kautsky, M. P., Thurman, G. W., and Hagerman, D. D., *J. Chromatogr.*, 114, 473, 1975. With permission.)

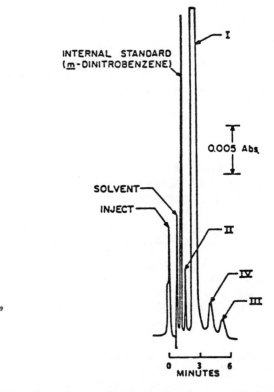

FIGURE HPLC 3. Separation of synthetic androgens. Instrument: Varian Model 4100; column: LiChrosorb SI-60 (25 cm × 2.1 mm); mobile phase: 3% ethylene chloride in 15% 2-propanol; and detection: 254 nm. (I) 2.5 μg methandrostenolone, (II) 25 ng methyltestosterone, (III) 17.5 ng 6α-hydroxymethandrostenolone, (IV) 6β-hydroxymethandrostenolone. (From Butterfield, A. G., Lodge, B. A., Pound, N. J., and Sears, R. W., *Pharm. Anal.*, 64, 441, 1975. With permission.)

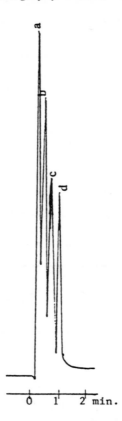

FIGURE HPLC 4.   Separation of synthetic androgens. Instrument: Dupont 830; column: Permaphase ODS (1 m × 2.1 mm); mobile phase: methanol-water (1:1); temperature: 60°C; and flow: 2 mℓ/min. (a) norethindrone, (b) norethindrone oxime, (c) norethindrone acetate, (d) 17-ethynyl-17β-acetoxy-19-norandrost-4-en-3-one-oxime. (From Huettermann, R. E. and Schroff, A. P., *J. Chromatogr. Sci.*, 13, 357, 1957. With permission.)

Table HPLC 1
SEPARATION OF 17-HYDROXY ANDROSTANES
(TIME IN MINUTES)[5]

	C_{18} column		CN-column	
	1 Time	2 Time	3 Time	4 Time
Methandienone	2.55	11.70	15.1	7.85
Testosterone (2α)	2.85	13.30	15.2	7.80
Epitestosterone (2β)		13.60	15.2	7.90
"Isotestosterone (5α)"	3.40	14.00	19.0	
Methyltestosterone	3.60	15.40	19.0	8.40
"Isomethyltestosterone (5α)"	4.30	16.75	24.3	9.55
17-Methyldihydrotestosterone (5α)	4.95	18.95	28.1	
Dihydrotestosterone (5α)	3.40	15.70	22.0	
Androstan-4-ene-3,17-dione	1.97	9.48	11.3	
17β-Hydroxyandrosta-1,4-dien-3-one	1.95	9.45	11.3	

Table HPLC 1 (continued)
SEPARATION OF 17-HYDROXY ANDROSTANES
(TIME IN MINUTES)[5]

Instrument: LDC Constametric Pump, Spectromonitor III

Column: Pelligard-LC-18 (Supelco) for guard column
Supelcosil-LC-C$_{18}$ 5 μm (250 $\times$ 4.6 mm)

Mobile phase: 1. Methanol-water (69:31)
2. Acetonitrile-water (49:51)
3. Methanol-acetonitrile-water (33:26:41)

Flow: 1 mℓ/min for each

Detection: Variable wavelength

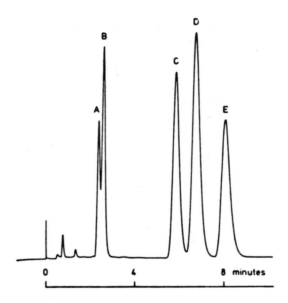

FIGURE HPLC 5. Separation of androgens. Instrument: Hewlett-Packard 1010A; column: LiChrosorb RP8 (0.25 m $\times$ 4.6 mm); mobile phase: Acetonitrile-water (1:1); flow: 2.8 mℓ/min; detection: 260 nm. (A) 19-nor-4,6-androstadien-3,17-dione, (B) 19-nor-4-androstene-3,17-dione, (C) 17α-hydroxy-19-nor-4,6-androstadien-3-one 17-acetate, (D) 17α-hydroxy-19-nor-4-androsten-3-one 17-acetate, (E) 17α-hydroxy-4-androsten-3-one 17-acetate. (From Higgins, J. W., *J. Chromatogr.*, 148, 335, 1978. With permission.)

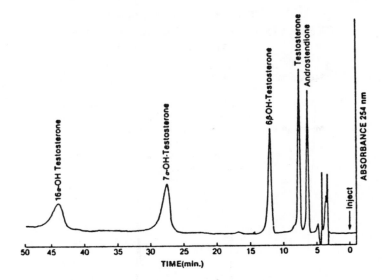

FIGURE HPLC 6.   Separation of testosterone derivatives. Instrument: LDC Constametric II Pump and Chrometronix detector; column: 10 μm Partisil 10 (26 cm × 4.6 mm); mobile phase: hexane-tetrahydrofuran-isopropanol (85:15:5); and detection: 254 nm. (From Shaikh, B. et al., *J. Liquid Chromatogr.*, 21, 943, 1979. With permission.)

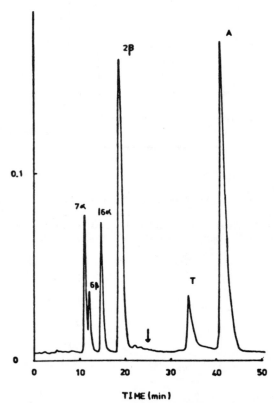

FIGURE HPLC 7.   Separation of testosterone derivatives. Instrument: Milton Roy minipump, Isco Model 1840 detector; column: 10 μm C_{18} reversed phase (guard column); mobile phase: methanol-water (55:45) then methanol-water-acetonitrile (55:35:10) (at arrow); flow: 0.7 mℓ/min; detector: 240 nm. (2β) 2β-Hydroxytestosterone, (6β) 6β-hydroxytestosterone, (7α) 7α-hydroxytestosterone, (From Vander Hoeven, T. A., *J. Chromatogr.*, 196, 494, 1980. With permission.)

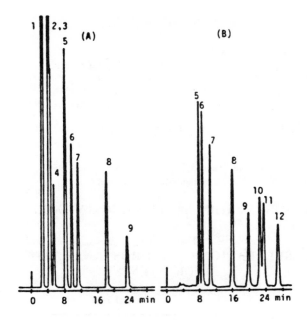

FIGURE HPLC 8. Separation of dansyl derivatives of androgens. Instrument: Hitachi Model 634; column: (A) Hitachi Gel # 3042 (250 × 4 mm) and (B) Zorbax S. 1 (250 × 4.6 mm); mobile phase: dichloromethane-ethanol-water (400:1:2) 1 mℓ/min; detection: fluorescence (A) excitation 350 nm, emission 505 nm and (B) excitation 365 nm, emission 505 nm; dansyl hydranones: peaks: (1) androsta-3,5-diene-17-one, (2) androstanedione, (3) androstadiene-dione, (4) androst-e-ene-3,17-dione, (5) androsterone, (6) dehydroepiandrosterone, (7) etiocholanolone, (8) 11-oxoandrosterone, (9) 11-oxoetiocholanolone, (10) 16α-hydroxy-dehydroepiandrosterone, (11) 11β-hydroxyandrosterone, (12) 16α-hydroxyandrosterone. (From Kawasaki, T., Maeda, M., and Tsugi, A., *J. Chromatogr.*, 226, 1, 1981. With permission.)

Table HPLC 2
SEPARATION OF ANDROGEN
DERIVATIVES[10]

	Elution volume (mℓ)		
Androgen	1	2	3
17β-Hydroxy-4-androsten-3-one	7.0	7.6	11.0
5α-Androstane-3β,17β-diol	9.8	7.6	12.5
5α-Androstane-3,17-dione	10.4	16.8	19.4
3β-Hydroxy-5α-androstan-17-one	11.0	11.2	16.0
17β-Hydroxy-5α-androstan-3-one	12.0	12.2	17.6
5α-Androstane-3α,17β-diol	15.8	10.8	20.8
3α-Hydroxy-5α-androstan-17-one	17.4	16.8	28.5

Instrument:	LDC Constametric Pump Spectromonitor III
Column:	Pelligard-LC-18 (Supelco) for guard column
	Supelcosil-LC-C$_{18}$ 5 μm (250 × 4.6 mm)
Mobile phase:	1. Methanol-water (69:31)
	2. Acetonitrile-water (49:51)
	3. Methanol-acetonitrile-water (33:26:41)
Flow:	1 mℓ/min for each
Detection:	Variable wavelength

REFERENCES (HPLC)

1. **Fitzpatrick, F. A., Siggia, S., and Dingmaryh, J.,** *Anal. Chem.,* 44, 2211, 1972.
2. **Kautsky, M. P., Thurman, G. W., and Hagerman, D. D.,** *J. Chromatogr.,* 114, 473, 1975.
3. **Butterfield, A. G., Lodge, B. A., Pound, N. J., and Sears, R. W.,** *Pharm. Anal.,* 64, 441, 1975.
4. **Huettermann, R. E. and Shroff, A. P.,** *J. Chromatogr. Sci.,* 131, 357, 1975.
5. **Frischkorn, C. G. B. and Frischkorn, H. E.,** *J. Chromatogr.,* 151, 331, 1978.
6. **Higgins, J. W.,** *J. Chromatogr.,* 148, 335, 1978.
7. **Shaikh, B., Hallmark, M. R., Issaq, H., Risser, N. H., and Kawalek, J. C.,** *J. Liquid Chromatogr.,* 21, 943, 1979.
8. **Van der Hoeven, T. A.,** *J. Chromatogr.,* 196, 494, 1980.
9. **Kawasaki, T., Maeda, M., and Tsugi, A.,** *J. Chromatogr.,* 226, 1, 1981.
10. **Sunde, A. and Lundmo, P. J.,** *J. Chromatogr.,* 242, 381, 1982.

SECTION IV. THIN LAYER CHROMATOGRAPHY

Table TLC 1
SEPARATION OF ANDROSTANES AND ANDROSTENES[1]

Steroid	Free steroid				Acetate
	1	2	3	4	5
5α-A-3,17-dione	13.6	8.5	—	15.3	—
5β-A3,17-dione	12.8	7.2	—	14.9	—
4-A-3,17-dione	9.0	4.0	2.3	12.5	—
17β-Hydroxy-5α-A-3-one	10.6	5.0	2.9	11.8	8.0
17β-Hydroxy-5β-A-3-one	8.7	3.2	2.0	—	7.6
17β-Hydroxy-4-A-3-one	6.7	2.3	1.4	8.7	5.0
17α-Hydroxy-4-A-3-one	6.7	—	—	—	—
3α-Hydroxy-4-A-17-one	8.1	—	—	9.2	—
3β-Hydroxy-4-A-17-one	10.2	—	—	11.5	—
3α-Hydroxy-5α-A-17-one	10.3	4.7	2.7	11.2	9.3
3β-Hydroxy-5α-A-17-one	9.0	4.0	2.4	—	9.5
3α-Hydroxy-5β-A-17-one	7.7	—	1.6	8.7	9.8
3β-Hydroxy-5β-A-17-one	10.9	—	3.2	11.9	9.3
5α-A-3α,17β-diol	7.9	3.2	—	—	—
5α-A-3β,17β-diol	7.6	2.9	—	8.2	—
5β-A-3α,17β-diol	4.7	1.2	—	5.3	—
5β-A-3β,17β-diol	9.0	3.7	—	9.4	—
5α-A-3α,17α-diol	5.7	—	—	5.8	—
5α-A-3β,17α-diol	7.0	—	—	—	—
4-A-3α,17β-diol	5.6	1.6	—	—	—
4-A-3β,17β-diol	8.5	3.0	—	—	—
5-A-3β,17β-diol	8.1	—	—	—	—
5-A-3β,17α-diol	8.1	—	—	—	—

Layer: Silica Gel 60 F254 (Merck)
Mobile phase: 1. Cyclohexane-ethyl acetate (1:1) (2 hr over-run)
2. *n*-Hexane-ethyl acetate (3:1) (90 min)
3. Cyclohexane-ethyl acetate (4:1) (90 min)
4. Benzene-ethyl acetate (1:1) (90 min)
5. Same as 3
Visualization: UV quench

A = Androstane
4-A = Androstene-4en
5-A = Androstene-5en

Table TLC 2
R_f VALUES FOR ANDROGEN DERIVATIVES[2]

	$R_f \times 100$		
Androgen	1	2	Color
3β,17β-Dihydroxy-17α-picolyl-5-androstene	34	37	Gray
3β-Acetoxy-17β-hydroxy-17α-picolyl-5-androstene	78	75	Brown
3β-Acetoxy-17-picolinyliden-5-androstene	83	93	Dark red
3β-Hydroxy-17-picolinyliden-5-androstene	40	49	Dark red
3β,17β-Dihydroxy-16-oximino-17α-picolyl-5-androstene	6	4	Red
3β-Hydroxy-16-oximino-5-androsten-17-one	17	27	Pink
3β-Acetoxy-5-androsten-17-one	80	94	Violet
Lactone of 3β,17β-dihydroxy-16,17-seco-androsten-16-oic acid	24	30	Gray-Violet
3β,17β-Dihydroxy-16,17-seco-5-androsteno-16-nitrile	13	18	Red
3β,17β-Dihydroxy-16-oximino-5-androstene	0	0	Crimson
3β-Hydroxy-5-androsten-17-one	42	53	Violet

Layer: Silica Gel G (Merck) 20 × 20 cm activated 1 hr at 110°C
Mobile phase: 1. Benzene-acetone (7:1) (1st dimension)
                    2. Benzene-ethyl acetate (2nd dimension)
Visualization: 50% H_2SO_4 in MeOH spray and 10—15 min at 100—100°C

Table TLC 3
TLC OF TESTOSTERONE AND $C_{19}O_3$ STEROIDS[3]

	Mobility (relative to testosterone)						
Steroid	1	2	3	4	5	6	7
Testosterone	1.00	1.00	1.00	1.00	1.00	1.00	1.00
Androst-5-ene-3β,16α,17β-triol	0.03	0.07	0.06	0.25	0.56	0.68	0.35
Androst-5-ene-3β,16β,17β-triol	0.35	0.40	0.42	0.77	0.80	0.80	0.51
5α-Androstane-3,6,17-trione	1.06	1.00	1.10	0.90	0.91	1.08	1.32
5α-Androstane-3,7,17-trione	1.28	1.11	1.08	0.90	0.90	1.12	1.31
3β-Hydroxy-5α-androstane-6,17-dione	0.31	0.30	0.23	0.34	0.69	0.75	0.68
3β-Hydroxy-5α-androstane-7,17-dione	0.48	0.50	0.34	0.51	0.71	0.80	0.71
17β-Hydroxy-5α-androstane-3,6-dione	0.60	0.60	0.51	0.64	0.79	0.83	0.80
3α,17β-Dihydroxy-5α-androstan-6-one	0.17	0.19	0.16	0.37	0.72	0.74	0.56
3β,6α-Dihydroxy-5α-androstan-17-one	0.04	0.06	0.02	0.10	0.45	0.60	0.48
3β,7α-Dihydroxy-5α-androstan-17-one	0.10	0.12	0.05	0.21	0.61	0.64	0.49
3β,7β-Dihydroxy-5α-androstan-17-one	0.11	0.13	0.07	0.19	0.54	0.61	0.46
3β,11β-Dihydroxy-5α-androstan-17-one	0.32	0.34	0.43	0.80	0.93	0.74	0.64
3β,17β-Dihydroxy-5α-androstan-6-one	0.17	0.19	0.13	0.30	0.65	0.65	0.54
3β,17β-Dihydroxy-5α-androstan-7-one	0.21	0.23	0.20	0.40	0.68	0.69	0.58
6α,17β-Dihydroxy-5α-androstan-3-one	0.18	0.21	0.17	0.43	0.73	0.63	0.55
5α-Androstane-2α,3α,17β-triol	0.10	0.12	0.11	0.38	0.65	0.55	0.53
5α-Androstane-2β,3α,17β-triol	0.06	0.08	0.09	0.33	0.67	0.45	0.48
5α-Androstane-2β,3β,17β-triol	0.17	0.20	0.19	0.57	0.83	0.63	0.56
5α-Androstane-3α,6β,17β-triol	0.14	0.13	0.12	0.35	0.69	0.63	0.51
5α-Androstane-3α,11β,17β-triol	0.06	0.08	0.09	0.36	0.76	0.56	0.50
5α-Androstane-3α,16α,17β-triol	0.02	0.03	0.03	0.14	0.50	0.41	0.39
5α-Androstane-3β,6α,17β-triol	0.04	0.06	0.02	0.13	0.46	0.43	0.42
5α-Androstane-3β,6β,17β-triol	0.05	0.07	0.04	0.25	0.60	0.50	0.47
5α-Androstane-3β,7α,17β-triol	0.06	0.05	0.03	0.17	0.48	0.40	0.38
5α-Androstane-3β,7β,17β-triol	0.06	0.06	0.04	0.21	0.50	0.46	0.42
5α-Androstane-3β,11β,17β-triol	0.05	0.06	0.07	0.35	0.70	0.47	0.53
5α-Androstane-3β,15α,17β-triol	0.09	0.06	0.04	0.15	0.46	0.43	0.39
5α-Androstane-3β,16α,17β-triol	0.05	0.07	0.07	0.21	0.57	0.54	0.42

Table TLC 3 (continued)
TLC OF TESTOSTERONE AND $C_{19}O_3$ STEROIDS[3]

Layer:	Silica Gel 60 F_{254} (Merck)
Mobile phase:	1. Chloroform-ethyl acetate (4:1, v/v) once
	2. Chloroform-ethyl acetate (4:1, v/v) twice
	3. Benzene-ethyl acetate (3:1, v/v) twice
	4. Cyclohexane-ethyl acetate (3:2, v/v) twice
	5. Cyclohexane-ethyl acetate-ethanol (45:45:10, v/v/v) once
	6. Chloroform-ethanol (9:1, v/v) once
	7. Benzene-ethanol (9:1, v/v) three times
Visualization:	H_2SO_4/MeOH (7:3) spray and heat 120°C for 15 min

Table TLC 4
R_f VALUES OBTAINED FOR THE MAJOR ANDROGENS AND ESTROGENS[4]

	Mobile phase				
Steroid	**1**	**2**	**3**	**4**	**5**
Estriol	0.01	0.05	0.19	0.09	0.04
Androstanediol	0.23	0.37	0.66	0.41	0.29
Testosterone	0.33	0.48	0.68	0.46	0.35
Estradiol-17β	0.33	0.53	0.77	0.43	0.50
5α-Dihydrotestosterone	0.40	0.60	0.79	0.47	0.49
Estrone	0.49	0.75	0.92	0.49	0.72

Layer:	Whatman K-5 Silica Gel
Mobile phase:	1. Chloroform-ethyl acetate (90:10)
	2. Chloroform-ethyl acetate (80:20)
	3. Chloroform-ethyl acetate (50:50)
	4. Chloroform-methanol (98:2)
	5. Benzene-ethyl acetate (60:20)
Detection:	Iodine vapor

Table TLC 5
SEPARATION OF KETONIC ANDROGENS AND ACETATES[5]

	R_f value			
Compound	**1**	**2**	**3**	**4**
Cholest-4-ene-3,6-dione	0.66	0.62	0.70	0.62
3-Oxoandrost-4-en-17β-yl acetate	0.40	0.34	0.58	0.38
7-Oxocholest-5-en-3β-yl acetate	0.70	0.61	0.70	0.65
(25R)-7-Oxospirost-5-en-3β-yl acetate	0.45	0.34	0.65	0.32
17-Oxo-5-androstan-3β-yl acetate	0.63	0.56	0.66	0.61
17-Oxoandrost-5-en-3β-yl acetate	0.62	0.55	0.65	0.61
3β-Hydroxy-5α-androstan-17-one	0.23	0.17	0.29	0.18
3β-Hydroxyandrost-5-en-17-one	0.23	0.19	0.29	0.20
3β-Hydroxypregn-5-en-20-one	0.23	0.22	0.34	0.22

Layer:	Silica Gel G (Merck)
Mobile phases:	1. Chloroform-methanol (99:1)
	2. Benzene-ethyl acetate (17:3)
	3. Benzene-methanol (19:1)
	4. Chloroform-ethylacetate (1:1)

Table TLC 6
SEPARATION OF ACETOXIMES OF ANDROGENS[5]

Compound	R_f value			
	1	2	3	4
3-Hydroxyiminoandrost-4-en-17β-yl acetate	0.26	0.36	0.39	0.30
3-Hydroxyiminoandrost-4-en-17-one	0.15	0.34	0.38	0.23
17,17-Ethylenedioxy-3-hydroxyiminoandrost-4-ene	0.19	0.38	0.39	0.22
3-Hydroxyiminocholest-4-en-6-one	0.34	0.41	0.39	0.36
3β-Methoxy-17-hydroxyimino-5α-androstane	0.20	0.36	0.45	0.25
7-Hydroxyiminocholest-5-en-3β-yl acetate	0.74	0.65	0.88	0.72
(25R)-7-Hydroxyiminospirost-5-en-3β-yl acetate	0.46	0.57	0.71	0.55
7-Hydroxyiminopregn-5-ene-3β,20β-diol diacetate	0.50	0.53	0.76	0.62
7,20-Dihydroxyiminopregna-5,16-dien-3β-yl acetate	0.48	0.38	0.57	0.55

Layer: Silica Gel G (Merck)
Mobile phase:
1. Benzene-ethylacetate (17:3)
2. Benzene-methanol (19:1)
3. Chloroform-methanol (39:1)
4. Chloroform-ethyl acetate (8:2)

Table TLC 7
SEPARATION OF ANDROSTANES[6]

No.	Abbreviation	Mobile phase	
		1	2
1	3α-ol-5α-A-17-one	1.35	2.58
2	3β-ol-5α-A-17-one	1.25	1.88
3	17β-ol-5α-A-3-one	1.37	2.03
4	17β-ol-5β-A-3-one	1.11	2.22
5	5α-A-3,17-one	1.89	4.22
6	5β-A-3,17-one	1.76	4.43
7	Δ^4-A-3,17-one	1.54	3.66
8	3β-ol-Δ^5-A-17-one	1.27	1.67
9	3α-ol-Δ^5-A-17-one	1.23	2.43
10	3α,17β-ol-Δ^4-A	0.69	0.33
11	3β,17β-olΔ^5-A	0.88	0.00
12	3β,17β-ol-5α-A	0.87	0.00
13	17β-ol-Δ^4-A-3-one	1.00	1.00

Layer:
1. Silica Gel G (Merck)
2. Cellulose impregnated with 1,2-propanediol

Mobile phase:
1. Chloroform-ethyl acetate-petroleum ether (50:45:5)
2. Benzene-cyclohexane (50:50)

13 = Testosterone
A = Androstane

Table TLC 8
SEPARATION OF ANDROGENS[7]

Steroid	R_f	
5α-Androstane-3α,17β-diol	1.00	1.00
5α-Androstane-3β,17β-diol	1.30	1.27
4-Androstene-3α,17β-diol	0.57	0.45
4-Androstene-3β,17β-diol	1.45	1.42
3α-Hydroxy-5α-androstan-17-one	1.63	
3α-Hydroxy-5α-androstan-17-one	1.80	
17β-Hydroxy-5α-androstan-3-one	2.00	
17β-Hydroxy-4-androsten-3-one	1.48	1.50
5α-Androstane-3,17-dione	2.57	

Layer: Schleicher and Schull F-1500
Mobile phase: 1. Dichloromethane-ethyl acetate, 2 developments
                2. Same as 1 but 5 developments at 4°C

Table TLC 9
SEPARATION OF ANDROGENS[8]

Androgen	System						
	1	2	3	4	5	6	7
3α,11β-ol-5αA-17-one	0.49	0.17	0.51	0.66	0.57	0.34	0.29
3β,11β-ol-5αA-17-one	0.45	0.13	0.43	0.68	0.53	0.28	0.22
3α,11β-ol-5βA-17-one	0.44	0.12	0.45	0.64	0.52	0.30	0.24
2α,17β-ol-A⁴-3-one	0.34	0.11	0.46	0.53	0.44	0.25	0.27
4,17β-ol-A⁴-3-one	0.52	0.28	0.58	0.70	0.61	0.47	0.46
11β,17β-ol-A⁴-3-one	0.31	0.05	0.34	0.46	0.45	0.22	0.11
11α,17β-ol-A⁴-3-one	0.18	0.02	0.26	0.27	0.33	0.13	0.05
14α,17β-ol-A⁴-3-one	0.27	0.04	0.31	0.39	0.39	0.19	0.11
16α,17β-ol-A⁴-3-one	0.19	0.02	0.26	0.33	0.38	0.18	0.09
17β,19-ol-A⁴-3-one	0.28	0.06	0.31	0.39	0.37	0.16	0.12
3α,11β-ol-5αA-17-one	0.49	0.17	0.51	0.66	0.57	0.34	0.29
3β,11β-ol-5αA-17-one	0.45	0.13	0.43	0.68	0.53	0.28	0.22
3α,11β-ol-5βA-17-one	0.44	0.12	0.45	0.64	0.52	0.30	0.24
2α,17β-ol-A⁴-3-one	0.34	0.11	0.46	0.53	0.44	0.25	0.27
4,17β-ol-A⁴-3-one	0.52	0.28	0.58	0.70	0.61	0.47	0.46
11β,17β-ol-A⁴-3-one	0.31	0.05	0.34	0.46	0.45	0.22	0.11
11α,17β-ol-A⁴-3-one	0.18	0.02	0.26	0.27	0.33	0.13	0.05
14α,17β-ol-A⁴-3-one	0.27	0.04	0.31	0.39	0.39	0.19	0.11
16α,17β-ol-A⁴-3-one	0.19	0.02	0.26	0.33	0.38	0.18	0.09
17β,19-ol-A⁴-3-one	0.28	.06	0.31	0.39	0.37	0.16	0.12
3β,7α-ol-A⁵-17-one	0.22	0.02	0.27	0.35	0.36	0.14	0.09
3β,11β-ol-A⁵-17-one	0.47	0.17	0.48	0.62	0.50	0.29	0.25
3β,16α-ol-A⁵-17-one	0.43	0.16	0.46	0.67	0.51	0.19	0.30
3β,17β-ol-A⁵-11-one	0.37	0.10	0.38	0.50	0.40	0.20	0.17
3α-ol-5αA-7,17-one	0.40	0.12	0.53	0.54	0.47	0.27	0.32
3β-ol-5αA-7,17-one	0.33	0.07	0.50	0.49	0.45	0.23	0.27
3α-ol-5αA-11,17-one	0.44	0.14	0.59	0.60	0.59	0.37	0.40
3α-ol-5βA-11,17-one	0.38	0.09	0.52	0.61	0.55	0.33	0.31
11β-ol-5αA-3,17-one	0.57	0.27	0.63	0.75	0.64	0.44	0.46
4-ol-A⁴-3,17-one	0.59	0.41	0.72	0.76	0.70	0.62	0.67
11β-ol-A⁴-3,17-one	0.47	0.17	0.58	0.64	0.55	0.41	0.42

Table TLC 9 (continued)
SEPARATION OF ANDROGENS[8]

Androgen	System						
	1	2	3	4	5	6	7
11α-ol-A^4-3,17-one	0.37	0.08	0.49	0.49	0.50	0.29	0.26
14α-ol-A^4-3,17-one	0.39	0.10	0.54	0.54	0.52	0.33	0.33
15α-ol-A^4-3,17-one	0.32	0.06	0.47	0.45	0.46	0.29	0.24
16α-ol-A^4-3,17-one	0.38	0.11	0.59	0.58	0.53	0.36	0.42
19-ol-A^4-3,17-one	0.31	0.06	0.44	0.43	0.51	0.26	0.23
17β-ol-19-al-A^4-3-one	0.39	0.10	0.53	0.56	0.54	0.38	0.35
3β-ol-A^5-7,17-one	0.35	0.08	0.49	0.49	0.45	0.30	0.28
3β-ol-A^5-11,17-one	0.47	0.19	0.59	0.62	0.51	0.37	0.41
17β-ol-5αA-3-one	0.56	0.35	0.61	0.73	0.57	0.42	0.49
17β-ol-5βA-3-one	0.55	0.30	0.61	0.71	0.59	0.42	0.49
17β-ol-A^4-3-one	0.50	0.22	0.59	0.64	0.52	0.37	0.43
17α-ol-A^4-3-one	0.51	0.23	0.60	0.64	0.52	0.39	0.43
17β-ol-A^{1,4}-3-one	0.41	0.14	0.55	0.58	0.49	0.31	0.34
3α-ol-5αA-17-one	0.56	0.33	0.63	0.74	0.58	0.44	0.50
3β-ol-5αA-17-one	0.52	0.30	0.58	0.70	0.56	0.40	0.44
3α-ol-5βA-17-one	0.53	0.24	0.58	0.71	0.55	0.39	0.43
3β-ol-5βA-17-one	0.57	0.36	0.62	0.75	0.59	0.44	0.50
3β-ol-A^5-17-one	0.53	0.36	0.57	0.73	0.55	0.39	0.43
3α-ol-A^5-17-one	0.54	0.26	0.67	0.68	0.64	0.49	0.52
5αA-17β-ol	0.64	0.56	0.64	0.78	0.62	0.57	0.57
A^5-3β-ol	0.65	0.49	0.62	0.75	0.60	0.44	0.52
5αA-3α,17β-ol	0.52	0.26	0.50	0.70	0.48	0.33	0.32
5αA-3β,17β-ol	0.51	0.25	0.47	0.68	0.49	0.29	0.29
5βA-3α,17β-ol	0.46	0.17	0.45	0.63	0.48	0.25	0.23
5βA-3β,17β-ol	0.53	0.30	0.50	0.70	0.50	0.34	0.32
A^4-3β,17β-ol	0.50	0.26	0.47	0.70	0.50	0.30	0.32
A^5-3β,17β-ol	0.49	0.26	0.48	0.69	0.50	0.29	0.31
A^5-3β,17α-ol	0.50	0.26	0.49	0.71	0.50	0.30	0.32
A^5-3β,7β,17β-ol	0.28	0.04	0.23	0.43	0.29	0.09	0.07
A^5-3β,11β,17β-ol	0.31	0.06	0.22	0.47	0.33	0.13	0.07
5αA-3,17-one	0.66	0.46	0.74	0.78	0.66	0.64	0.69
5βA-3,17-one	0.64	0.43	0.74	0.78	0.67	0.63	0.67
5αA^1-3,17-one	0.61	0.38	0.76	0.75	0.66	0.58	0.67
5αA^2-7,17-one	0.67	0.52	0.77	0.81	0.70	0.61	0.70
A^4-3,17-one	0.53	0.30	0.73	0.68	0.64	0.55	0.65
A^{1,4}-3,17-one	0.49	0.20	0.71	0.62	0.58	0.47	0.58
5βA-3,11,17-one	0.52	0.25	0.69	0.70	0.66	0.50	0.58
19-al-A^4-3,17-one	0.48	0.19	0.70	0.63	0.64	0.54	0.61
A^4-3,11,17-one	0.39	0.14	0.67	0.60	0.56	0.48	0.53

Layer: Silica Gel G (Merck)

Mobile phase:
1. Cyclohexane-ethyl acetate-ethanol (45:45:10)
2. Cyclohexane-ethyl acetate (50:50)
3. Chloroform-ethanol (90:10)
4. Ethyl acetate-n-hexane-acetic acid-ethanol (72:13.5:10:4.5)
5. Benzene-ethanol (80:20)
6. Benzene-ethanol (90:10)
7. Chloroform-ethanol (95:5)

A = Androstane
A^4 = 4-Androsten
A^5 = 5-Androsten

Table TLC 10
SEPARATION OF TESTOSTERONE DERIVATIVES[9]

	Developing solvent system			
	1	**2**	**3**	**4**
Testosterone-3-CMO	0.00	0.62	—	0.54
Cysteamine	0.00	—	—	—
1,3-diaminopropane	—	0.00	—	0.00
1,7-diaminoheptane	—	0.00	—	0.00
Testosterone-3-cysteamine	0.60	—	1.00	—
Testosterone-3-DAP	0.00	0.65	—	0.05
Testosterone-3-DAH	0.00	0.65	—	0.10
IAEDANS	—	—	0.25	—
FITC	0.74	—	—	0.91
Testosterone-cysteamine-DANS	—	—	0.45	—
Testosterone-DAP-fluorescein	0.37	—	—	0.68
Testosterone-DAH-fluorescein	0.32	—	—	0.60

Layer: Silica Gel 60 (Merck)
Mobile phase:
1. Benzene:ethyl acetate:acetone (1:8:1)
2. Chloroform:methanol:water (60:39:1)
3. Chloroform:ethanol (1:1)
4. Chloroform:ethanol (7:3)

REFERENCES (TLC)

1. **Lisboa, B. P. and Hoffman, V.,** *J. Chromatogr.,* 115, 177, 1975.
2. **Petrovic, J. A. and Petrovic, S. M.,** *J. Chromatogr.,* 119, 625, 1976.
3. **Kerber, A., Morfin, R. F., Berthou, F. L., Picart, D., Bardou, L. G., and Floch, H. H.,** *J. Chromatogr.,* 140, 229, 1977.
4. **Ruh, T. S.,** *J. Chromatogr.,* 121, 82, 1976.
5. **Singh, H., Paul, H., Bhardwaj, T. R., Bhutani, K. K., and Ram, J.,** *J. Chromatogr.,* 137, 202, 1977.
6. **Tinschert, W. and Trager, L.,** *J. Chromatogr.,* 152, 447, 1978.
7. **Sunde, A., Stenstad, P., and Eik-Nes, K. B.,** *J. Chromatogr.,* 175, 219, 1979.
8. **Mattox, V. R., Litiviller, R. D., and Carpenter, P. C.,** *J. Chromatogr.,* 175, 243, 1979.
9. **Evrain, C., Rajkowski, K. M., Cittanova, N., and Jayle, M. F.,** *Steroids,* 36, 611, 1980.

Chapter 7

BILE ACIDS AND ALCOHOLS

SECTION I. PREPARATIONS

The cholanic acids are a group of substances that are products of bile acid metabolism in the GI tract and are found in the feces and portal blood. They are intermediates in the synthesis of bile acids. Derivatives are used in the treatment of liver disorders related to bile acid metabolism. It has been suggested that large bowel cancer may be related to the concentration and composition of the fecal bile acids. The bile acids in the feces are secondary bile acids which are formed from the primary compound via bacterial 7-dehydration.

A current technique for isolation and characterization of the bile acids is some form of chromatography. This section presents this method as well as extraction and detection methods currently in use.

Figure 1 gives the structures for the more common bile acids and their metabolites. Table 1 provides a glossary and explanation of abbreviations used.

There are few methods presently available for extraction and analysis of bile acid conjugates. Many methods hydrolyze and derivatize before analysis by gas chromatography (GC). Representative methods are described below. More recent methodology involves separation without hydrolysis. This section is divided into (1) preparation for direct TLC or HPLC without hydrolysis and (2) preparation including hydrolysis and derivatization for GC.

Extraction Methods for Bile Acids
Free, Conjugated, or Total Sample

Methods described in this section for the extraction of bile acids from various sources vary widely. The method of choice depends on the source. Often, it is possible to apply the sample directly to the thin layer as was done by Levin and Touchstone[1] with bile samples. Bile aliquots were diluted (1:1) with water or saline and 2 to 5 $\mu\ell$ diluted bile was streaked on Whatman C_{18} layers and developed for separation. This separation worked well for conjugated bile acids.

Separation of bile acids from jejunal juice (method of O'Moore and Percy-Robb[2]) — Aspirates of jejunal juice were centrifuged at 100,000 × g for 1.8 hr (MSE Superspeed 50) at 4°C. This separated the samples into three phases: pellet, aqueous, and lipid. The whole of the aqueous phase was carefully removed by piercing the centrifuge tube with a needle and withdrawing the fluids into a syringe. This is considered to be the "micellar" phase. Aliquots of jejunal juice, bile, or the micellar phase were extracted into 10 mℓ ethanol at 80°C and the precipitate removed by decanting the supernatant after centrifuging for 10 min. This was repeated 3 times and the ethanolic phases combined, reduced in volume *in vacuo*, and made up to a known volume with ethanol. Some samples of micellar phase were chromatographed without prior extraction.

Use of the ion exchange column to separate free, glyco- and tauro-bile acids (method of Alme et al.[3]) — A suspension of chlorohyroxypropyl Sephadex® LH-20 (27.2 g) in piperidine (100 to mℓ) was stirred at room temperature for 36 min. After addition of 5.74 g of KOH in methanol (302 mℓ), the mixture was heated at 20 to 50°C for 3 hr with occasional shaking. The product was collected by filtration, washed with 50% aqueous ethanol, 0.2 *M* acetic acid in 70% ethanol, and finally 90% ethanol until the washings were neutral. The PHP-LH-20 so prepared had an exchange capacity of 0.5 meq/g and was stored

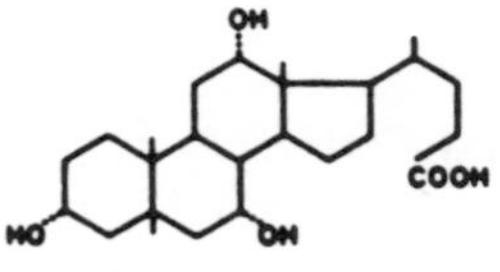

FIGURE 1.   Representative bile acids and conjugates: (1) cholic acid, (2) chenodeoxycholic acid, (3) deoxycholic acid, (4) hyodeoxycholic acid, (5) lithocholic acid, (6) glycocholic acid, (7) taurocholic acid, (8) glycolithocholic acid sulfate, (9) taurolithocholic acid sulfate.

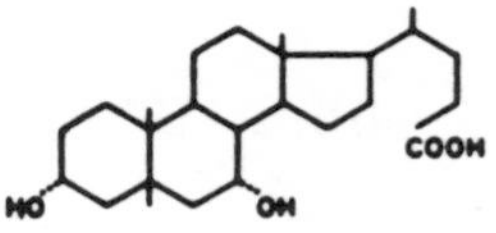

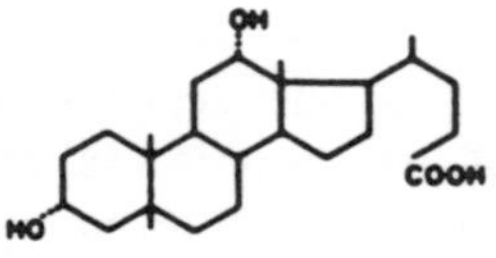

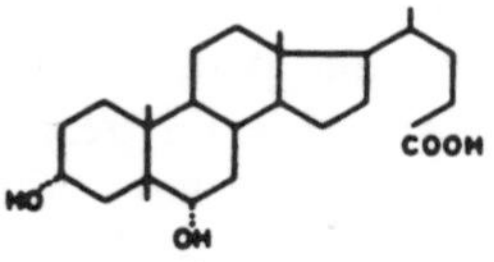

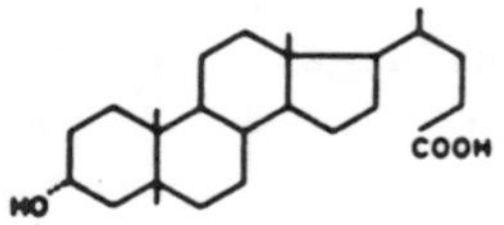

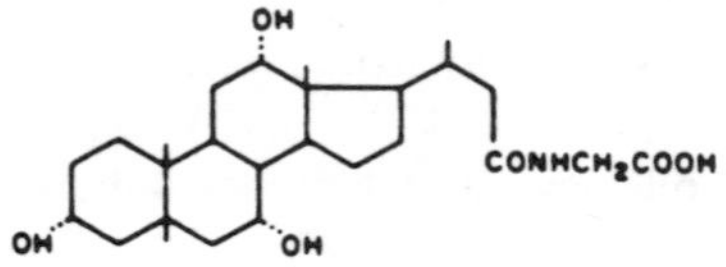

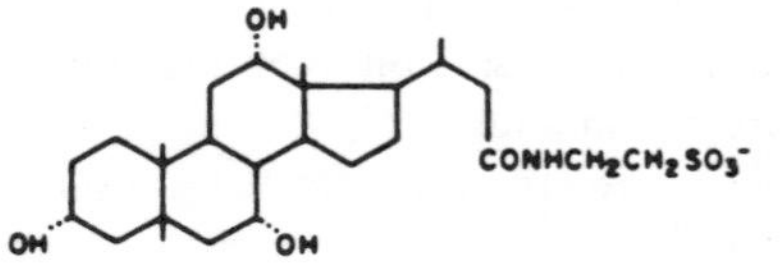

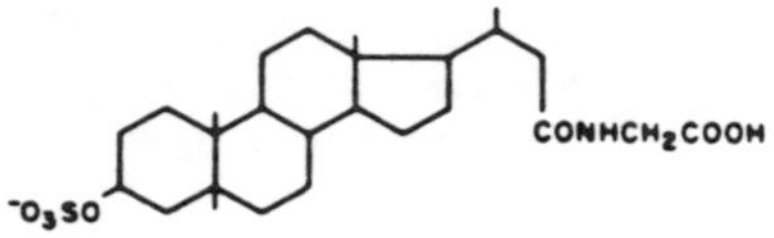

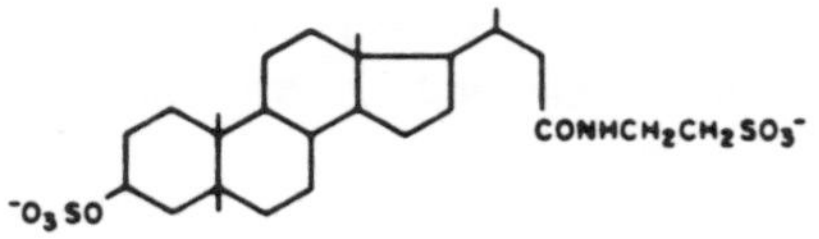

Table 1
GLOSSARY FOR BILE ACIDS

Cholic acid	CA
Chenodeoxycholic acid	CDCA
Deoxycholic acid	DCA
Lithocholic acid	LCA
Urosodeoxycholic acid	UDCA

Note: When preceded by G (glyco) the derivative is a conjugate with glycine; when preceded by T (tauro) the derivative is a conjugate with taurine; (TCA) Taurocholic acid, (GCA) glycocholic acid.

in the acetate form in ethanol. Plasma (usually 5 mℓ) was diluted with 9 vol 0.1 *N* NaOH and the sample was passed through a 1-g column of Amberlite® XAD-2. The column was washed with water and hexane, and the separation of bile acids was achieved by chromatography on Sephadex® LH-20 in chloroform-methanol 1:1 (v/v). This contained 0.01 mol/ ℓ sodium chloride, which yielded a nonsulfate fraction; a sulfate fraction was obtained by eluting the column with methanol. A fraction collected between the two fractions was found not to contain any bile acids.

These fractions can be further separated by TLC on silica gel G using the mobile phase ethyl acetate-ethanol-ammonia (5:5:1) or ethylene chloride-acetic acid-water (10:10:1). The sulfates have lower R_f values than the nonsulfated acids in both systems. Tauro-bile acids showed higher R_f values than the glycine conjugates in the alkaline system. With the acidic mobile phase, the respective mobilities were reversed.

Sample preparation for HPLC (method of Shaw et al.[5]) — Bile was deproteinized by the addition of 5 ethanol. This was kept overnight at −20°C and then centrifuged. The supernatant and combined washings were evaporated to dryness at room temperature. The residue was redissolved in the mobile phase used for HPLC or in ethanol and filtered through a Swinney filter before injection of an aliquot into the HPLC column. Bloch and Wattons[6] prepared bile and duodenal fluids for HPLC multiple procedures. Gallbladder bile samples were diluted (1 mℓ bile + 1.86 mℓ methanol). The duodenal samples (1 mℓ) were evaporated to dryness under nitrogen and dissolved in the mobile phase. The samples were vortexed and centrifuged at 1000 rpm for 5 min. An aliquot was then injected into HPLC.

Extraction Methods Including Hydrolysis[7]

Serum Extraction

The serum (1 to 2 mℓ) was extracted with 5 vol isopropanol at 65°C for 30 min. The supernatant was decanted and the precipitated protein extracted twice with 1 to 2 vol isopropanol. The pooled isopropanol extracts were evaporated to dryness under N and the residue completely redissolved in 2.5 M sodium hydroxide (10 mℓ) with gentle warming. The nonpolar lipids were removed from this solution by extraction with 3 vol petroleum ether (bp 60 to 80).

The bile acid conjugates that remained in solution were hydrolyzed in nickel crucibles in an autoclave for 3 hr at 110°C under a pressure of 15 lb. for 2 min. The hydrolyzate was then acidified to pH 4 using 1 N HCl and extracted with petroleum ether (bp 60 to 80) to remove other lipids.

Bile acids are removed by extracting three times with equal volumes of freshly distilled peroxide-free diethyl ether. The ether extract was washed to neutrality with distilled water, which in turn was washed with ether. The pooled ether extracts were taken to dryness under N and redissolved in a small volume of ethanol. The ethanolic solution was supplied as a narrow band to a thin layer plate coated with silica gel 600-μm thick.

Duodenal Contents Extraction

The duodenal contents (1 to 2 mℓ) were extracted at room temperature with 9 vol chloroform-methanol (9:1), and the protein filtered from the organic phase. An aliquot was taken from this chloroform-methanol extract and applied directly to the thin layer plate.

Use of Enzymatic Reactions to Hydrolyze Conjugated Bile Acid Prior to Extraction[8]

Extraction from serum — Five milliliters of 0.2 M acetate buffer (pH 5.6) were added to 5 mℓ serum. This buffer also contained EDTA and β-mercaptoethanol (10 to 20 μmol each) to activate the enzymatic hydrolysis. The pH 5.6 required for optimal enzymatic acitivity was attained by adding about three drops of 1 M HCl, to which an excess of N-cholylglycine hydrolase was added (0.1 to 0.2 mℓ) with a specific activity of 12,000 U/mℓ.

This mixture was incubated at 37°C for 3 hr; this period was found necessary for maximal hydrolysis of the conjugated bile acids present. Next, 60 mℓ absolute ethanol was added, whereupon the mixture was heated in a boiling water bath for 1 to 2 min for denaturation of the proteins present. This mixture was then kept overnight in a refrigerator at 4°C. It was then centrifuged and filtered. The residue was washed several times with 96% ethanol. The filtrate was then evaporated to heat dryness with the aid of a film evaporator (Rotavapor), and dissolved in 5 mℓ toluene-isopropanol-methanol-30% aqueous NaOH (10:20:20:6). The mixture was then shaken 4 times with 5 mℓ petroleum ether (boiling range 40 to 60°C) after the addition of 1 mℓ water in order to eliminate the neutral sterols, chiefly cholesterol.

The combined substrates were acidified with 6 *M* HCl to pH 1 and extracted 4 times with diethyl ether with very vigorous shaking to extract the free bile acids in the ether layer. The petroleum ether extract was eluted with water 2 more times. The water layers were acidified and extracted once with diethyl ether because the layer may contain bile acids. The combined extracts of diethyl ether were eluted with water twice more until acid-free, and then evaporated to dryness under a stream of N.

Extraction from bile — To 0.1 mℓ bile, 2.5 mℓ 0.1 *M* acetate buffer (pH 5.6) was added and hydrolysis was carried out as described for isolation from serum. Next, 5 mℓ water was added and acidified with about 0.3 mℓ 6 *M* HCl, whereupon extraction with 10 mℓ ether was carried out 4 times. The combined extracts of diethyl ether were eluted with water 2 more times until acid-free, and evaporated dry under an N stream. The extracts prepared can be used directly for HPLC or TLC, or derivatives can be prepared for separation by GC.

Bile acid removal from serum (Method of Van Berge Henegouwen and Hofmann[10]) — For batch extraction, serum (1 or 2 mℓ) was added to 12 mℓ 0.1 *N* NaOH; after gentle mixing, 1 g XAD-7 (or XAD-2) was added. The mixture was then gently shaken in a rotator for 60 min to ensure exposure of all resin particles to bile acids.

The aqueous phase was then removed using a Pasteur pipette attached to a vacuum line and the resin washed once with 10 mℓ water. Bile acids were then eluted from the resin with methanol or isopropanol (4 mℓ) elutions 3 times. In experiments using bile acid sulfates, the eluted bile acids were treated first with a solvolysis step. After elution, samples were then hydrolyzed using nickel Parr-Bombs and 8 mℓ 2 *N* NaOH and 50% methanol in H_2O (1:1, v/v). Samples were hydrolyzed for 3 hr at 115°C, neutralized with Dowex 50H', and the supernatant removed and evaporated. This sample can then be esterified and treated for trimethylsilylation.

Isolation of bile alcohols (free neutral sterols) — Bile samples were diluted with 20-fold excess of methanol and kept at 4°C for 1 hr. The precipitated protein was removed by filtration and washed with 3 mℓ methanol. The methanol fractions were pooled and evaporated under N. The samples were suspended in 2 mℓ 1% ammonium hydroxide solution and the free neutral sterols were extracted from the ammonia solution with four 4-mℓ portions of ethyl acetate-methanol (19:1, v/v). The extracts were pooled, washed with 4 mℓ water, and evaporated under N. Samples were redissolved in methanol.

Solvolysis of bile alcohols — The aqueous layer left after removal of the free neutral sterols was evaporated in air at 50°C. Solvolysis was then carried out in 50 mℓ acetone-ethanol (9:1, v/v) which was made acidic with 2 *N* HCl (pH 1 to 2). After 3 days at room temperature, the mixture was adjusted to pH 7 to 8 with 2 *N* NaOH and evaporated to dryness. The residue was redissolved in 2 mℓ water and extracted with four 4-mℓ portions of ethyl acetate-methanol (19:1, v/v). The extracts were pooled and washed with water to neutrality, dried over sodium sulfate, and evaporated. The residue was dissolved in 1 mℓ methanol.

Extraction of bile for bile acids (method of Igimi[11]) — An aliquot of the gall bladder-bile collected was diluted exactly 10 times with water. To 0.2 mℓ of each diluted solution was added 5 mℓ 5% NaOH, and hydrolysis was carried out for 3 hr in an autoclave maintained at 121°C and 2 kg/cm². The free bile acids were extracted directly with 3 vol of ether sucessively 4 times from the hydrolysate acidified with 3 mℓ HCl. After evaporation of the combined ether extract to dryness, the residue obtained was dissolved with methanol and methylated with freshly distilled diazomethane in ether to produce the bile acid methyl esters. The standard solutions of bile acids prepared above were likewise methylated in this way.

To the dry residue, 200 $\mu\ell$ trimethylsilylimidazole were added to convert the methyl ester to trimethylsilyl ethers. After 30 min at room temperature an aliquot can be injected into the GC.

Extraction of bile from pregnant women (method of Laatikainer et al.[12]) — Bile (100 mℓ) was added to 10 mℓ 96% (v/v) ethanol in an ultrasonic bath. After centrifugation, the sediment was treated in the same manner. After evaporation of the pooled supernatant, the sample was dissolved in chloroform-methanol (1:1, v/v) containing) 0.01 M NaCl and applied onto a 4-g column of Sephadex® LH-20. The fraction of 10 to 70 mℓ contained nonsulfated bile acid conjugates, then sulfated bile acids were eluted with 70 mℓ methanol. The bile salt in the first fraction was subjected to alkaline hydrolysis for 5 hr in 15% NaOH in 50% aqueous ethanol at 120°C under pressure in Teflon® containers. The second fraction was subjected to solvolysis for 2 days in acidified ethanol-acetone (1:9 v/v) and thereafter to alkaline hydrolysis. Bile acid methyl ester derivatives were prepared with diazomethane in ether-methanol (5:1 v/v). Before GC, a suitable amount of stigmasterol was added as an internal standard and trimethylsilyl ether derivatives of bile acid and methyl esters were prepared. Recoveries of deoxycholyl and chenodeoxycholyl taurine and cholic acid-3-sulfate added to bile samples varied from 71 to 95%.

Serum bile acids (method of Karlaganis and Paumgartner[13]) — Hydrodeoxycholic acid 2.5 μg (6.36 nmol; internal standard) in 50 $\mu\ell$ methanol was pipetted into a glass tube and evaporated to dryness; 2 mℓ serum and 18 mℓ 0.1 M sodium hydroxide in 0.9% sodium chloride were added. The mixture was kept in an ultrasonic bath for 15 min and then passed over a column (35/5 mm) of approximately 0.5 g Amberlite® XAD-2 in water, which had been washed with methanol, acetone, and water. This procedure was carried out at 4°C at a flow rate of about 0.2 to 0.3 mℓ/min.

The column was subsequently washed with 1 mℓ 0.1 M sodium hydroxide in 0.9% sodium chloride and 5 mℓ water and then eluted with 5 mℓ methanol. The eluate was evaporated to dryness at 50°C under a stream of N.

Solvolysis — Methanolic hydrogen chloride was prepared by bubbling dry gaseous hydrogen chloride through cold methanol (dried over a 0.3-nm molecular sieve). One milliliter dry 1 M HCl in methanol and 9 mℓ acetone (dried over molecular sieve 0.3 nm) were added to the dry sample, placed in an ultrasonic bath for several seconds, and kept at room temperature for 3 hr. The mixture was then neutralized by 1 mℓ 1 M dry methanolic sodium hydroxide and evaporated to dryness at 50°C under a stream of N.

Enzymatic hydrolysis — The residue was dissolved in 4 mℓ glass-distilled water, and the pH was adjusted to 5.8 to 6.0 with a few drops of 0.2 M acetic acid, followed by 0.4 mℓ 0.1 M sodium acetate buffer (pH 5.6). The mixture was placed in an ultrasonic bath for a few seconds; 20 $\mu\ell$ (+ 5 units) cholylglycine hydrolase solution, 0.1 mℓ freshly prepared solutions of the disodium salt of EDTA (0.2 M), and 2-mercaptoethanol (0.2 M) were added. The mixture was incubated at 37°C overnight for at least 12 hr as suggested, cooled to room temperature, and acidified to pH 1.0 with 6 M HCl. The aqueous solution was extracted 3 times with 10 mℓ diethyl ether. The ether extract was evaporated to dryness with N at 35°C after the addition of 0.5 mℓ methanol.

Extraction of urine for bile acids (method of Makins et al.[14]) — Urine (30 mℓ) was percolated through 10 g Amberlite® XAD-2 resin in a column (1.0 × 20.0 cm) at a rate of about 1 drop every 2 sec. The column was washed with 60 mℓ water, and bile acids were eluted with 60 mℓ methanol. The elute from the Amberlite® XAD-2 column was evaporated to dryness. The residue was dissolved with ultrasonic agitation in 2 mℓ chloroform-methanol 1:1 (v/v) containing 0.01 M sodium chloride. This solution was then added to a column (1.0 × 25.0 cm) containing 4 g Sephadex® LH-20, which had been packed and equilibrated overnight with the same solvent. The glycine or taurine conjugates (non-sulfated fraction) were eluted with 70 mℓ solvent. Elution was carried out with 50 mℓ methanol to obtain the sulfated fraction.

Solvolysis, hydrolysis, and methylation of bile acids — The sulfated fraction was subjected to solvolysis in 30 mℓ ethanol-acetone 1:9 (v/v) and acidified with a few drops of 2 N HCl to pH 1. After 2 days at room temperature, the mixture was neutralized by addition of 1 N NaOH, and the solvent evaporated.[16] The residue was hydrolyzed with 15 mℓ 15% NaOH for 4 hr at 120°C. After acidification to pH 1 with 6 N HCl, the free bile acids were extracted with 3 × 50 mℓ of ether. The pooled extracts were washed to neutrality with 5 mℓ water and then re-extracted with 20 mℓ ether. The ether phases were pooled and the solvent was evaporated. The residue was dissolved in 5 mℓ ether-methanol 9:1 (v/v), and diazomethane was added in excess. This mixture was allowed to stand 15 min, and then the solution was evaporated to dryness. The nonsulfated bile acid fraction was treated in the same manner except for initial solvolysis.

Separation of bile acids from serum (method of Dinh and Penner[15]) — Serum (2 mℓ) was diluted with 9 vol 0.1 N NaOH in 0.9% NaCl. The sample was percolated through a small column of Amberlite® XAD-2 resin (0.5 × 20.0 cm) at a rate of about 1 drop every 2 sec. The column was washed with water until the effluent was neutral, and then bile acids were eluted with 10 mℓ ethanol. The separation of sulfated and nonsulfated bile acids on Sephadex® LH-20 and the solvolysis, hydrolysis, and methylation procedure were as described above for urinary bile acids.

Bile acids (Method of Barnes and Clintrenukooh[16]) — Bile was collected by placing the gall bladder of rabbits into ethanol. After separation and evaporation of the ethanol the crude bile salts were obtained. This material (3.7 g from 60 rabbits) was dissolved in 500 mℓ water, acidified with 2 N HCl, and finally extracted with three 300 mℓ portions of ethyl acetate. The ethyl acetate was washed with 5% sodium carbonate to remove acidic materials (glycine conjugated bile acids) and also with water. The washed extract was dried over anhydrous Na_2SO_4, and the solvent was evaporated to dryness, leaving a residue (211 mg) consisting of neutral lipids, i.e., biliary sterols and unconjugated bile alcohol. The residue was dissolved in 1 mℓ benzene. This solution was then added to a column (1.6 × 11 cm) of silica gel (10 g, Merck) made up with benzene. The biliary sterols were eluted from the silica gel column with 100 mℓ ethyl acetate. The column was then eluted with 100 mℓ acetone to yield unconjugated bile alcohols. Evaporation of the solvent from the acetone eluate gave 34.9 mℓ extract.

Acid hydrolysis of sulfated bile alcohols — The aqueous layer left from ethyl acetate extract was adjusted to pH 7 with 1 N NaOH and percolated through 450 g Amberlite® XAD-2 resin in a column (3.8 × 63 cm) at, a rate of about 1 drop every 2 sec. The column was washed with 1.5 ℓ water, and then with 1.5 ℓ methanol. The methanol eluate was evaporated to dryness, leaving a residue (114 mg) that was subjected to acid hydrolysis. The residue was dissolved in 30 mℓ 0.1 N HCl and NaCl (1.6) and ethyl acetate (12 mℓ) was added. After 2 days at 37°C the mixture was extracted with two 200 mℓ portions of ethyl acetate. The combined ethyl acetate extract was washed with water and 5% Na_2CO_3 solution and water, successively. Evaporation of the solvent from the washed extract left a residue of 22.1 mg.

Preparation of bile samples for analysis of bile acids (method of Haeffner et al.[17]) — Human duodenal juice (1 mℓ) was extracted with 20 mℓ chloroform-methanol (2:1). A 2-mℓ aliquot was evaporated and 10 mℓ 15% aqueous sodium hydroxide added. The solution was heated at 120°C for 4 hr and acidified to pH 3. It was extracted twice with 20 mℓ ether, the ether was dried, evaporated to dryness, and the residue dissolved in 100 μℓ 1,1,1,3,3,3,hexafluoroisopropanol and 200 μℓ trifluoroacetic anhydride. The solution was heated at 37°C for 30 min, evaporated, dissolved in 100 μℓ acetonitrile, and an aliquot injected into the GC.

Bile acids in serum (method of Haeffner et al.[17]) — XAD-7 resin (1 g) was added to 10 mℓ glass tubes fitted with a screw cap lined with Teflon® spacer. Serum (0.2 to 2.0 mℓ) was added to the tubes followed by sodium hydroxide solution to give a final volume of 9 mℓ. The tubes were then mixed on a Rollermix at room temperature (21°C) for 60 min. Tubes were placed upright and the supernatant removed with a Pasteur pipette connected to a vacuum line. In order to maintain a pH of at least 10 at all stages during the washing procedure, dilute sodium hydroxide solution (2 mℓ, 0.004 mol/ℓ) was added. The tubes were mixed briefly by hand on a Whirlimixer, and the supernatant removed. Methanol (6 mℓ) was added to the resin, and the tubes were mixed for a further 20 min on the Rollermix. The methanolic supernatant was transferred to 50 mℓ-stoppered boiling tubes. The resin was extracted with a second 6-mℓ portion of methanol, and the supernatant solutions were combined. Methanol was removed *in vacuo* using a rotary evaporator at 50°C. This extract can be derivatized and subjected to GC/MS or applied directly for TLC.

Bile acids in intestinal aspirates (this method of Goto et al.[18] gives a total bile acid and does not separate the conjugates) — Intestinal samples for determination of bile acid distribution were processed as follows: 20 mℓ absolute alcohol were added to a 2.0 mℓ aliquot of aspirate. The sample was placed in a boiling water bath for 10 min, cooled, and centrifuged. After the addition of 4.6 mℓ 1 *N* NaOH, the supernatant was extracted 3 times with 20 mℓ petroleum ether. The combined ether extracts were washed with a mixture of 3.5 mℓ water and 6.5 mℓ 1 *N* NaOH in 50% ethanol; this wash was added to the sample. The sample was then mixed with 4 mℓ 10 *N* NaOH, transferred into nickel crucibles, and the bile acids were hydrolyzed for 3 hr in an autoclave (15 psi, 122°C). After hydrolysis, the pH was adjusted to 1.1, and the sample was stored at 4°C overnight. The sample was then sequentially extracted with 35, 50, 35, and 50 mℓ chloroform, passed through anhydrous sodium sulfate, and evaporated to dryness. The residue was dissolved in 0.5 mℓ of methanol and 4.5 mℓ of diethyl ether. The methyl esters of the bile acids were prepared by adding fresh diazomethane until a yellow color persisted and were dried under nitrogen. Trifluoroacetic anhydride, 0.5 mℓ, was added to each sample and allowed to react for 30 min at 38°C. The sample was then dissolved in 0.3 mℓ of acetone immediately prior to GLC.

REFERENCES

1. **Levin, S. S. and Touchstone, J. C.,** in *Advances in Thin Layer Chromatography,* Touchstone, J. C., Ed., Wiley, New York, 1982, 229.
2. **O'Moore, R. R. L. and Percy-Robb, W.,** *Clin. Chim. Acta,* 43, 39, 1973.
3. **Alme, B., Bremmelgaard, A., Sjovall, J., and Thomassen, P.,** *J. Lipid Res.,* 18, 339, 1977.
4. **Back, P., Sjovall, J., and Sjoval, K.,** *Med. Biol.,* 52, 31, 1979.
5. **Shaw, R., Smith, J. A., and Elliott, W. H.,** *Anal. Biochem.,* 86, 1450, 1978.
6. **Bloch, C. A. and Wattons, J. B.,** *J. Lipid Res.,* 19, 510, 1978.
7. **Eastwood, M., Hamilton, D., and Mobray, L.,** *J. Chromatogr.,* 65, 407, 1972.
8. **Van Berge Henegouwen, G. P., Ruben, A., and Bramet, K. H.,** *Clin. Chim. Acta,* 54, 249, 1974.
9. **Kuramoto, T., Cohen, B., and Morbach, E. N.,** *Anal. Biochem.,* 71, 481, 1974.
10. **Van Berge Henegouwen, G. P. and Hofmann, A. F.,** *Clin. Chim. Acta,* 73, 469, 1976.
11. **Igimi, H.,** *Life Sci.,* 18, 993, 1976.
12. **Laatikainer, T., Lehtoner, P., and Hasco, A.,** *Clin. Chim. Acta,* 85, 145, 1978.

13. **Karlaganis, P. and Paumgartner, G.,** *Clin. Chim. Acta,* 92, 19, 1979.
14. **Makins, I., Shinozaki, K., Makagawa, S., and Mashimo, K.,** *J. Lipid Res.,* 15, 132, 1974.
15. **Dinh, D. M. and Penner, J. W.,** *Steroids,* 29, 193, 1977.
16. **Barnes, S. and Chitranukroh, A.,** *Ann. Clin. Biochem.,* 14, 1235, 1971.
17. **Haeffner, J., Gordon, S. J., Strum, S., Elliot, M., and Kowlessar, D. O.,** *Ann. Clin. Sci.,* 6, 11, 1976.
18. **Goto, J., Hasagawa, M., Kato, H., and Nambara, T.,** *Clin. Chim. Acta,* 87, 141, 1978.

SECTION II. GAS CHROMATOGRAPHY

GC of bile acids and alcohols requires the preparation of a derivative. The methods for this are described in the first part of this section. Usually the carboxyl group is esterified with diazomethane or methanol saturated with hydrogen chloride. Precautions for these reactions are described. The acetate trimethylsilyl derivative can also be prepared to block the hydroxyl groups. The examples given in this section are selective, and by no means are all possibilities described. The data given are taken from the literature and have not been verified by the author of this handbook.

Table GC 1
RETENTION TIMES FOR BILE ACID METHYL ESTERS AND ACETATES[1]

Steroid	Retention time (min)
Cholestrol	4
Lithocholate	7
3β-Cholanate	8
Deoxycholate	14
Chenodeoxycholate	18
Hyodeoxycholate	22
Cholate	28
Hyocholate	36
α-Muricholate	42
β-Muricholate	50

Instrument:	Packard 7401 with flame detector
Column:	1% OV-225 on Gas Chrom Q (100—200 mesh), (3 ft × 2 mm)
Temperatures:	Column, detector, and oven 250°C
Flow:	Helium 20 mℓ/min

Table GC 2
SEPARATION OF BILE ACID METHYL ESTERS[2]

Bile acid[a]	RRT[b]	Bile acid[a]	RRT[b]
$C_{24}5\beta,12\alpha,OAc,\Delta3$	0.31	$C_{24}5\beta,3\alpha OAc,7=0$	1.95
$C_{24}5\beta,3\alpha OAc,\Delta11$	0.70	$C_{24}5\beta,3\alpha OAc,12=0$	2.02
$C_{24}5\beta,3\alpha OAc$	0.71	$C_{24}5\beta,3\alpha OAc,12=0,\Delta9(11)$	2.11
$C_{24}5\beta,3=0$	0.75	$C_{24}5\beta,3\alpha OAc,6\beta OAc,7\alpha OAc$	2.26
$C_{24}5\beta,3\alpha OAc,12\alpha OAc,\Delta8(14)$	0.86	$C_{24}5\beta,3\alpha OAc,6\alpha OAc,7\alpha OAc$	2.36
$C_{24}5\beta,3\alpha OEt,7\alpha OAc,12\alpha OAc$	0.92	$C_{24}5\beta,3\alpha OAc,7=0,12\alpha OAc$	2.40
$C_{24}5\beta,3\alpha OMe,7\alpha OAc,12\alpha OAc$	0.92	$C_{24}5\alpha3=0,12=0$	2.42
$C_{24}5\beta,3\alpha OAc,12\alpha OAc$	0.93	$C_{24}5\beta,3=0,6=0$	2.80
$C_{24}5\alpha,3=0$	0.93	$C_{24}5\beta,3\alpha OAc,6\beta OAc,7\beta OAc$	3.10
$C_{24}5\beta,3\alpha OAc,12\alpha OAc$	1.00	$C_{24}5\beta,3\alpha OAc,7\alpha OAc,12=0$	3.12
$C_{24}5\alpha,3\alpha OAc,12\alpha OAc$	1.14	$C_{24}5\beta,3\alpha OAc,7\alpha OAc,12\alpha OH$	(3.59)
$C_{24}5\beta,3=0,12\alpha OAc$	1.14	$C_{24}5\beta,3=0,7\alpha OAc,12=0$	3.62
$C_{24}5\beta,3\alpha OAc,7\alpha OAc,\Delta11$	1.26	$C_{24}5\beta,3\alpha OAc,7=0,12=0$	4.20
$C_{24}5\beta,3\alpha OAc,7\alpha OAc$	1.30	$C_{22}3\beta,OAc,\Delta5$	0.38
$C_{24}5\alpha,3\beta OAc,12\alpha OAc$	1.45	$C_{22}5\beta,3\alpha OAc,12\alpha OAc$	0.46
$C_{24}5\beta,3\alpha OAc,7\alpha OAc,12\alpha OAc$	1.62	$C_{22}5\beta,3\alpha OAc,11\alpha OAc$	0.64
$C_{24}5\beta,3=0,7\alpha OAc$	1.62	$C_{22}5\beta,3\alpha OAc,7\alpha OAc,12\alpha OAc$	0.75
$C_{24}5\alpha,3=0,12\alpha OAc$	1.62	$C_{23}5\beta,3\alpha OAc,12\alpha OAc$	0.73
$C_{24}5\beta,3\beta OAc,7\alpha OAc,12\alpha OAc$	1.65	$C_{23}5\beta,3\alpha OAc,11\alpha OAc$	0.98
$C_{24}5\beta,3\alpha OAc,6\beta OAc$	1.69	$C_{27}3\alpha OAc,\Delta5$	0.37
$C_{24}5\beta,3\alpha OAc,6\alpha OAc$	1.82	$C_{27}5\beta,3\alpha OAc,7\alpha OAc,12\alpha OAc,25 OAc$	0.76
$C_{24}5\beta,3=0,7\alpha OAc,12\alpha OAc$	1.82	$C_{27}5\beta,3\alpha OAc,7\alpha OAc,12\alpha OAc,24 OAc$	0.98
$C_{24}5\alpha,3\alpha OAc,7\alpha OAc,12\alpha OAc$	1.83	$C_{27}5\beta,3\alpha OAc,7\alpha OAc$	2.03
$C_{24}5\beta,3\alpha OAc,7\beta OAc$	1.93	$C_{27}5\beta,3\alpha OAc,7\alpha OAc,12\alpha OAc$	2.50

Instrument: Varian 1400 with flame detector
Column: 0.5% P-525 on Gas Chrom Q (100—200 mesh)
Temperatures: Column 240°C, detector and injector, 280°C
Flow: Nitrogen at 15 mℓ/min

Note: C_{24}-5β = 5β-cholan-24-oic acid, C_{22}-5β = 24,23-dinor-5β-cholan-22-oic acid, C_{23}-5β = 24 nor-5β-cholan-23-oic acid, C_{27}-5β = -cholestane.

[a] Derivatives of cholanic acid.
[b] Time relative to deoxycholate at 665 sec.

= O is keto.
− OAC is acetate.
Δ is double bond.

Table GC 3
SEPARATION OF FECAL BILE ACIDS[3]

Bile acid[a]	Retention times[b]	
	Standard	Fecal component
Lithocholic	0.64	0.64
3β,12α-Dihydroxycholanic	0.85	0.84

Table GC 3 (continued)
SEPARATION OF FECAL BILE ACIDS[3]

	Retention times[b]	
Bile acid[a]	**Standard**	**Fecal component**
Deoxycholic	1.00	1.00
3α,12β-Dihydroxycholanic	1.12	1.10
Chenodeoxycholic	1.26	1.22
Ursodeoxycholic	1.48	1.43
Cholic	2.10	2.09
7-Keto,α-lithocholic	2.44	2.43
3-Keto,7α-hydroxycholanic	2.66	2.69
3β,12-Keto-cholanic	1.97	1.98
3α,12-Keto-cholanic	2.23	2.20

Instrument:	Varian 2700 with flame detector
Column:	3% QF-1 on Chromosorb W (100—120 mesh)
Temperature:	Column 225°C, detector 245°C, injector 245°C
Flow:	Helium 50 mℓ/min

a Trifluoroacetates of the methyl esters.
b Retention times relative to deoxycholate.

Table GC 4
RETENTION TIMES OF BILE ACID METHYL ESTER ACETATES RELATIVE TO DEOXYCHOLATE[4]

Group on cholanic acid nucleus	1[a]	2[b]
$C_{24}5\beta,12\alpha OAc,\Delta3$[c]	0.31	0.28
$C_{24}5\beta,12 = O$		0.52
$C_{24}5\beta,7 = O$		0.59
$C_{24}5\beta,3\alpha OAc,\Delta11$	0.70	0.65
$C_{24}5\beta,3\alpha OAc$ (lithocholic acid)	0.71	0.67
$C_{24}3\beta,\Delta5$		0.83
$C_{24}5\beta,3\alpha OAc7\alpha OAc12\alpha OAc$	0.915	0.92
$C_{24}5\beta,7 = O,12\alpha OAc$		0.93
$C_{24}5\beta,3\alpha OAc,\Delta8(14)$	0.86	0.97
$C_{24}5\beta,3 = 0$	0.75	0.99
$C_{24}5\beta,3\alpha OAc,12\alpha OAc$ (deoxycholic acid)	1.00	1.00
$C_{24}5\beta,3\alpha OMe,7\alpha OAc,12\alpha OAc$	0.925	1.02
$C_{24}5\alpha,3 = O$	0.93	1.14
$C_{24}5\alpha,3\alpha OAc,12\alpha OAc$ (allodeoxycholic acid)	1.14	1.22
$C_{24}5\beta,3\alpha OAc,7\alpha OAc,\Delta11$	1.26	1.36
$C_{24}5\beta,3\alpha OAc,7\alpha OAc$ (chenodeoxycholic acid)	1.30	1.38
$C_{24}5\beta,3 = O,12\alpha OAc$	1.14	1.46
$C_{24}5\alpha,3\beta OAc,12\alpha OAc$	1.45	1.52
$C_{24}5\beta,7 = O,12 = 0$		1.62
$C_{24}5\beta,7 = O,12\alpha OH$		1.85
$C_{24}5\beta,3\alpha OAc,7\alpha OAc,12\alpha OAc$ (cholic acid)	1.62	1.87
$C_{24}5\beta,3\alpha OAc,6\alpha OAc$ (hyodeoxycholic acid)	1.82	1.87
$C_{24}5\beta,3\beta OAc,7\alpha OAc,12\alpha OAc$	1.65	1.88
$C_{24}5\beta,3\alpha OAc,6\beta OAc$	1.69	1.91
$C_{24}5\beta,3\alpha OAc,7\beta OAc$ (ursodeoxycholic acid)	1.93	2.02
$C_{24}5\alpha,3 = O,12\alpha OAc$	1.62	2.14
$C_{24}5\alpha,3\alpha OAc,7\alpha OAc,12\alpha OAc$ (allocholic acid)	1.83	2.15
$C_{24}5\beta,3 = O,7\alpha OAc$	1.62	2.16

Table GC 4 (continued)
RETENTION TIMES OF BILE ACID METHYL ESTER ACETATES RELATIVE TO DEOXYCHOLATE[4]

Group on cholanic acid nucleus	1[a]	2[b]
$C_{24}5\beta,3\alpha OAc,12 = 0$	2.02	2.29
$C_{24}5\beta,3\alpha OAc,6\alpha OAc,7\ OAc$ (hyocholic acid)	2.36	2.45
$C_{24}5\beta,3\alpha OAc,7 = O$	1.95	2.47
$C_{24}5\beta,3\alpha OAc,12 = 0,\Delta 9(11)$	2.11	2.62
$C_{24}5\beta,3\alpha OAc,6\beta OAc,7\alpha OAc$ (α-muricholic acid)	2.26	2.66
$C_{24}5\beta,3\alpha OH,7\alpha OH$		2.74
$C_{24}5\beta,3 = O,7\alpha OAc,12\alpha OAc$	1.82	2.89
$C_{24}5\beta,3\alpha OAc,7 = O,12\alpha OAc$	2.40	3.24
$C_{24}5\beta,3OAc,6\beta OAc$ (β-muricholic acid)	3.10	3.85
$C_{24}5\beta,3\alpha OAc,7\alpha OAc,12\alpha OH$	3.59	3.87
$C_{24}5\beta,3\alpha OAc,7\alpha OAc,12 = O$	3.12	4.18
$C_{24}5\beta,3\alpha OAc,7\alpha OAc,12 = O$	2.80	4.95

Instrument: Varian 1400
Column: 1, 0.5% SP-525, 2, 1% Poly S-175 both on Gas Chrom Q (100—120 mesh), (18.3 cm × 1 mm)
Temperatures: Column 250°C, detector and flash 280°C (both)
Flow: Nitrogen 15 mℓ/min

[a]   Deoxycholate = 665 sec.
[b]   Deoxycholate = 825 sec.
[c]   Derivatives of cholamic acid.

= O is keto.
OAc is acetate.
Δ is double bond.

Table GC 5
SEPARATION OF BILE ACIDS AS THEIR HEPTAFLUOROBUTYRATES[5]

Derivative from	Column/retention time				
	1	2	3	4	5
LC-acid[a]	1.00	1.00	0.72, 1.48	1.00	
LC-gly-Na	1.00	1.04	0.72, 1.48	1.00	
LC-tau-Na	1.01	1.00	0.71, 1.48	0.99	
DC-acid	1.18	1.57	0.79, 1.25	1.31	1.22
DC-gly-Na	1.17	1.56	0.78, 1.24	1.32	1.25
DC-tau-Na	1.18	1.59	0.79, 1.24	1.31	1.24
CDC-acid	1.42	1.77	1.46	1.00	1.25
CDC-gly-Na	1.41	1.78	1.45	1.01	1.26
CDC-tau-Na	1.41	1.77	1.46	1.10	1.26
C-acid	2.55	3.53			2.85
C-gly-Na	2.56	3.55			2.84
C-tau-Na	2.55	3.53			2.85

Instrument: Hewlett-Packard, 5830A with [63]Ni electron, capture detector; columns, temperatures, and flow rates: detector at 300°C.

Table GC 5 (continued)
SEPARATION OF BILE ACIDS AS THEIR HEPTAFLUOROBUTYRATES[5]

Coating	Support	Column temp. (°C)	Injection temp. (°C)	Flow (mℓ/min)	Absolute RT of LC-HFB/min
1. OV-225 3%	GC-Q,AW-DMCS 100—120 mesh	230	245	27	3.40
2. QF-1 0.5%	GC-QAW-DMCS 100—120 mesh	215	250	26	1.52
3. SE-30 0.75%	GC-QAW-DMCS 100—120 mesh	240	265	25	3.73
4. SE-30/QF-1 4%/6%	Chromosorb W-HP 80—100 mesh	260	275	24	6.46
5. OV-17 1%	Chromosorb W-HP 80—100 mesh	235	260	25	2.36

[a]   See glossary in Section I of this chapter.

Table GC 6
SEPARATION OF TRIMETHYL SILYL ETHERS OF BILE ALCOHOLS[6]

Compound	RRT (column) 1	2	3	4	5
5β-Cholestane-3α,4α,12α-triol	2.33	2.26	2.43	2.92	3.11
5β-Cholestane-3α,7α,25-triol	4.13	1.75	4.88	2.99	3.05
5β-Cholestane-3α,7α,26-triol	4.79	2.29	5.67	3.39	3.47
5β-Cholestane-3α,7α,12α,26-tetrol	4.19	1.19	4.99	3.08	3.00
5β-Cholestane-3α,7α,12α,26-tetrol	4.80	1.50	5.68	3.30	3.28
5β-Cholestane-3α,7α,12α,22e,25-pentol	6.32	1.45	7.22	3.66	3.60
5β-Cholestane-3α,7α,12α,23e,25-pentol	6.49	1.52	7.55	3.85	3.72
5β-Cholestane-3α,7α,12α,24e,25-pentol	6.62	1.67	8.05	4.48	4.27
5β-Cholestane-3α,7α,12α,25,26-pentol	7.57	1.94	9.36	4.91	4.74

Instrument: Hewlett-Packard, 7610A with flame detector
Columns: U-shaped columns in all cases
    1.   3% SE-30 on Supelcoport (100—120 mesh)
    2.   1% HiEff-3Bp Supelcoport (100—120 mesh)
    3.   1.5% OV-1, Gas Chrom Q (100—120 mesh)
    4.   2% OV-210, Gas Chrom Q (100—120 mesh)
    5.   3% QF-1, Gas Chrom Q (100—120 mesh)
Temperatures: 1.   Column, 250°C, detector and injector 280°C
    2.   Column, 235°C, detectors and injector 270°C
    3—5. Column 240°C, detector and injector 270°C
Flow: Nitrogen, rate not given

Table GC 7
RETENTION TIMES OF BILE ACID DERIVATIVES[7]

Component	RT TFA-HFIP	TFA-methyl	HFB
Cholesterol	0.49	0.32	0.47
Cholanoic	0.20	0.25	0.19
3α	0.60	0.69	0.54
3β	0.59	0.68	0.56
3β-Δ⁵	0.66	0.73	0.66
3β,Δ⁵-22,23-bisnor	0.35	0.36	—
5α,3β	0.74	0.81	0.73
3α,6α	1.48	1.37	1.22
3α,6β	1.38	1.24	1.21
3β,7α	1.05	1.02	1.25
3α,7α	1.29	1.22	1.21
3α,7β	1.49	1.40	1.20
3β,12α	0.86	0.88	0.76
3α,12α	1.00	1.00	1.00
7α,12α	0.62	0.63	0.60
3α,12α,Δ	0.86	0.85	0.85
3-keto	1.37	1.49	1.19
3α,6-keto	2.69	2.69	2.33
3α,7-keto	2.30	2.24	1.93
3α,12-keto	1.93	2.03	1.68
3-keto,7α	2.62	2.37	2.10
3-keto,7β	2.74	2.84	2.32
3-keto,12α	2.07	2.00	1.66
3α,12-keto,Δ⁹⁽¹¹⁾	2.12	2.55	1.82
5α-3α,6-keto	2.46	2.63	2.01
3,6-diketo	5.89	5.96	4.99
3,7-diketo	4.01	4.34	3.40
3,12-diketo	3.63	4.29	3.10
7,12-diketo	1.72	2.18	1.46
5α-3,6-diketo	5.91	6.06	5.08
3α,6α,7α	2.00	1.84	2.31
3α,7α,12α	2.22	1.90	2.58
5α-3α,7α,12α	2.44	2.02	2.80
3α,7α,12-keto	3.77	3.56	3.31
3α,12α,7-keto	3.28	2.93	3.02
3-keto,7α,12α	4.30	3.59	3.71
3α,7,12-diketo	5.50	6.35	4.70
3,7-diketo,12α	5.16	5.28	3.96
3,7,12-triketo	8.24	9.70	6.89

Instrument:	Varian 1700 with flame detector
Column:	QF-1 on Chromosorb W, AWDNCS (100—120 mesh) (1.8 m × 3 mm)
Temperatures:	Column and detector 230°C, injection port 240°C
Flow:	Nitrogen 30 mℓ/min

TFA-HFIP = trifluoroacetyl-hexafluoro-isopropyl ester, TFA-methyl = trifluoroacetyl-methyl ester, and HFB = heptafluorobutyrate. Relative retention times refer to the corresponding deoxycholate derivative. The deoxycholate was injected simultaneously with each of the derivatives.

Note: α and β are configurations of the hydroxyls except at 5 which is the hydrogen.

REFERENCES (GC)

1. **Yousef, I. M., Fisher, M. M., Myher, J. J., and Kukis, A.,** *Anal. Biochem.,* 75, 538, 1976.
2. **Szczepanik, P. A., Hachey, D. L., and Klein, P. D.,** *J. Lipid Res.,* 17, 314, 1976.
3. **Subbiah, M. T. R., Tyler, N. E., Buscoglia, M. D., and Mari, L.,** *J. Lipid. Res.,* 17, 78, 1976.
4. **Szczepanik, P. A., Hachey, D. L., and Klein, P. D.,** *J. Lipid. Res.,* 19, 280, 1978.
5. **Musial, B. C. and Williams, C. N.,** *J. Lipid Res.,* 20, 78, 1979.
6. **Kuramotow, T., Cohen, B. D., and Mosbach, E. H.,** *Ann. Biochem.,* 71, 481, 1976.
7. **Edeharder, R. and Slemr, J.,** *J. Chromatogr.,* 222, 1, 1981.

SECTION III. HIGH PERFORMANCE LIQUID CHROMATOGRAPHY

The separation of bile acids by HPLC has not been fully developed. Due to the similarity and number of the hydroxyl functions, separation has been difficult to achieve. The molecules do not have strong UV absorbance, and therefore, some investigators have resorted to derivatization. In this section some of the methods will be described. For the sake of clarity, the actual chromatograms are shown in most cases instead of retention times.

Table HPLC 1
SEPARATION OF FREE BILE ACIDS[1]

	RT		
Bile acid	1	2	3
5β-Cholanic acid	1	1	1
3-Keto-5β-cholanic acid	2.0	2.2	2.9
3β-Hydroxy-5β-cholanic acid	3.6	5.1	7.8
3α-Hydroxy-5β-cholanic acid	5.6	7.7	12.6

Instrument:	LDC Constametric 2 pump and Chromatronix detector
Column:	Porasil-10 (10 μm) (25 cm × 4.6 mm)
Mobile phase:	1. 2% Isopropanol in isooctane
	2. 1.5% Isopropanol in isooctane
	3. 1% Isopropanol in isooctane
Flow:	1.5 mℓ/min
Detection:	254 nm

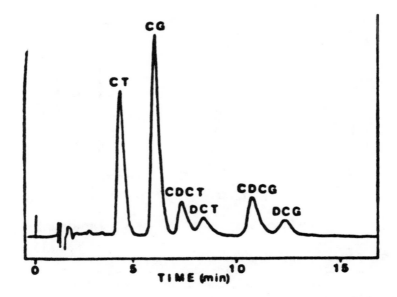

FIGURE HPLC 1. Separation of conjugated bile acids.[2] Instrument: Waters 6000A solvent delivery and model V6K detector; column: μBondapak C_{18} (30 cm × 4 mm); mobile phase: methanol-water (26:14), add buffer to pH 4.7 (buffer: 10 *M* NaOH added to 375 mℓ acetic acid to pH 4.7); detection: 254 nm. CT: cholyl taurine, CG: cholyl glycine, CDCT: chenodeoxycholyl taurine, DCT: deoxycholyl taurine, CDCG: chenodeoxycholyl glycine, and DCG: deoxycholyl glycine.

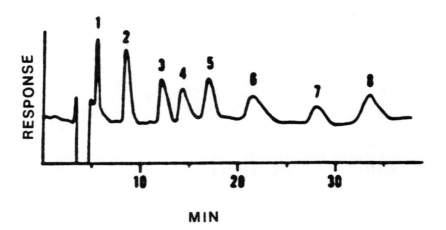

FIGURE HPLC 2. Separation of conjugated bile acids.[3] Instrument: Waters Model ALC 201 with refractive index detector; column: Waters "fatty acid" column (30 cm × 4 mm); mobile phase: 2-propanol-8.8 m*M* phosphate (pH 2.5) (16:34). (1) TMCA, (2) TCA, (3) TCDCA, (4) TDCA, (5) GCA, (6) TLCA, (7) GCDCA, (8) GCA. See Table 1 (Chapter 7, Section I) for glossary.

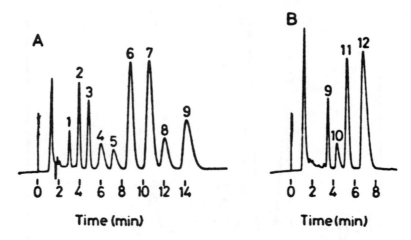

FIGURE HPLC 3.   Separation of sulfated bile acids.[4] Instrument: Nikon Bunko model Tri-Rotar with 9 UV detector; column: ODS SC-02 (25 cm × 4.6 mm); mobile phase: (A) 0.5% ammonium carbonate-acetonitrile (28:8), (B) 0.5% ammonium carbonate-acetonitrile (20:8); flow: 2.1 mℓ/min; detection: 210 nm. Cholanic acids as sulfate: (1) CA, (2) GCA, (3) TCA, (4) CDCA, (5) DCA, (6) GCDCA, (7) GDCA, (8) TCDCA, (9) TDCA, (10) LC, (11) GLC, and (12) TLC. See Table 1 (Chapter 7, Section I) for glossary.

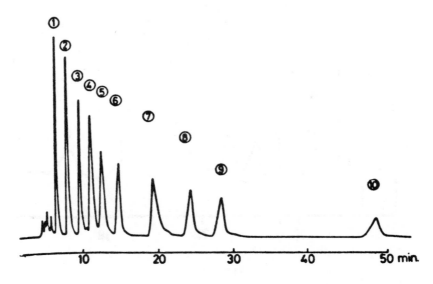

FIGURE HPLC 4.   Separation of conjugated bile acids.[5] Instrument: Hitachi 635 with variable wavelength detector; column: LiChrosorb RP18 (5 μm) (300 × 8.0 mm); mobile phase: methanol-water 75:25 adjusted to pH 2 with phosphoric acid; flow: 2 mℓ/min; detector: 210 nm. (1) TUDCA, (2) TCA, (3) TCDCA, (4) TDCA, (5) TLCA, (6) GUDCA, (7) GCA, (8) GCDCA, (9) GCA, and (10) GLCA.

Table HPLC 2
SEPARATION OF CONJUGATED BILE
ACIDS[6]

Bile acid	Relative elution vol
Tauroursodeoxycholic acid	0.46
Glycoursodeoxycholic acid	0.53
Taurocholic acid	0.60
Glycocholic acid	0.71
Taurochenodeoxycholic acid	0.90
Taurodeoxycholic acid	1.00
Glycochenodeoxycholic acid	1.12
Glycodeoxycholic acid	1.26
Taurolithocholic acid	1.59
Glycolithocholic acid	2.07
Testosterone acetate (internal standard)	1.45

Instrument:	Waters ALC/GP C202 with Shimadzu Model SPD-1 detector
Column:	μBondapak C_{18} (30 cm × 3.9 mm) guard column of 350 mg Bondapak C_{18}/Corasil
Mobile phase:	Acetonitrile-methanol-0.03 M phosphate (pH 3.4) (10:60:30)
Flow:	0.5 mℓ/min
Detector:	200 nm

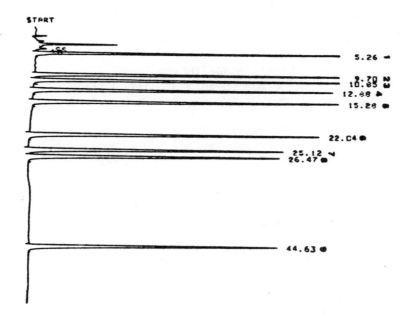

FIGURE HPLC 5.   Separation of bile acids as *p*-bromophenacyl esters.[7] Instrument: Hewlett-Packard 1084B with UV detector; column: Brownlee RP-18 (5 μm) (25 cm × 4 mm); mobile phase: methanol-water (to pH 3.1 with phosphoric acid) (70:30); temperature: 40°C; flow: 0.5 mℓ/min; detection: 200 nm. (1) GCA, (2) GCDCA, (3) GDCA, (4) CA, (5) UDCA, (6) GLCA, (7) CDCA, (8) DCA, and (9) LCA.

Table HPLC 3
SEPARATION OF CONJUGATED BILE ACIDS[8]

| | 1 | | 2 | |
Bile salt	A $RRT_{(TDC)} =$ 12.80 min	B $RRT_{(TDC)} =$ 12.84 min	A $RRT_{(TDC)} =$ 18.20 min	B $RRT_{(TDC)} =$ 16.44 min
TUDCA	0.48	0.52		0.56
TCA	0.53	0.54	0.67	0.52
TCDCA	0.94	0.90	0.96	0.86
TDCA	1.00	1.00	1.00	1.00
TLCA	1.76	1.72		1.75
GUDCA	1.08	1.06		1.06
GCA	0.64	0.64	1.06	0.74
GCDCA	1.22	1.20	1.65	1.44
GDCA	1.27	1.32	1.82	1.70
GLCA	2.17	2.14		2.82

Instrument: Du Pont 840 with a Cecil 2012 detector
Column: 1.   Partisil-10 ODS (25 cm × 4.6 mm) precolumn CO:Pell ODS
                    (60 × 2 mm)
               2.   μBondapak fatty acid column
Mobile phase: A.   Methanol-water (55:45) (pH 2.5 with phosphoric acid)
               B.   Isopropanol-8.8 mmol/ℓ phosphate pH 2.5 (32:68)
Detection: 1.   200 nm
               2.   193 nm
Flow: 1.   1.0 mℓ/min
               2.   1.2 mℓ/min

Note: See Table 1 (Chapter 7, Section I) for glossary.

Table HPLC 4
SEPARATION OF BILE METHYL ESTER
TRIMETHYLSILYL ESTERS[9]

| | Elution vol (mℓ) | |
Bile acid ester	A	B
5β-Cholestane-3α,7α,12,25α,25-pentol	8.42	5.54
Methyl ursodeoxycholate	10.13	6.62
Methyl cholate	11.12	7.23
Methyl 7-keto-3α-hydroxy-5β-cholanoate	12.54	7.74
Methyl 5α-cholate	12.64	7.77
5β-Cholestane-3α,7α,12α,25-tetrol	14.13	7.76
Methyl 7-keto-3α,12α-dihydroxy-5β-cholestanoate	15.14	7.98
Methyl 3-keto-7α,12α-dihydroxy-5β-cholestanoate	17.72	9.92
Methyl chenodeoxycholate	19.87	11.09
Methyl deoxycholate	19.98	11.23
Methyl 3α,7α,12α-trihydroxy-5β-cholestanoate	23.30	12.19
Methyl lithocholate	40.10	20.31
Methyl 3α,7α,-dihydroxy-5β-cholestanoate	46.63	21.11
5β-Cholestane-3α,7α,12α-triol		23.96

Instrument: Waters ALC 201 with refractive index detector
Column: μBondapak C_{18} (30 cm × 4 mm)
Mobile phase: 1. Methanol-water-chloroform (80:25:3)
               2. Methanol-water-chloroform (85:17:3)
Detection: Refractive index

REFERENCES (HPLC)

1. **Shaikh, B., Pontzer, N. J., Molina, J. E., and Kelsey, M. I.,** *Anal. Biochem.,* 85, 1978.
2. **Laatikainen, T., Lehtonea, P., and Hesso, A.,** *Clin. Chem.,* 85, 145, 1978.
3. **Shaw, R., Smith, J. A., and Elliott, W. H.,** *Anal. Biochem.,* 86, 450, 1978.
4. **Goto, J., Kato, H., and Namara, T.,** *Lipids,* 13, 1109, 1978.
5. **Marujama, M., Tanimura, H., and Kikasa, Y.,** *Clin. Chim. Acta,* 100, 47, 1980.
6. **Nakayama, F. and Kakayaki, M.,** *J. Chromatogr.,* 183, 287, 1980.
7. **Mingrone, G. and Greco, A. V.,** *J. Chromatogr.,* 183, 277, 1980.
8. **Sian, M. S. and Rains, A. J. H.,** *Clin. Chem. Acta,* 98, 243, 1979.
9. **Tint, G. S., Dayal, B., Batta, A. K., Shefer, S., Joanen, T., McNease, L., and Salen, G.,** *J. Lipid Res.,* 21, 110, 1980.

SECTION IV. THIN LAYER CHROMATOGRAPHY

Most of the methods in the literature are designed for TLC systems to separate the "free" bile acids and their derivatives. More recently, since the advent of reversed phase layers, a number of methods for resolution of the conjugated steroids have appeared. There is, at present, no chromatographic system that will separate all of the free and conjugated bile acids in one system. Group separations as described in the section add sample preparation are the rule. Also it is possible, as discussed, to apply the biological sample without treatment directly onto the starting line of the chromatogram.

Table TLC 1
SEPARATION OF BILE ACIDS ($R_f \times 100$)[1]

Mobile phase	LC	DOC	CDC	C	GLC	GDOC	GCDC	GC	TLC	TDOC	TCDC	TC
1	75	86	53	38	18	12	12	5	18	12	12	8
2	59	54	49	39	32	20	20	15	8	5	5	1
3	67	64	64	62	61	57	56	44	34	27	26	16
4	52	38	33	16	22	9	9	2	0	0	0	0
5	38	29	23	11	16	6	6	1	0	0	0	0
6	54	37	31	13	21	8	7	2	1	0	0	0
7	64	43	35	10	30	6	5	1	0	0	0	0
8	88	85	81	73	74	63	63	44	29	19	19	9
9	51	33	27	9	17	6	6	1	0	0	0	0
10	43	26	20	6	12	4	4	1	0	0	0	0
11	77	63	55	33	47	25	24	9	6	2	2	0
12	52	36	30	10	26	7	7	0	0	0	0	0
13	74	66	61	46	60	41	41	18	12	6	6	1
14	46	29	22	6	17	5	5	0	0	0	0	0
15	53	39	31	9	30	8	8	1	0	0	0	0

Mobile phase	Composition	Ratio (v/v)
1	$CHCl_3$-Methanol-H_2O	70:25:3
2	Isooctane-isopropanol-HAC	60:40:1
3	*n*-Butanol-HAC-H_2O	100:7:5
4	Isooctane-isopropyl ether-HAC	50:25:40
5	Toluene-HAC-H_2O	50:50:10
6	Isooctane-HAC-isopropyl ether-isopropanol	10:6:5:1
7	HAC-CCl_4-isopropyl ether-isoamyl acetate-*n*-propanol-benzene	5:20:30:40:10:10
8	Ethyl acetate-methanol-HAC	70:20:10
9	Isooctane-isopropyl ether-HAC	100:50:70
10	Isooctane-ethylene chloride-HAC	60:30:30

Table TLC 1 (continued)
SEPARATION OF BILE ACIDS (R$_f$ × 100)[1]

Mobile phase	Composition	Ratio (v/v)
11	Isooctane-ispropyl ether-HAC-isopropanol	2:1:1:1
12	Isooctane-ethyl acetate-HAC-*n*-butanol	20:10:3:3
13	Isooctane-isopropyl-ether-isopropanol-HAC	1:1:1:1
14	Isooctane-ethyl acetate-HAC	5:5:1
15	CHCl$_3$-ethyl acetate	45:45:10

Note: LC = Lithocholic acid; DOC = deoxycholic acid; CDC = chenodeoxycholic acid; C = cholic acid; UrsoDoc = ursodeoxycholic acid; 7-ketoDoc = 7 ketodeoxycholic acid; HyoC = hyocholic acid; 12-ketoLC = 12-ketolithocholic acid. For conjugated bile acids: GLC = glycolithocholic acid; GDOC = glycodeoxycholic acid; GCDC = glycochenodeoxycholic acid; GC = glycocholic acid; TLC = taurolithocholic acid; TDOC = taurodeoxycholic acid; TCDC = taurochenodeoxycholic acid; TC = taurocholic acid; Layer: Silica Gel; HAC: acetic acid; CHCl$_5$: chloroform.

Table TLC 2
SEPARATION OF BILE ACIDS (R$_f$ VALUES)[2]

Steroid	Mobile phase 1	2
TCA	0.08	0.11
TDCA	0.21	0.24
TCDCA	0.21	0.24
—	—	—
GCA	0.35	0.53
GDCA	0.50	0.80
GCDCA	0.50	0.79

Note: Layer: Silica Gel G (Merck); mobile phase (1) butanol-acetic acid-water (10:1:1), (2) isoamyl acetate-propionic acid-propanol-water (4:3:2:1). See Table 1 (Chapter 7, Section I) for glossary.

Table TLC 3
SEPARATION OF FREE AND CONJUGATED BILE ACIDS (R$_f$)[3]

Compound	1	2	3	4	5	6	7
Free acids							
Lithocholic	0.84	0.85	0.84	0.72	0.74	0.57	0.46
Deoxycholic	0.74	0.78	0.76	0.61	0.54	0.42	0.33
Chenodeoxycholic	0.74	0.77	0.77	0.58	0.52	0.33	0.25
Cholic	0.61	0.70	0.67	0.42	0.21	0.13	0.10
Glycine conjugates							
Lithocholic	0.65	0.71	0.68	0.53	0.42	0.22	0.15
Deoxycholic	0.53	0.66	0.53	0.34	0.13	0.03	0.03
Chenodeoxycholic	0.54	0.67	0.53	0.32	0.13	0.03	0.03
Cholic	0.37	0.50	0.34	0.13	0.06	0.00	0.00

Table TLC 3 (continued)
SEPARATION OF FREE AND CONJUGATED BILE ACIDS (R$_f$)[3]

	Mobile phase						
Compound	1	2	3	4	5	6	7
Taurine conjugates							
Lithocholic	0.23	0.35	0.21	0.01	0.00	0.00	0.00
Deoxycholic	0.16	0.26	0.11	0.00	0.00	0.00	0.00
Chenodeoxycholic	0.15	0.26	0.12	0.00	0.00	0.00	0.00
Cholic	0.05	0.11	0.05	0.00	0.00	0.00	0.00

Note: Layer: Silica Gel G (Analtech); mobile phase: (1) benzene-acetic acid-water (10:10:1), (2) isopentanol-acetic acid-water (18:5:3), (3) isopentyl acetate-propionic acid-propanol-water (4:3:2:1), (4) isooctane-diisopropyl ether-isopropanol-acetic acid (2:1:1:10), (5) ethyl acetate-cyclohexane-acetic acid (23:7:3), (6) isooctane-ethyl acetate-acetic acid (5:5:1), (7) isooctane-diisopropyl ether-acetic acid (2:1:1).

Table TLC 4
R$_f$ VALUES FOR BILE ACIDS AND CONJUGATES[4]

	Mobile phase (R$_f$) relative to TC						
Compound	1	2	3	4	5	6	7
Taurolithocholic acid	4.40	1.47	1.60	1.52	1.22	1.17	1.19
Taurochenodeoxycholic acid	2.50	1.28	1.36	1.36	1.12	1.11	1.12
Taurodeoxycholic acid	2.20	1.28	1.36	1.32	1.12	1.11	1.12
Glycolithocholic acid	13.2	2.10	1.86	1.14	0.84	0.88	0.84
Glycochenodeoxycholic acid	10.5	1.76	1.62	0.94	0.67	0.79	0.73
Glycodeoxycholic acid	9.5	1.76	1.62	0.90	0.67	0.79	0.73
Glycocholic acid	5.20	1.55	1.26	0.64	0.51	0.70	0.56
Mobility of taurocholic acid (cm)	1.0	5.8	8.3	5.0	10.7	8.1	11.2

Note: Layer: Silica Gel G (Merck); mobile phase: (1) chloroform-isopropanol-acetic acid-water (30:30:4:1), (2) n-butanol-acetic acid-water (20:4:3), (3) ethyl acetate-methanol-acetic acid-water (35:12:2:2), (4) n-butanol-water (20:30), (5) ethyl acetate-methanol-water (35:12:2), (6) n-butanol-pyridine-water (20:4:3), (7) ethyl acetate-methanol-pyridine-water (35:12:5:2).

Table TLC 5
R$_f$ VALUES FOR CONJUGATED
BILE ACIDS ON REVERSE PHASE
THIN LAYERS[5]

Bile acid	R$_f$
Taurocholic acid (TC)	0.38
Glycocholic acid (GC)	0.31
Taurochenodeoxycholic acid (TCDC)	0.26
Taurodeoxycholic acid (TDC)	0.23
Glycochenodeoxycholic acid (GCDC)	0.20
Glycodeoxycholic acid (CDC)	0.17
Glycolithocholic acid (GLC)	0.10

Note: Layer: Whatman LHPC$_{18}$; mobile phase: chloroform 0.3 M CaCl$_2$ (pH 3.5) (50:50); detection: spray with 10% sulfuric acid in ethanol, heat at 170°C for 2 min, and observe fluorescence.

Table TLC 6
MOBILITIES OF BILE ACID SULFATES $(R_f)^6$

Bile acid	Mobile phase 1	Mobile phase 2
Lithocholic acid	0.93	0.96
Lithocholic acid sulfate	0.70	0.79
Taurolithocholic acid	0.41	0.44
Taurolithocholic acid sulfate	0.26	0.19
Glycolithocholic acid	0.59	0.85
Glycolithocholic acid sulfate	0.48	0.42
Deoxycholic acid	0.91	0.92
Dimethyl deoxycholate 3-hemisuccinate-12-sulfate	0.72	0.66
Deoxycholic acid 12-sulfate	0.67	0.60
Deoxycholic acid disulfate	0.28	0.23
Chenodeoxycholic acid	0.90	0.91
Dimethyl chenodeoxycholate 3-hemisuccinate 7-sulfate	0.69	0.58
Chenodeoxycholic acid 7-sulfate	0.65	0.54
Chenodeoxycholic acid disulfate	0.27	0.22
Taurocholic acid trisulfate	0.03	0.00

Note: Layer: Silica Gel G (Merck); mobile phase: (1) butanol-acetic acid-water (10:1:1), (2) methyl ethyl ketone-chloroform-methanol isopropanol-acetic acid-water (20:6:2:2:4:1).

Table TLC 7
SEPARATION OF BILE ACID CONJUGATES[7]

R_f Values of Bile Salts and Their Sulfates

Compound	Mobile phase 1	2	3	4	5	6
Lithocholic acid	0.95	0.98	0.93	0.95	0.96	0.98
3β-Hydroxy-5-cholenoic acid	0.95	0.98	0.92	0.95	0.96	0.98
Deoxycholic acid	0.93	0.96	0.91	0.92	0.94	0.98
Chenodeoxycholic acid	0.92	0.96	0.91	0.92	0.94	0.97
Cholic acid	0.85	0.93	0.86	0.89	0.92	0.95
Lithocholate 3-sulfate	0.75	0.73	0.73	0.83	0.95	0.96
3β-Hydroxy-5-cholenoate 3-sulfate	0.75	0.71	0.71	0.81	0.93	0.94
Deoxycholate 3-sulfate	0.66	0.69	0.73	0.82	0.91	0.94
Chenodeoxycholate 3-sulfate	0.63	0.68	0.71	0.79	0.88	0.92
Deoxycholate 12-sulfate	0.59	0.66	0.71	0.79	0.87	0.92
Chenodeoxycholate 7-sulfate	0.54	0.64	0.68	0.77	0.84	0.90
Cholate 3-sulfate	0.50	0.58	0.65	0.77	0.81	0.89
Cholate 7-sulfate	0.39	0.53	0.60	0.73	0.73	0.84
Cholate 12-sulfate	0.37	0.51	0.58	0.72	0.73	0.84
Deoxycholate disulfate	0.23	0.35	0.39	0.63	0.60	0.77
Chenodeoxycholate disulfate	0.22	0.35	0.37	0.61	0.58	0.76
Cholate 3,7-disulfate	0.17	0.25	0.31	0.56	0.47	0.68
Cholate 3,12-disulfate	0.16	0.25	0.31	0.55	0.46	0.66
Cholate 7,12-disulfate	0.14	0.23	0.26	0.49	0.41	0.57
Cholate trisulfate	0.06	0.11	0.14	0.34	0.26	0.48
Glycolithocholate	0.84	0.88	0.89	0.88	0.88	0.94
Glycodeoxycholate	0.68	0.79	0.75	0.82	0.81	0.88
Glycochenodeoxycholate	0.67	0.78	0.75	0.82	0.80	0.88

Table TLC 7 (continued)
SEPARATION OF BILE ACID CONJUGATES[7]

R_f Values of Bile Salts and Their Sulfates

Compound	Mobile phase					
	1	2	3	4	5	6
Glycocholate	0.40	0.62	0.61	0.74	0.65	0.78
Glycolithocholate 3-sulfate	0.45	0.53	0.57	0.73	0.65	0.78
Glycodeoxycholate 3-sulfate	0.31	0.42	0.48	0.67	0.59	0.76
Glycochenodeoxycholate 3-sulfate	0.28	0.42	0.48	0.67	0.59	0.75
Glycochenodeoxycholate 7-sulfate	0.20	0.38	0.42	0.62	0.47	0.68
Glycodeoxycholate 12-sulfate	0.18	0.38	0.41	0.62	0.44	0.65
Glycocholate 3-sulfate	0.18	0.30	0.32	0.56	0.41	0.64
Glycochenodeoxycholate disulfate	0.09	0.16	0.18	0.42	0.24	0.48
Glycodeoxycholate disulfate	0.08	0.15	0.16	0.39	0.21	0.46
Glycocholate trisulfate	0.02	0.04	0.05	0.19	0.05	0.26
Taurolithocholate	0.43	0.55	0.62	0.74	0.73	0.83
Taurodeoxycholate	0.29	0.48	0.54	0.67	0.65	0.75
Taurochenodeoxycholate	0.28	0.47	0.54	0.67	0.64	0.74
Taurocholate	0.13	0.36	0.38	0.55	0.41	0.55
Taurolithocholate 3-sulfate	0.17	0.29	0.34	0.55	0.48	0.65
Taurodeoxycholate 3-sulfate	0.11	0.20	0.27	0.47	0.41	0.57
Taurochenodeoxycholate 3-sulfate	0.10	0.20	0.27	0.47	0.41	0.56

Note: Layer: Silica Gel G (Merck); mobile phase: (1) methyl ethyl ketone-chloroform-methanol-propan-2-ol-acetic acid-water (20:6:2:2:4:1), (2) ethyl acetate-*n*-butanol-acetic acid-water (8:6:3:3), (3) *n*-butanol-acetic acid-water (10:2:1), (4) *n*-butanol-acetic acid-water (10:5:1), (5) methyl ethyl ketone-methanol-propan-2-ol-acetic acid-water (10:1:1:1:1), (6) methyl ethyl ketone-methanol-propan-2-ol-acetic acid-water (10:1:1:3:1).

Table TLC 8
SEPARATION OF BILE ACIDS AND CONJUGATES
(R_f)[8]

Bile acid	Free	Glyco-	Tauro-
Chenodeoxycholic acid	0.48—0.49	0.48—0.49	0.53—0.55
Deoxycholic acid	0.48—0.49	0.46—0.47	0.51—0.53
Cholic acid	0.42—0.43	0.40—0.41	0.49—0.50
Lithocholic acid	0.50—0.52	—	—
Hyodeoxycholic acid	0.48—0.49	—	—

Note: Layer: Silca Gel G (Merck); mobile phase: isopropanol-ethyl acetate-water-ammonia (20:25:6:4); glyco- : glycine conjugate; tauro- : taurine conjugate.

Table TLC 9
SEPARATION OF BILE
ACIDS AND CONJUGATES[9]

Bile acid	R_f
Lithocholic acid	0.99
Lithocholylglycine	0.94
Sulfolithocholic acid	0.48

Table TLC 9 (continued)
SEPARATION OF BILE
ACIDS AND CONJUGATES[9]

Bile acid	R$_f$
Lithocholyltaurine	0.44
Sulfolithocholylglycine	0.41
Sulfolithocholyltaurine	0.17
Deoxycholylglycine	0.84
Chenodeoxycholylglycine	0.83
Cholylglycine	0.65
Deoxycholyltaurine	0.38
Chenodeoxycholyltaurine	0.37
Cholyltaurine	0.31

Note: Layer: Silica Gel H with 10% calcium sulfate; mobile phase: chloroform-methanol-acetic acid-water
(65:24:15:19).

Table TLC 10
SEPARATION OF BILE ACIDS AND
CONJUGATES[10]

Bile acid	Mobile phases					
	1	2	3	4	5	6
Unconjugated						
L[a]	0.86	0.87	1.00	0.97	0.56	0.30
CD	0.81	0.82	0.90	0.92	0.49	0.20
D	0.82	0.82	0.90	0.92	0.47	0.20
U	0.81	0.82	0.91	0.92	0.49	0.20
HD	0.77	0.80	0.84	0.88	0.49	0.20
C	0.74	0.72	0.67	0.84	0.34	0.09
Conjugated						
GL	0.78	0.85	0.85	0.94	0.29	0.25
NaGCD	0.70	0.73	0.63	0.84	0.21	0.16
GHD	0.65	0.72	0.60	0.77	0.23	0.16
NaGC	0.50	0.57	0.33	0.69	0.13	0.08
NaTL	0.32	0.37	0.07	0.47	0.53	0.58
NaTCD	0.21	0.24	0.03	0.33	0.43	0.50
NaTD	0.21	0.24	0.03	0.33	0.43	0.50
NaTU	0.21	0.25	0.03	0.34	0.46	0.50
NaTC	0.10	0.12	0.01	0.17	0.35	0.36
Sulfate						
NaLS	—	0.40	—	0.65	—	0.32
NaTLS	—	0.03	—	0.12	—	0.12

Note: Layer: Silica Gel G (Merck); mobile phases: (1) ethyl acetate-
ethanol-acetic acid (7:3:1), (2) chloroform-ethanol-ethyl acetate-
acetic acid-water (8:6:5:4:1), (3) chloroform-ethanol-acetic acid-
water (17:4:3:1), (4) chloroform-ethanol-acetic acid-water
(12:8:4:1), (5) chloroform-methanol-ethanol (5:4:1), and (6)
chloroform-ethanol-28% aqueous ammonia (25:35:1).

[a] The following abbreviations are used: L, lithocholic acid; CD, chenodeoxycholic acid; D, deoxycholic acid; U, ursodeoxycholic acid;
HD, hyodeoxycholic acid; G, glycine conjugates; T, taurine conjugates; S, sulfate.

Table TLC 11
SEPARATION OF BILE ACIDS[11]

R_f Values for Free and Conjugated Bile Acids

	Mobile phase	
	1	2
Bile acids class		
Free bile acids	0.86	
Taurolithocholic	0.47	
Taurochenodeoxy with deoxycholic	0.38	
Taurocholic	0.29	
Glycolithocholic	0.71	
Glycochenodeoxy with deoxycholic	0.67	
Glycocholic	0.51	
Free bile acids		
Ketolithocholic		0.93
Lithocholic		0.78
Deoxycholic		0.67
Chenodeoxycholic		0.43
Cholic		0.21

Note: Layer: Silica Gel G (Merck); mobile phase: (1) isopropanol-acetic acid (93:7), (2) hexane-methyl ethyl ketone-acetic acid (56:36:8).

Table TLC 12
SEPARATION OF 3 Oxo
BILE ACIDS[12]

5β-Cholan-24-oic acid	R_f
3-Oxo-[a]	0.64
3,6-Dioxo-	0.48
3,12-Dioxo-	0.44
3,7,12-Trioxo-	0.24
6α-Hydroxy-3-oxo-	0.32
7α-Hydroxy-3-oxo-	0.42
12α-Hydroxy-3-oxo-	0.39
7α,12α-Dihydroxy-3-oxo-	0.16
3α,7α-Dihydroxy-	0.24
3α,12α-Dihydroxy-	0.29
3α,7α,12α-Trihydroxy-	0.07

Note: Layer: Silica Gel G (Merck); mobile phase: isooctane-ethyl acetate-acetic acid (10:10:2).

[a]   Substituent on 5β-cholan-24-oic acid.

Table TLC 13
SEPARATION OF FREE AND SULFATED BILE ACIDS[13]

Bile acid	Mobile phase (R_f)		
	1	2	3
NSBA			
Lithocholic acid	0.82	0.91	0.98
Nordeoxycholic acid	0.53	0.76	0.89
Deoxycholic acid	0.40	0.74	0.96
Chenodeoxycholic acid	0.42	0.76	0.96
Cholic acid	0.16	0.75	0.93
Glycolithocholic acid	0.40	0.72	0.88
Glycodeoxycholic acid	0.08	0.70	0.84
Glycochenodeoxycholic acid	0.09	0.61	0.88
Glycocholic acid	0.00	0.48	0.70
Taurolithocholic acid	0.00	0.30	0.51
Taurochenodeoxycholic acid	0.00	0.28	0.47
SBA			
Lithocholic acid 3-sulfate	0.01	0.31	0.70
Nordeoxycholic acid 3-sulfate	0.04	0.30	0.61
Deoxycholic acid 3-sulfate	0.03	0.31	0.68
Chenodeoxycholic acid 3-sulfate	0.03	0.31	0.68
Chenodeoxycholic acid 7-sulfate	0.01	0.27	0.67
Cholic acid 3-sulfate	0.01	0.26	0.65
Glycolithocholic acid 3-sulfate	0.01	0.24	0.60
Glycodeoxycholic acid 3-sulfate	0.00	0.46	
Glycodeoxycholic acid 3-sulfate	0.00	0.11	0.46
Glycochenodeoxycholic acid 3-sulfate	0.00	0.19	0.48
Glycochenodeoxycholic acid 7-sulfate	0.00	0.18	0.45
Glycocholic acid 3-sulfate	0.00	0.06	0.34
Taurolithocholic acid 3-sulfate	0.00	0.02	0.20
Taurochenodeoxycholic acid 3-sulfate	0.00	0.01	0.16
Taurochenodeoxycholic acid 7-sulfate	0.00	0.00	0.14

Note: Layer: Silica Gel HR (Macherey-Nagel); Mobile phases: (1) acetic acid-carbon tetrachloride-diisopropyl ether-isoamyl-acetate-*n*-propanol-benzene (5:20:30:40:10:10), (2) propionic acid-isoamylacetate-water-*n*-propanol (15:20:5:10), (3) ethyl acetate-*n*-butanol-acetic-acid-water (40:30:15:15); detection: spray with 5% phosphomolybdic acid in ethanol; heat at 120°C for 10—20 min.

Table TLC 14
SEPARATION OF CONJUGATED BILE ACIDS[14]

Compound	R_f
TCA	0.18
TDCA	0.34
TCDCA	0.37
TUDCA	0.43
TLCA	0.53
GCA	0.47
GDCA	0.73

Table TLC 14 (continued)
SEPARATION OF
CONJUGATED BILE
ACIDS[14]

Compound	R_f
GCDCA	0.76
GUDCA	0.81
GLA	0.87

Note: Layer: Silica Gel G (Merck); mobile phase: (1st development) butanol-water (20:3) 1st development to 5 cm, dry then develop in (2nd development) chloroform-isopropanol-acetic acid-water (30:30:4:1) to 17 cm. See Table 1 (Chapter 7, Section I) for glossary.

Table TLC 15
R_f VALUES FOR BILE ACIDS[15]

Bile acid	Mobile phase 1	2
Free bile acids		
CA	0.11	0.04
CDCA	0.46	0.20
DCA	0.52	0.20
LCA	0.77	0.68
3-Keto derivatives of		
CA	0.31	0.11
CDCA	0.68	0.61
DCA	0.68	0.41
LCA	0.99	0.96

Note: Layer: Silica Gel G (Merck); mobile phase: (1) hexane-methyl ethyl ketone-glacial-acetic acid (58:36:8, v/v), (2) glacial acetic acid-diethyl ether (1.99, v/v). See Table 1 (Chapter 7, Section 1) for glossary.

REFERENCES (TLC)

1. **Goswami, S. K. and Frey, C.,** *J. Chromatogr.,* 89, 87, 1974.
2. **Rocchi, E. and Salivioli, G.,** *Minerva Gastroenterol.,* 22, 130, 1976.
3. **Kim, H. K. and Kritchenevsky,** *J. Chromatogr.,* 117, 222, 1976.
4. **Batta, G. K., Salen, G., and Shefer, S.,** *J. Chromatogr.,* 168, 557, 1979.
5. **Touchstone, J. C., Levitt, R. E., Levin, S. S., and Soloway, R.,** *Lipids,* 15, 386, 1980.
6. **Pageanx, J. F., Duperray, B., Anker, D., and Dubois, M.,** *Steroids,* 34, 73, 1979.
7. **Parmentier, G. and Eyssen, H.,** *J. Chromatogr.,* 152, 285, 1978.

8. **Goswami, S. K. and Frey, C.,** *J. Chromatogr.,* 100, 200, 1974.
9. **Cass, O. W., Cowen, A. E., Hoffman, A. F., and Coffin, S. B.,** *J. Lipid Res.,* 16, 159, 1975.
10. **Ikawa, S. and Goto, M.,** *J. Chromatogr.,* 114, 237, 1975.
11. **Chavez, M. N. and Krone, C. L.,** *J. Lipid Res.,* 17, 545, 1976.
12. **Beher, W. T., Sanfield, J., Stradniko, S., and Lin, G. L.,** *J. Chromatogr.,* 155, 421, 1978.
13. **Lepage, G., Roy, C. C., and Weber, A. M.,** *J. Lipid Res.,* 22, 705, 1981.
14. **Batta, A. K., Shefer, S., and Salen, G.,** *J. Lipid Res.,* 22, 712, 1981.
15. **Chavez, M. N.,** *J. Chromatogr.,* 162, 71, 1979.

Chapter 8

CARDENOLIDES

SECTION I. PREPARATIONS

Introduction

Cardenolides occur in the plant families Apocyanaceae and Asclepiadaceae, and also in the families Moraceae, Scrophulariaceae, Liliaceae, and Ranunculaceae. Useful cardenolides include digoxin, digitoxin and related glycosides from *Digitalis purpurea* and *D. lanata* (Scrophulariaceae), and ouabain and strophanthidin from *Strophanthus* spp. (Apocyanaceae). The cardenolides (Asclepiadaceae) are associated biologically with livestock poisoning and ecological defense mechanisms in insects which sequester these compounds from milkweed plants. These compounds differ in the stereochemistry of the steroid A/B ring junction, being *trans* in the former and *cis* in the latter.

Cardenolides may be produced by chrysomelid beetles and larvae, and are also present in some toad species. Cardenolides from toad sources often occur conjugated to sulfate, dicarboxylic acid, or dicarboxylic acid arginine moieties.

Bufadienolides have been isolated from plants, amphibians, and arthropods. Structural features vary with the source. Bufadienolides from plant isolations are glycosides and aglycones, while no glycosides have been found among the toad poisons. The latter are genins or dicarboxylic acid arginine esters or sulfate conjugates at C_3. The discovery of bufadienolides in arthropods is relatively recent. The five bufadienolides so far isolated contain acetate esters in the 2 and/or 3 positions.

The cardiac glycosides and their genins are a therapeutically important group of plant constituents. There is a continually increasing need for micromethods of quantitation and identification in therapeutics, pharmacology, and phytochemistry as well as analytical chemistry. These compounds are steroids possessing the specific ability to increase contractility of the cardiac muscle in mammals and certain amphibia. Small amounts can be beneficial to a weakened heart, but large amounts can be detrimental, leading to cardiac arrhythmia and even death. Plant and animal extracts containing these steroids have been used as heart tonics, emetics, diuretics, arrow poisons, and rodenticides.[1,2] These steroids are divided into classes according to the nature of the aglycones. Digitoxigenin (C_{23}) and bufalin (C_{24}) having a side chain with an unsaturated 5- or 6-membered ring are characteristic of this class of compounds, as illustrated in Figure 1. Other features include hydroxyl groups at C_3 and C_{14}, the former being the usual site of conjugation with one or more sugar molecules. Additional oxygenation, principally at C_2, C_{11}, C_{12}, C_{16}, and C_{19}, differentiates individual cardenolide and bufadienolide genins. Other structures are illustrated in Figure 2. The flow sheet in Figure 3 provides a general scheme for extraction of digoxin from *D. lanata* leaf prior to separation by HPLC.

Detection Reagents

The reagents for TLC detection of digoxin and dihydrodigoxin[1] are

1. A solution of 25% trichloroacetic acid in chloroform containing 4 drops of hydrogen peroxide in each 50 mℓ is sprayed on the chromatograms and heated at 90 to 100°C for 2 min.
2. Acetic anhydride-sulfuric acid-abs. ethanol 5:5:100 can be sprayed.
3. 0.05 mℓ *p*-anisaldehyde, 0.2 mℓ sulfuric acid, and 10 mℓ glacial acetic acid are mixed. The plates are sprayed with fresh reagents, and heated at 110°C for 4 to 6 min.

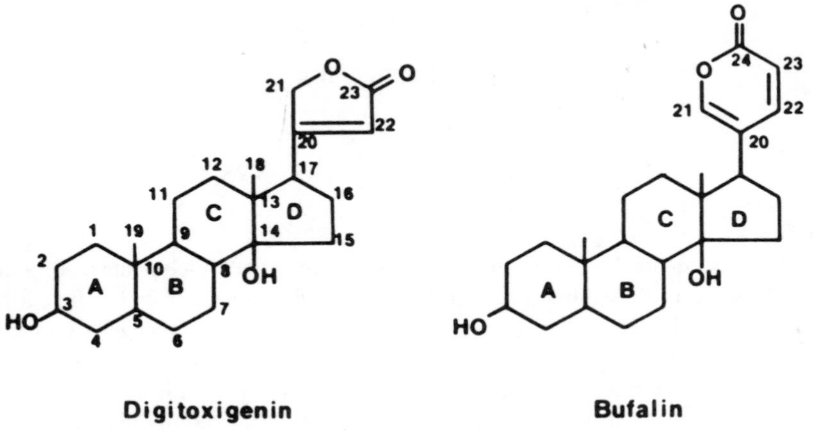

Digitoxigenin   **Bufalin**

FIGURE 1.   Structures of digitalis.

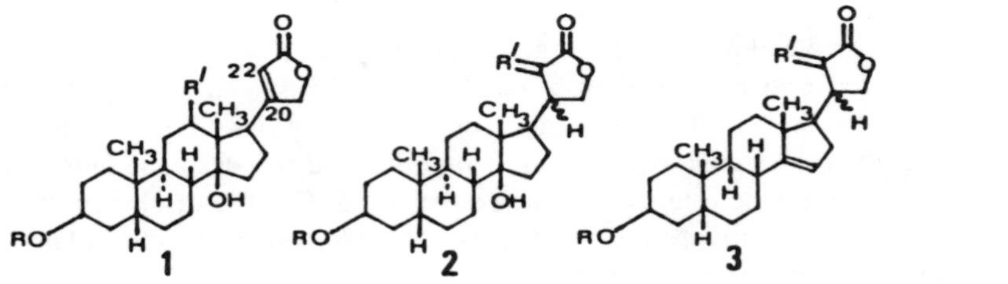

	R	R'	C(20)*
1a	Digitoxose$_3$	OH	sp^2
1b	Digitoxose$_3$	H	sp^2
1c	H	H	sp^2
2a	H	H,H	---H (R)
2b	H	H,H	—H (S)
2c	THP	H,H	---H (R)
2d	THP	H,H	—H (S)
2e	H	CH$_2$	---H (S)
2f	H	CH$_2$	—H (R)
2g	THP	CH$_2$	---H (S)
2h	THP	CH$_2$	—H (R)
3a	H	CH$_2$	—H (R)
3b	H	CH$_2$	---H (S)
4	3β-hydroxy-5β-carda-14,20(22)-dienolide (14-Anhydrodigitoxigenin)		

THP = 2'(R,S)-tetrahydropyran-2'-yl)

*(R) and (S) designations change for compounds with the same C(20)
stereochemistry because of sequence rule priorities.

FIGURE 2.   Structures of the digitoxin derivatives.

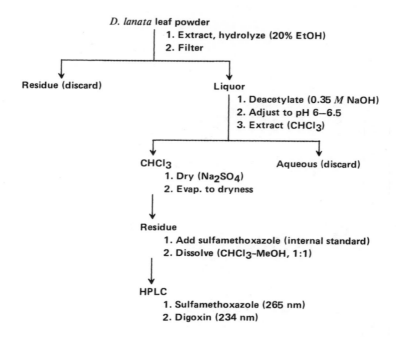

FIGURE 3.   Extraction of digoxin from *D. lanata* leaves for HPLC.

4.   20 mg ascorbic acid, 19 mℓ methanol, 30 mℓ concentrated HCl, 2.1 μℓ 30% hydrogen peroxide are mixed together. The plates are sprayed and heated at 110°C for 4 min.

5.   10 mℓ freshly prepared 3% aqueous solution of chloramine T and 40 mℓ 25% trichloroacetic acid in ethanol are mixed prior to spraying, and the plates are heated at 110°C for 8 min.

The sprayed chromatograms are viewed under both visible and long wave UV light (see Table 1).

Derivatives of Cardiac Glycosides for HPLC

Precolumn

The formation of 4-nitrobenzoates of digitalis glycosides has been a useful means to derivatize for better separation by HPLC. This derivative is described in the section on derivatization and follows the method of Nachtman et al.[3] The acid chloride and the glycoside are dissolved in pyridine and allowed to react at room temperature for 10 min. The pyridine is removed *in vacuo*. The tubes are then flushed with nitrogen, sodium carbonate solution and 4-dimethylaminopyridine are added, and the reaction mixture is allowed to stand 5 min. The reaction product is extracted with chloroform, washed with 0.5% bicarbonate and 0.05 N HCl containing 5% NaCl. This reaction is quantitative and all but the 14-hydroxyl are esterified. Derivatives of all the compounds can be adequately separated isocratically, which has not been possible with underivatized cardenolides. Because of the nature of the mobile phase, partition processes may be responsible for compound separation. The nitrobenzoates do not always emerge in the expected order based on polarity.

The major advantage of the procedure is the increased absorptivity of the nitrobenzoyl derivatives. Limits of detectability of selected glycosides are in the 1- to 2-ng range. This is an increase in sensitivity of at least tenfold over that attainable with underivatized compounds. Absorption of the derivatives at 260 nm makes possible the use of a wider variety of mobile phase solvent systems. Derivatization may increase the selectivity of analysis of cardenolide mixtures isolated from biological matrixes or pharmaceutical preparations in two

Table 1
DETECTION OF ALKALOIDS[2]

Alkaloid	Daylight					UV light				
	Starting		Final color	Detection (µg)		Starting		Final color	Detection (µg)	
	Time (sec)	Color		80°C	300°C	Time (sec)	Color		80°C	300°C
Tomatidine	130	PK	BE-GN	0.20	0.10	0	BT-BE	GN	0.04	0.04
Tomatidenol	86	PK	RE	0.25	0.20	76	LT-BE	BT-BE	0.05	0.05
Soladulcidine	300	LT-GN	BE-GN	1.00	0.25	210	BE-GN	GN	0.05	0.05
Solasodine	110	PK	PU	0.25	0.10	111	LT-BE	BT-BE	0.03	0.03
Veramine	100	GY	GY	0.85	0.10	23	PK	TAN	0.07	0.05
9-Hydroxytomatidene	250	PK	GY	0.90	0.25	131	GY	TAN	0.08	0.08
7-Hydroxytomatidene	80	BE	BE-GY	1.00	0.25	96	PK-BE	GY-GN	0.05	0.05
7,11-Dihydroxytomatidine	270	LT-BE	BE	8.00	0.50	322	PK-BE	PK	0.30	0.30
9,11-Dihydroxytomatidine	NIL	NIL	NIL	1.00	0.20	322	PK-BE	PK	0.20	0.10
Demissidine	132	PK	PE-LC	40.00	1.00	118	BE	BE	0.25	0.25
Solanidine	74	RE	DK-RE	0.025	0.10	0	LT-BE	LT-BE	0.10	0.10
5-Solanidan-3-one	NIL	NIL	NIL	14.00	0.25	NIL	NIL	NIL	2.00	0.05
Solanid-4-en-3-one	437	LT-PK	NIL	80.00	0.25	300	BE-GY	GY-GN	10.00	0.07
Rubijervine	60	TAN	LC	0.50	0.25	119	DK-BE	PK	0.06	0.04
Isorubijervine	53	RE	LC	0.90	0.35	89	LT-BE	LT-BE	0.08	0.06
Veralobine	NIL	NIL	NIL	80.00	0.25	360	VY-LT-BE	VY-PE-BE	2.00	0.07
Verarine	166	YW	OE	0.15	0.10	106	YW	BT-YW	0.02	0.02
Cyclopamine	NIL	GY	GY	0.25	0.10	0	PK-BE	TAN	0.05	0.10
Jervine	330	LT-YW	GY	0.50	0.30	151	BE	BT-GN	0.05	0.05
Veratramine	139	YW	LT-OE	0.40	0.01	59	LT-BE	BY-YW	0.02	0.02
Veramarine	70	PK	OE	1.20	0.07	64	YW	BT-BE	0.02	0.02
Solanocapsine	NIL	NIL	NIL	4.00	0.01	NIL	NIL	NIL	0.10	0.01
Verazine	120	PK	LC	2.00	0.90	86	TAN	LT-LC	0.40	0.07
Tomatillidine	80	PK	LC	0.20	0.10	92	TAN	LT-LC	0.10	0.05
Etioline	60	PK	LC	0.30	0.20	45	LT-BE	BT-BE	0.05	0.05
Veralkamine	118	IC	PU-GY	0.35	0.25	64	TAN	TAN	0.01	0.01

Note: Alkaloids can be detected by spraying thin layers of Silica Gel G (Analtech) with 50% sulfuric acid, followed by heating on a hot plate at 80°C and up to 300°C. Observation of color development is in visible light and under UV light in a darkroom. Abbreviations: BE = blue; BT = bright; DK = dark; GN = green; GY = gray; LC = lilac; LT = light; MDQ = minimum detected; NIL = no response; OE = olive; PE = pale; PK = pink; PU = purple; RE = rose; UV = ultraviolet radiation of 366 nm; VY = very; YL = yellow.

ways: (1) co-extracted, potentially interfering materials may not form nitrobenzoyl esters with chromatographic properties similar to those of cardenolide derivatives; (2) interference of non-derivatized co-extracted materials is less likely at 260 nm than at the less selective wavelength of 220 nm.

Post-Column

A procedure for post-column reaction of cardenolides to form highly fluorescent derivatives was described by Gfeller et al.[4] The fluorogenic reaction was carried out in a miniaturized Technicon AutoAnalyzer system, employing the air segmentation principle and based on the reaction of HCl with the steroid portion of the cardiac glycosides. Fluorescence of the derivatives was further enhanced by the addition of a hydrogen peroxide/ascorbic acid mixture. Reaction times and temperature were optimized for reproducible derivatization conditions.

The derivatization step was reproducible within a relative standard deviation of 1.2%. Calibration graphs for 2 to 20 ng of Ianatoside C were linear. Some peak broadening did occur as a result of derivatization.

The method was reported to increase the cardenolide detection limit at least 100-fold over that attainable by UV detection (220 nm). For a nonretained glycoside, the detection limit was approximately 0.5 ng. The method of post-column derivatization has several drawbacks. First, it is not obvious that the post-column derivatization method is compatible with the use of other reverse phase or adsorption chromatography mobile phases. Second, the nature of the fluorescing moiety formed during derivatization is not known and it is postulated to be the product of a dehydration reaction acting upon the steroid nucleus. Fluorescence yield of the molecules thus, will remain the same, regardless of the size of cardenolide molecule, and limits of detectability of higher molecular weight and more strongly retained compounds may be greater than for lower molecular weight compounds. In the pre-column nitrobenzoylation of cardenolides, the number of UV-absorbing nitrobenzoyl esters increases as the number of sugars on the genin increases. Third, post-column derivatization does not have the potential to enhance the chromatographic properties of compounds to be separated by modifying their polarity.

Digitoxin extraction from serum (method of Farber et al.[6]) — 2 mℓ serum were shaken for 2 min with 8 mℓ chloroform and 2 drops of 4 N NaOH in a 10-mℓ glass stoppered centrifuge tube in a Vibromixer, then centrifuged for 15 to 20 min at $14 \times g$; 7 mℓ supernatant were transferred into a 10-mℓ centrifuge tube, washed with 3 mℓ 1% ammonia solution, and finally centrifuged. Six milliliters of supernatant are transferred to a 10-mℓ centrifuge tube with a tapered base of 0.2 mℓ vol. The chloroform extract was evaporated to dryness at 40°C in the tube by passing dried compressed air for 20 to 30 min to prevent absorption. The residue was then taken up in 200 μℓ chloroform by mixing vigorously for 2 min.

Digitoxin and cardioactive metabolite extraction from urine and serum (method of Storstein[7]) — Serum or urine (5.0 mℓ) was shaken with 15 mℓ dichloromethane for 10 min; after separation of the phases by centrifugation, 10 mℓ dichloromethane extract were evaporated to dryness at 50°C on a water bath, then 3 mℓ 70% ethanol were added to the residue and the solution was washed twice with 0.7 mℓ portions of light petroleum (bp 40 to 60°C). A portion (2.5 mℓ) of the ethanol extract was subsequently transferred to glass-stoppered conical tubes and evaporated on a water bath at 50°C in a stream of air. The residue was dissolved in 25 μℓ chloroform-methanol (50:50), and 15 μℓ of this solution were applied to thin layer plates. Stock solutions of standards were dissolved in absolute ethanol.

Diosgenin extraction from *Dioscorea composita* (method of Sanchez et al.[8]) — Roots were dried at 60°C for 2 hr in a vacuum oven and quantitatively transferred to a Soxhlet extractor. They were extracted with 80 mℓ chloroform for 2 hr, and the extract was transferred into a 100-mℓ calibrated flask and made up to a volume of 100 mℓ with chloroform.

Extraction of cardiac glycosides from milkweed plants and monarch butterflies (method of Enson and Seiber[9]) — Cardenolides were extracted from plant and butterfly material by incubating 0.1g dried, powdered sample in 4.5 mℓ 95% ethanol for 1 hr at 70 to 78°C in a shaker water bath. Extracts were cooled to room temperature and the volume was brought to 5 mℓ with ethanol; 3 mℓ extracts were cleaned up for HPLC analysis with lead acetate for precipitating pigments and fats.

Sample preparation for analysis of digitalis glycoside tablets (method of Erni and Frei[10]) — Tablets were pulverized, the powder being made into a slurry with mobile phase, and the clear solution obtained after centrifugation or filtration used for injection.

Extraction of digoxigenin from serum (method of Loo et al.[11]) — Serum (1.0 mℓ) was spiked with 7000 d.p.m of the internal standard and was extracted with 6 mℓ ether-methylene chloride (60:40) on a constant speed shaker. Following centrifugation at 7000 g at 0°C, 55 mℓ organic phase was rapidly washed with 1 mℓ 0.01 N HCl. 5 mℓ organic phase were evaporated to dryness at 50°C under a stream of N, and the residue was dissolved in 100 μℓ mobile phase (0.2% acetic acid, 7% ethanol, and 30% methylene chloride in *n*-hexane).

Extraction of an ampoule solution of desacetyl lanatoside C (method of Nachtmann et al.[3]) — To the contents of ampoule (2 mℓ) were added 15 mℓ 2% sodium hydrogen carbonate solution. The solution was extracted 5 times with 10 mℓ chloroform-2-propanol (3:2), and each extract was washed with the same 10 mℓ water starting with the first extract. This washing procedure was repeated with another 10 mℓ water. The combined fractions were filtered into a 100-mℓ round-bottom flask with some of the solvent mixture. After evaporation to dryness, the residue was transferred into a 10 mℓ centrifuge tube with two 1-mℓ vol of chloroform/pyridine (10:1) and the solvent evaporated at 50°C on a sand bath. The residue was then taken up directly into the reagent suggested for derivatization (200 μℓ vol).

Extraction of digoxin and its metabolites from urine (method of Sugden et al.[12]) — 20 mℓ urine, were extracted with 3X40-mℓ vol chloroform/ethanol (50:50), reconstituted in chloroform/methanol (85:15) after solvent evaporation. Extraction of tritiated digoxin and dihyrodigoxin from urine using chloroform/ethanol (50:50) was complete.

Extraction of digoxin and its metabolites from urine (method of Gault et al.[13]) — Tritiated digoxin and its by-products were added to urine in concentrations approximating physiological levels. Ten milliliter aliquots of urine were extracted 3 times with 20 mℓ chloroform/ethanol (50:50) and shaken vigorously for 5 min, centrifuged at 1600 rpm for 10 min, and the organic solvent phase separated. This was repeated twice. The three organic extracts were then pooled and the total radioactivity determined.

Extraction of cardiotonic steroid conjugates from Japanese toad venom[14] (method of Shimada et al.[14]) — Fresh venom secreted from the parotid glands of the Japanese toad was diluted with chloroform/methanol (1:1), filtered or centrifuged, concentrated *in vacuo*, and then submitted to partition chromatography on silica gel columns employing chloroform/methanol/water (80:20:2.5) as a solvent. Bufotoxins can thus be divided into three fractions.

Preparation of hydrolysate from plant material for the analysis of sapogenins from Agave (method of Higgins[15]) — Approximately 1 g dried acid-hydrolysis product of the plant juice was accurately weighed into a 22 × 80 mm thimble and placed in a suitable Soxhlet extractor. A 70-mℓ vol of heptane was added to a 100-mℓ round-bottomed flask of known tare weight. The sample was extracted for 16 hr. The heptane solution was evaporated to dryness on a rotary evaporator and the flask with the residue was then dried for 3 hr at 100°C in an oven. After determining the weight of the dry heptane extract, approximately 30 mg was weighed into a 40-mℓ screw-capped test tube. A 2.0-mℓ vol of pyridine was added. The tube was shaken in an attempt to dissolve the sample completely. Then, 0.1 mℓ benzoyl chloride was added to the mixture and the tube was sealed with a PTFE lined screw cap. The sample was heated in a water bath at 80°C for 30 min, and after cooling 10.0 mℓ distilled dichloromethane as added, followed by 10 mℓ water and 2.0 mℓ concentrated HCl. The sealed tube was shaken for 15 sec. After phase separation, the aqueous layer was aspirated off and the washing was repeated once with 10 mℓ water.

REFERENCES

1. **Sabatka, J. J., Brent, D. A., and Murphy, I.,** *J. Chromatogr.,* 125, 535, 1976.
2. **Hunter, I. R., Walden, M. K., Wagner, J. R., and Heftman, E.,** *Analyst,* 97, 1973.
3. **Nachtmann, F., Spitzy, H., and Frei, R. W.,** *Anal. Chem.,* 48, 1576, 1976.
4. **Gfeller, J. C., Frey, G., and Frei, R. W.,** *J. Chromatogr.,* 142, 271, 1977.
5. **Cobb, P. H.,** *Analyst,* 101, 768, 1976.
6. **Farber, D. B., Dokolc, A., and Brinkman, U. A.,** *J. Chromatogr.,* 143, 95, 1977.
7. **Storstein, L.,** *J. Chromatogr.,* 117, 87, 1976.
8. **Sanchez, G. L., Acevedo, J. C. M., and Soto, R. R.,** *Analyst,* 97, 973, 1973.
9. **Enson, J. M. and Seiber, J. N.,** *J. Chromatogr.,* 148, 521, 1978.
10. **Erni, F. and Frei, R. W.,** *J. Chromatogr.,* 130, 160, 1977.
11. **Loo, J. C. K., McGilveray, I. J., and Jordan, N.,** *Res. Commun. Chem. Pathol. Pharmacol.,* 17, 497, 1977.
12. **Sugden, D., Ahmed, M., and Gault, M. H.,** *J. Chromatogr.,* 121, 401, 1976.
13. **Gault, M. H., Ahmed, M., Symes, A., and Vana, J.,** *Clin. Biochem.,* 9, 46, 1976.
14. **Shimada, K., Hasegawa, M., Itasebe, K., Fujic, Y., and Nambara, T.,** *Chem. Pharm. Bull.,* 24, 2995, 1976.
15. **Higgins, I. W.,** *J. Chromatogr.,* 121, 329, 1976.

SECTION II. GAS CHROMATOGRAPHY

GC of the intact cardenolides is more involved due to the highly polar nature of these derivatives. The results shown here are for the acid hydrolysis products. (See sample preparation methodology at the beginning of this section.) The following figures are representative of results described by various investigators for GC of these materials.

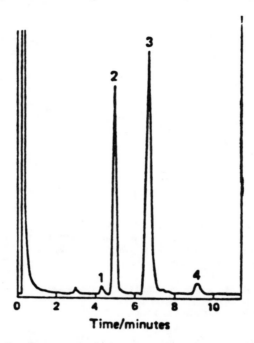

Time/minutes

FIGURE GC 1.   Gas chromatogram of xylene extract of hydrolysate of the tuber of *Dioscorea comporita*. Instrument: HP-700; column: (241 cm × 3 mm) 2% SE-30 on Gas Chrom Q 100 mesh with flame ionization detector; temperatures: column 278°C, injector 280°C, detector 300°C; flow: helium 60 mℓ/min. (1) 25R-Spirost-3,5-diene, (2) 5α-cholestan-3β-ol, (3) diosgenin, (4) pennogenin. (From Rozanski, A., *Analyst*, 97, 968, 1972. With permission.)

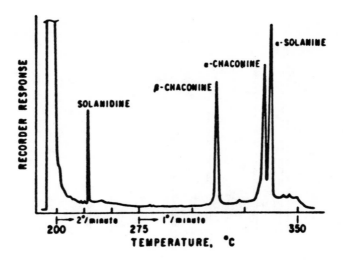

FIGURE GC 2.   Gas chromatogram of a mixture of glycoalkaloids. These compounds are permethylated. Instrument: Varian 2100 with flame ionization; column: 90 cm × 2 mm glass, 3% Dexsil 300 on Supelcoport (100—120 mesh), temperatures: 200° isothermal for 2 min, 2°C/min to 275°C, then 10/min to 350°C; flow: Helium 55 mℓ/min. (From Herb, S. F., Fitzpatrick, T. J., and Osman, S. F., *J. Agric. Food Chem.*, 23, 520, 1975. With permission.)

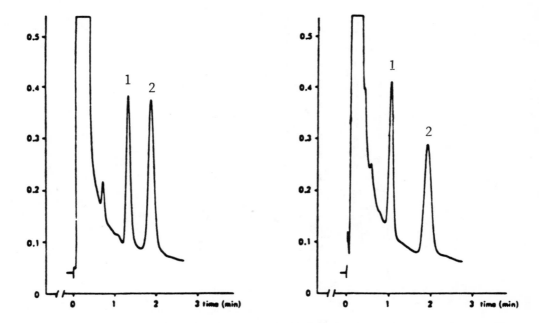

FIGURE GC 3.   Separation of digitoxin acetates. Instrument: Pye Unicam 104 with flame ionization; column: 50 cm × 4 mm glass (left) 2% OV-1, (right) 0.5% PEG 20M on silanized Chromosorb WAW (70—80 mesh); temperature: Oven 250°C. (1) Digitoxigenin, (2) Digoxigenin as acetates of the epoxigeninic methyl ester. (From Meilink, J. W., Lenstra, J. B., and Svendsen, A. B., *J. Chromatogr.*, 170, 35, 1979. With permission.)

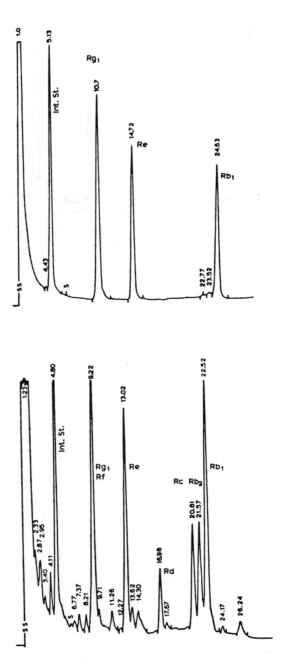

FIGURE GC 4. Gas chromatograms of ginsing derivatives. Instrument: H.P. 5830 A with flame detector, 30 cm × 2 mm column of 0.5% OV-101 on Chromosorb WHD (100—120 mesh); temperatures: column program 250—350°C at 5°C/min hold at 350°C for 10 min; injector: 350°C; detector: 350°C; ginsing saponins (see Figure 5) were separated as trimethylsilyl esters (TMS) for the ginsenosides (upper panel) or extract of Panax ginsing (lower panel). (From Bombardelli, E., Bonati, A., Gabetta, B., and Martinelli, E. M., *J. Chromatogr.*, 196, 121, 1980. With permission.)

A

	R	R′
20(S)-Protopanaxadiol	H	H
Rc	glc²-¹glc	glc⁶-¹ara (fu)
Rd	glc²-¹glc	glc
Rb₁	glc²-¹glc	glc⁶-¹glc
Rb₂	glc²-¹glc	glc⁶-¹ara (py)
Rb₃	glc²-¹glc	glc⁶-xyl (py)
F₂	glc	glc

B

	R	R′
20(S)-Protopanaxatriol	H	H
Rg₁	glc	glc
Rg₂	glc²-¹rham	H
Rf	glc²-¹glc	H
Re	glc²-¹rham	glc
Rh₁	glc	H
F₁	H	glc
F₃	H	glc⁶-¹ara (py)

FIGURE GC 5. Structures of ginsing sapogenins. Glc = β-D-Glucopyranosyl, rham = α-L-rhamnopyranosyl, xyl = β-D-xylopyranosyl, ara (py) = α-L-arabinopyranosyl, and ara (fu) = α-L-arabinofuranosyl.

REFERENCES (GC)

1. **Rozanski, A.,** *Analyst,* 97, 968, 1972.
2. **Herb, S. F., Fitzpatrick, T. J., and Osman, S. F.,** *J. Agric. Food Chem.,* 23, 520, 1975.
3. **Meilink, J. W., Lenstra, J. B., and Svendsen, A. B.,** *J. Chromatogr.,* 170, 35, 1979.
4. **Bombardelli, E., Bonati, A., Gabetta, B., and Martinelli, E. M.,** *J. Chromatogr.,* 196, 121, 1980.

SECTION III. HIGH PERFORMANCE LIQUID CHROMATOGRAPHY

The steroidal glycosides, being highly polar, are well separated by HPLC. Since separation can be performed without derivatization some advantages are inherent in the use of this technique over other methods (GC or TLC).

Table HPLC 1
SEPARATION OF DIGITOXINS[1]

Compound	Retention times (min)		Min. detectable amt. (ng)	
	Isocratic	Gradient[a]	Isocratic	Gradient
Digoxigenin	2.8[b]	4.0	4	10
Digoxin monodigitoxoside	3.8	5.0	8	13
Digoxin bisdigitoxoside	7.1	8.0	16	10
Digoxin	13.2	10.5	39	14
Digitoxigenin	7.9[c]	14.8	9	4

Table HPLC 1 (continued)
SEPARATION OF DIGITOXINS[1]

Compound	Retention times (min)		Min. detectable amt. (ng)	
	Isocratic	Gradient[a]	Isocratic	Gradient
Digitoxin monodigitoxoside	9.6	16.6	25	8
Digitoxin bisdigitoxoside	14.4	19.4	48	10
Digitoxin	23.0	21.5	76	11

Instrument:	Waters 202 with variable wavelength detector
Column:	μBondapak C_8 (30 cm × 4 mm)
Mobile phase:	[a] Linear gradient of 25% acetonitrile in water to 40% acetonitrile in water at 5%/min at flow of 2.2 mℓ/min
	[b] Acetonitrile-water (25:75)
	[c] Acetonitrile-water (33:67)
Detection:	220 nm

Table HPLC 2
SEPARATION OF CARDENOLIDES AS 4-NB[a]
DERIVATIVES[2]

Alkaloid as derivative	Mobile phase (retention time)		
	1	2	3
Digitoxigenin	4.68	3.87	3.07
Gitoxigenin	3.55	3.09	2.72
Digoxigenin	7.41	5.66	4.77
Diginatigenin	4.42	3.73	3.52
Gitaloxigenin	8.06	6.19	5.01
Digitoxigenin monodigitoxoside	6.55	5.19	4.31
Digitoxigenin bisdigitoxoside	9.17	6.92	5.93
Digoxigenin monodigitoxoside	9.36	6.96	6.40
Digoxigenin bisdigitoxoside	13.0	9.08	8.42
Acetyldigitoxin	11.6	8.13	6.93
Digitoxin	12.9	9.14	8.15
Gitoxin	10.6	7.75	7.51
Acetyldigoxin	16.5	10.7	9.90
Digoxin	18.3	11.9	11.4
Diginatin	12.4	8.77	9.16
Gitaloxin	21.1	13.9	12.9
Lanatoside A	32.7	19.1	18.4
Lanatoside B	27.6	16.0	16.9
Lanatoside C	46.0	24.6	25.6
Lanatoside D	32.1	18.0	20.1
Desacetyl lanatoside A	39.5	22.8	23.1
Desacetyl lanatoside B	32.3	19.3	21.0
Desacetyl lanatoside C	55.5	28.9	32.2

Instrument:	Hewlett-Packard UFC 1000; Dupont 842
Column:	Si-60, 5 μm (15 cm × 3 mm)
Mobile phase:	1. Hexane-methylene chloride-acetonitrile (10:3.5:2.5)
	2. Hexane-methylene chloride-acetonitrile (10:3:3)
	3. Hexane-chloroform-acetonitrile (30:10:9)
Flow:	1.55—1.4 mℓ/min
Detection:	254 nm

[a] 4-NB: Nitrobenzoyl.

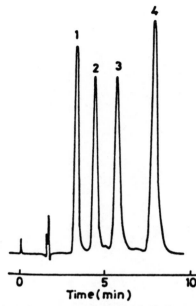

FIGURE HPLC 1. Separation of bufotalin esters. Instrument: Waters ALC/GPC, μBondapak C_{18} (1 ft × 1/4 in.); mobile phase: methanol-water (5:4); flow: 1.5 mℓ/min; detection. 280 nm. (1) Gamabufotalin 3-succinoylarginine ester, (2) Gamabufotalin 3-adipolyarginine ester, (3) Gamabufotalin 3-pimeloylarginine ester, (4) Gamabufotalitoxin.

Table HPLC 3
SEPARATION OF DIGITOXINS[4]

Mobile phase		Retention time (cm/min)		
		Digoxigenin	Digoxin	Lanatoside C
Methanol-isopropanol-acetonitrile-hexane-water	400:350:200:1000:10	—	—	5.9[a]
Isopropanol-acetonitrile-hexane-water	230:115:445:8.6	—	—	8.2[b]
n-Butanol-acetonitrile-heptane-water	230:100:700:10	—	3.5[a]	20.2[b]
tert-Butanol-acetonitrile-heptane-water	220:70:800:10	—	10.0[c]	—
	204:93:712:10.4	—	3.6[c]	9.6[c]
n-Pentanol-acetonitrile-isooctane-water	270:93:660:9.3	—	3.8[a]	14.1[a]
	230:100:700:10	—	5.2[b]	—
	175:60:620:10	3.7[a]	10.4[a]	—
				22.7[d]
	175:60:620:6	—	8.2[a]	41.0[a]
				19.2[d]

Instrument:	HP 1010 A, LiChrosorb Si-60 10 μm (25 cm × 3 mm) stainless steel column
Mobile phase:	As above
Flow rate:	[a] 1.3 mℓ/min
	[b] 1.4 mℓ/min
	[c] 2.2 mℓ/min
	[d] 2.5 mℓ/min
Detection:	254 nm

Table HPLC 4
SEPARATION OF BUFADIENOLIDES[5]

Compound	1			2	3	
	MeOH/H₂O (2:1) 1 ml/min	CH₃CN/H₂O (1:1) 1 ml/min	THF/H₂O (2:3) 1 ml/min	Hexane/THF (1:1) 1.5 ml/min	Hexane/THF (3:1) 1.5 ml/min	CHCl₃/AcOEt/cyclohexane (1:1:1) 1.5 ml/min
Gamabufotalin	0.46	0.46	0.41	1.79	3.44	4.55
Hellebrigenin	0.54	0.50	0.44	2.10	4.48	6.45
Arenobufagin	0.55	0.48	0.54	1.36	1.96	2.41
Desacetylcinobufotalin	0.58	0.51	0.41	4.87	17.44	13.95
Desacetylbufotalin	0.62	0.63	0.49	1.79	4.07	4.34
Resibufagin	0.66	0.82	0.59	1.68	2.22	1.55
Bufotalin	0.71	0.69	0.71	1.19	1.48	1.77
Telocinobufagin	0.72	0.68	0.61	1.78	2.82	4.02
Cinobufotalin	0.78	0.82	0.71	1.46	1.82	1.77
Marinobufagin	0.78	0.82	0.66	1.50	1.67	1.73
Desacetylcinobufagin	0.82	0.68	0.55	2.90	8.19	5.32
Cinobufagin	1.20	1.33	1.15	1.19	1.37	0.91
Resibufogenin	1.20	1.45	1.08	1.09	1.00	1.02
Bufalone	0.92	0.83	1.00	0.87	0.78	0.82
3-Epibufalin	0.99	0.95	0.92	1.18	1.33	1.32
Bufalin	1.00	1.00	1.00	1.00	1.00	1.00
	(8.2 min)	(9.4 min)	(14.0 min)	(3.3 min)	(2.7 min)	(2.2 min)
Separation of Cardenolides						
Sarmentogenin	0.50	0.33	0.37	1.96	4.13	3.45
Digoxigenin	0.54	0.36	0.38	1.92	4.00	3.45
Gitoxigenin	0.71	0.53	0.57	1.91	4.03	2.88
14α,15α-Epoxy-···β···-anhydro-digitoxigenin	1.02	1.36	1.08	0.90	0.73	0.80
14β,15β-Epoxy-···β···-anhydrodigitoxigenin	1.15	1.52	1.15	1.13	1.22	0.80
···β···-Anhydrodigitoxigenin	2.31	3.81	2.76	0.79	0.53	0.60
Digitoxigenone	0.96	1.28	0.94	0.85	0.83	0.80
3-Epidigitoxigenin	1.04	1.01	1.00	1.27	1.73	1.40
Digitoxigenin	1.00	1.00	1.00	1.00	1.00	1.00
	(6.9 min)	(12.6 min)	(11.1 min)	(3.9 min)	(3.0 min)	(2.0 min)

Instrument: Waters ALC/GPC 202
Column:
1. μBondapak C₁₈ (1 ft × 1/4 in)
2. μ Porasil (1 ft × 1/4 in)
3. Corasil J on (2 ft × 1/8 in)

Mobile phase and flow as in the table
Detection: 254 nm for cardenolides, 280 nm for bufadienolides

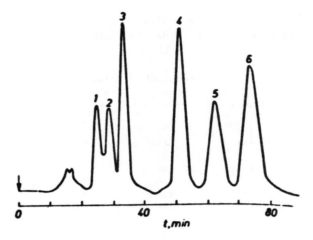

FIGURE HPLC 2. Separation of cardiac glycosides. Instrument: Spectra Physics 3500 B; column: LiChrosorb Si-60 (1 m × 2.1 mm); mobile phase: ethanol-water (35:65); flow rate: 0.12 cm³/min; temperature: 50°C; detection: variable wavelength, 217 nm. (1) G-Strophanthin, (2) K-strophanthoskle, (3) K-strophanthin-β, (4) cymarin, (5) isolanid, (6) dygoxin.

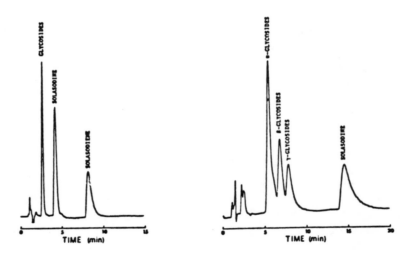

FIGURE HPLC 3. Separation of solasodines. Instrument: Waters 6000 A delivery system, Valco injector, and PE LC 55 detector; column: μBondapak (30 cm × 3.9 mm); mobile phase: Methanol-0.01 M Tris buffer, (left) 75:25, (right) 88:12; flow: 2 mℓ/min for both; detection: 205 nm; temperature: 25°C.

Table HPLC 5
SEPARATION OF CARDENOLIDES BY HPLC
(k' VALUES)[b]

Compound	Adsorption	Acetonitrile in water (37%)	Dioxane in water (45%)	Tetrahydrofuran dioxane (2:1 v/v); 33% water
Digitoxigenin	0.35	7.15	2.55	3.77
Gitoxigenin	0.91	2.65	1.10	2.20
Digoxigenin	1.56	1.30	0.30	1.14
Diginatigenin	3.9	0.85	0.17	0.63
Gitaloxigenin	0.35	4.60	1.70	2.64
Digitoxin	0.76	13.80	11.10	7.77
Gitoxin	1.55	4.90	4.56	4.21
Digoxin	1.90	1.95	1.43	1.65
Diginatin	3.40	1.10	0.80	1.10
Gitaloxin	0.86	8.40	7.20	8.00
Lanatoside A	5.50	8.60	9.00	9.52
Lanatoside B	8.71	3.10	3.85	4.89
Lanatoside C	11.27	1.30	1.18	1.99
Lanatoside D	13.40	0.80	0.70	1.16
Lanatoside E	8.60	3.10	3.70	4.54
Desacetyllanatoside C	15.40	0.90	1.40	1.41

Instrument:	Perkin-Elmer LC55
Column:	1. LiChrosorb Si 60, 5 μm
	2. Nucleosil C_{18}, 10 μm
Mobile Phase:	As in Table
Flow:	Not Given
Detection:	254 nm

Table HPLC 6
TOTAL RETENTION TIMES OF DIGITALIS GLYCOSIDES[9]

Compound	1	2	3	4	5
Digoxigenin	2.8		3.4	5.5	2.8
Digoxigenin monodigotoxoside	2.9		4.0	5.9	3.2
Digoxigenin bisdigitoxoside	3.6		6.1	7.1	4.4
Digoxin	4.5		9.8	9.0	6.3
Gitoxin	10.6				
Digitoxigenin	7.0	4.5		12.2	9.6
Digitoxigenin monodigitoxoside	9.8	5.0		14.9	11.4
Digitoxigenin bisdigitoxoside	14.4	6.3		15.5	12.4
Digitoxin	24.5	8.5		17.2	13.6
α-Acetyldigoxin			18.0		
β-Acetyldigoxin			23.4		

Instrument:	Beckman 322
Column:	Ultrasphere RPC$_{18}$ (25 × 0.46 cm)
Mobile phases, flow rates:	1. Isocratic, water, methanol-isopropanol-dichloromethane (47:40:9:4) at 1.2 mℓ/min
	2. Isocratic, water, methanol-isopropanol-dichloromethane (44:34:15:7)
	3. Isocratic, water, methanol-isopropanol-dichloromethane (51:42:5:2)
	4. Single step gradient, 0 time, water-methanol-isopropa-nol-dichloro-methane (49:41:7:5); solvent ratio at 5 min changed to (41:34:17:8)
	5. Linear gradient, at 0 time water-methanol-isopropanol-dichloromethane (49:41:7:3); solvent was linearly changed from time 2.5—3.0 min to (38:32:20:10) and was maintained here to completion with flow of 1.7 mℓ/min
Detection:	254 nm

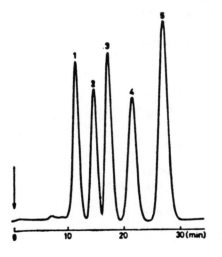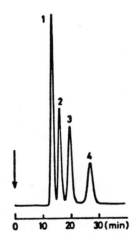

FIGURE HPLC 4.   Digitalis principles separated by HPLC. Instrument: Familic 100 and Uridec 100 UV detector; column: SC-01 (Japan Spectroscopic) (165 × 0.5 mm); mobile phase: (left) acetonitrile-methanol-water (1:1:1), (right) methanol-water (5:2); flow rate: 4 mℓ/min; detector: 220 nm. (Left) (1) Diogoxin, (2) lanatoside B, (3) gitoxin, (4) lanatoside, (5) digitoxin. (Right) (1) Digitoxin genin, (2) digitoxin genin monodigitoxoside, (3) digitoxigein biodigitoxoside, (4) digitoxin.

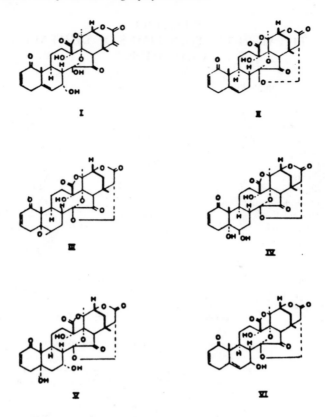

FIGURE HPLC 5. Structures of physalins (13,14-seco-16,24-cy-closteroids). (1) Physalin A, (2) physalin B, (3) B-epoxide, (4) D(5,6-dehydroxyphysalin B), (5) physalin E, (6) physalin H.

Table HPLC 7
SEPARATION OF PHYSALINS[a11]

Sample	Retention time (min)	k'
Physalin A	5.12	0.46
Physalin B	11.61	2.32
Epoxyphysalin B	6.3	0.8
Physalin D	4.53	0.29
Physalin E	4.63	0.32
Physalin H	6.01	0.72

Instrument:	Waters ALC/GPC 2440
Column:	μBondapak 98 (10 μm) (300 × 3.9 mm)
Mobile phase:	Water-methanol (2:3)
Flow rate:	1 mℓ/min
Detection:	254 nm

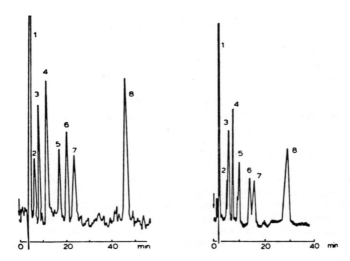

FIGURE HPLC 6. Separation of cardiac glycosides. Instrument: Spectra Physics; column: (left) LiChrosorb Si 60 (5 μm) modified with diphenylchlorosilane (125 × 4.8 mm) (temp 50°C), (right) LiChrosorb Si 100 (7 μm) (125 × 4.8 mm) (temp 40°C); mobile phase: (left) ethanol-water (40:60), (right) ethanol-water (30:70); flow rate: (left) 0.34 mℓ/min, (right) 0.7 mℓ/min; detection: not given; temperature: 50°C; peaks: (1) G-strophanthin, (2) K-strophanthoside, (3) K-strophanthin, (4) erysimin, (5) cymarin, (6) isoland, (7) digoxin, (8) neriolin. (From Davydov, Y., Gonzales, M. E., and Kiseler, A. V., *J. Chromatogr.*, 204, 293, 1981. With permission.)

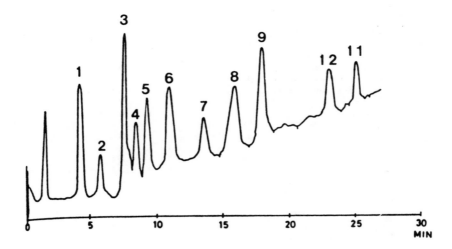

FIGURE HPLC 7. HPLC of digitalis extracts. Instrument: HP 1084B; column: LiChrosorb RP-18 (200 × 4 mm); mobile phase: acetonitrile-water (25:75); flow rate: 2.5 mℓ/min; detector: 230 nm. (1) Digitalinum verum, (2) glucoverodoxin, (3) glucogitoroside, (4) odorobisidl G, (5) lanatoside C, (6) glucoevatromonoside, (7) glucodigitoxingeninbisdigitoxoside, (8) lanatoside B, (9) acetylgitoxin, (10) lanatoside A, (11) digitoxin. (From Wichtl, M., Mangkadidjojo, and Wichtl-Bleier, W., *J. Chromatogr.*, 234, 503, 1980. With permission.)

REFERENCES (HPLC)

1. **Castle, M. C.,** *J. Chromatogr.,* 115, 437, 1975.
2. **Nachtmann, F., Spitzz, H., and Frei, R. W.,** *J. Chromatogr.,* 122, 293, 1976.
3. **Shimada, K., Hasegawa, M., Itasebe, K., Fujic, Y., and Nambara, T.,** *J. Chromatogr.,* 124, 79, 1976.
4. **Linder, W. and Frei, R. W.,** *J. Chromatogr.,* 117, 81, 1976.
5. **Shimada, K., Hasegawa, M., Hasebe, K., Fujic, Y., and Nambara, T.,** *Chem. Pharm. Bull.,* 24, 2995, 1976.
6. **Erni, F. and Frei, R. W.,** *J. Chromatogr.,* 130, 160, 1977.
7. **Davydov, V. Y., Kisler, A. V., Muionova, I. U., and Sapojnikov, Y. M.,** *J. Chromatogr.,* 290, 137, 1982.
8. **Crabbe, P. G. and Fryer, C.,** *J. Chromatogr.,* 187, 87, 1980.
9. **Desta, B., Kwong, E., and McErlane, K. M.,** *J. Chromatogr.,* 290, 137, 1982.
10. **Fujic, Y., Fukuda, H., Saito, Y., and Yamazski, M.,** *J. Chromatogr.,* 202, 1139, 1980.
11. **Sen, G., Mulchandany, N. B., and Patankar, A. V.,** *J. Chromatogr.,* 198, 203, 1980.
12. **Davydov, Y., Gonzalez, M. E., and Kiseler, A. V.,** *J. Chromatogr.,* 204, 293, 1981.
13. **Wichtl, M., Mangkudidjojo, M., and Wichtl-Bleier, W.,** *J. Chromatogr.,* 234, 503, 1980.

SECTION IV. THIN LAYER CHROMATOGRAPHY

Table TLC 1
$R_f \times 100$ Values for Alkaloids[1]

Alkaloid	Mobile phase							
	1	2	3	4	5	6	7	8
Tomatidine	47	90	49	66	37	87	73	94
Tomatidenol	47	92	46	66	40	89	73	96
Soladulcidine	26	87	40	53	23	85	71	97
Solasodine	27	85	41	59	23	82	70	96
Veramine	10	75	27	35	9	60	68	93
9α-Hydroxytomatidene	8	86	27	37	7	60	45	91
7α-Hydroxytomatidene	4	88	20	35	5	53	37	93
7α.11α-Dihydroxytomatidine	0	69	5	6	0	10	10	93
9α.11α-Dihydroxytomatidine	0	69	5	6	0	10	10	93
Demissidine	58	85	21	33	19	87	57	97
Solanidine	59	83	42	61	15	95	81	89
5β-Solanidan-3-one	75	85	61	50	43	0	91	100
Solanid-4-en-3-one	63	83	48	50	39	0	71	89
Rubijervine	15	73	17	13	7	56	55	82
Isorubijervine	19	81	27	33	11	60	43	91
Veralobine	11	75	43	43	17	64	55	100
Verarine	3	60	22	17	3	19	64	73
Cyclopamine	3	46	25	21	5	17	57	81
Jervine	0	35	26	19	3	14	53	83
Veratramine	5	76	28	17	11	36	37	91
Veramarine	3	29	12	8	5	17	27	60
Solanocapsine	0	2	0	0	0	0	5	12
Verazine	69	92	59	64	83	84	92	97
Tomatillidine	69	91	61	69	82	84	92	97
Etioline	1	47	12	14	3	9	47	77
Veralkamine	3	52	19	14	5	22	60	73

Layer: Silica Gel G
Mobile phase: 1. *n*-Hexane-ethyl acetate (1:1)
2. *n*-Hexane-ethanol (1:1)
3. Dichloromethane-methanol (23:2)
4. Dichloromethane-methanol (9:1)
5. Dichloromethane-acetone (4:1)
6. *n*-Hexane-acetone (1:1)
7. Dichloromethane-methanol-acetic acid (85:13:2)
8. Dichloromethane-methanol-ammonia (100:100:1)

Table TLC 2

HEART GLYCOSIDES AND CARDENOLIDES ($R_f \times 100$)[2]

Heart glycosides, genins	Color reaction	Mobile phase													
		I	II	III	IV	V	VI	VII	VIII	IX	X	XI	XII	XIII	XIV
Digitoxigenin (DT)	bl-gr → gr	38	44	55	35	43	39	43	48	42	62	38	38	45	61
epi-DT	gr	34	43		30	38	33	38	45	39	57	38	34	38	57
DT mono-S	bl-v	33			19	34	30	26						33	
DT bis-S		30	38	50	15	30	22	22	35	29	39	30	33	21	44
DT tris-S (digitoxin)		27	36	48	9	24	16	19	34	29	33	28	27	15	38
Digoxigenin (DG)	v	23	33	41	16	26	21	16	26	20	33	22	28	23	41
epi-DG	v	19	30	36	11	19	14	10	21	16	24	22	22	17	34
DG mono-S	v	21	32	42	11	23	18	12	24	20	29	20	24	19	38
DG bis-S	v	18	28	39	7	17	12	10	23	17	23	19	19	11	30
DG tris-S (digoxin)		16	28	36	4	14	9	9	23	16	21	20	17	8	25
DG tetra-S		11	21	28	2	9	6	6	18	18	18	20	14	6	21
neo-Digoxin	v	14	26	34	2	11	5	9	24	16	18	16	11	6	21
Gitoxigenin (GT)	y → y-gr → bl-gr	25	36	43	18	28	22	20	31	24	34	29	30	23	44
GT mono-S	ol → bl-gr	21	33	39	9	22	19	12	25	18	27	26	25	19	39
GT tris-S (gitoxin)		17	29	36	4	14	9	9	24	18	20	24	18	8	27
Gitaloxigenin	y → y-gr → bl-gr	31	42	48	32	42	40	41	47	40	34	36	40	44	
Acovenosigenin	bl				24	31		30	39			33	34	29	46
Periplogenin	gr				16	29		18	29			33	32	23	40
Sarmentogenin	bl								23			24	26	23	39
7β-ol-DT					23	37		23				34	36	31	
Strophanthidol	bl-gr		30	34				21	18		26	18	16		

Table TLC 2 (continued)
HEART GLYCOSIDES AND CARDENOLIDES ($R_f \times 100$)[2]

Heart glycosides, genins	Color reaction	Mobile phase													
		I	II	III	IV	V	VI	VII	VIII	IX	X	XI	XII	XIII	XIV
Strophanthidin	v → bl-v	23	35	41	12	23	17	12	25	21	25	29	26	12	26
g-Strophanthin	y	S.L.	S.L.	S.L.	S.L.	S.L.	S.L.	S.L.	S.L.	S.L.	S.L.	3	S.L.	S.L.	

Layer: Silica Gel G (merck)

Mobile phases: (I) Chloroform-acetone-methanol (56:36:7); (II) chloroform-acetone-methanol (55:35:10); (III) chloroform-methanol (40:50:10); (IV) chloroform-acetone (65:35); (V) diethyl ether-methanol (90:10); (VI) ethylacetate (water-saturated); (VII) chloroform-isopropanol (90:10); (VIII) chloroform-ethanol (90:10); (IX) chloroform-methanol (92:8); (X) pyridine-chloroform (1:6); (XI) toluene-methanol (80:20); (XII) cyclohexane-acetone-glacial acetic acid (49:49:2); (XIII) ethyl acetate-n-hexane-glacial acetic acid (80:10:10); (XIV) ethyl acetate (water saturated)-n-hexane-glacial acetic acid-ethanol (72:13.5:10:4.5); (XV) same as XIV, but the ethyl acetate was not water-saturated; (XVI) same as XIII, but the ethyl acetate was saturated with water

Detection: The spots (5—15 μg) were made visible after spraying with 0.5% anisaldehyde in acetic acid-sulfuric acid (98:2); the colors obtained by heating at 100°C for 6—10 min are tabulated as bl = blue; gr = green; v = violet; y = yellow; ol = olive

SL = Starting Line.

DT = 3β,14β-dihydroxy-5β-cardenolide; epi-DT = 3α,14β-dihydroxy-5β-cardenolide; S = digitoxoside; DG = 12β-ol-DT; GT = 16β-ol-DT; gitaloxigen = 16β-formyloxy-DT; acovenosigenin = 1β-ol-DT; periplogenin = 5β-ol-DT; sarmentogenin = 11α-ol-DT; strophanthidol = 5 β,19-dihydroxy-DT; g-strophanthin = 5 β-ol-19-oxo-DT; strophanthidin (ouabain) = ouabagenin rhamnoside (1β,3β,5,11α,14β,19-hexahydroxy-cardenolide rhamnoside.

Table TLC 3
SEPARATION OF DIGITALIS[3]

Solvent system, proportions	Conc. formamide solution in acetone used for impregnation (%)	R_f			
		DT-0	DT-3	DG-0	DG-3
Cyclohexane-acetone-acetic acid					
49:49:2	—	0.40	0.23	0.26	0.16
49:49:2	—	0.82	0.52	0.52	0.33
45:45:10	—	0.60	0.45	0.45	0.34
16:80:4	—	0.71	0.65	0.60	0.59
Chloroform-pyridine					
64:6	—	0.86	0.27	0.43	0.13
64:6	10	0.84	0.65	0.44	0.38
Ethyl methyl ketone-xylene-formamide					
50:50:0	10	0.55	0.28	0.27	0.12
50:50:4	10	0.47	0.20	0.20	0.09
50:50:4	15	0.52	0.29	0.21	0.10
50:50:4	20	0.54	0.33	0.19	0.09
70:30:0	10	0.72	0.59	0.50	0.36

Layer: Silica gel and silica gel impregnated with formamide
Mobile phase: As in table
DT-0 Digitoxigenin
DT-3 Digitoxin
DG-0 Digoxigenin
DG-3 Digoxin

Table TLC 4
SEPARATION OF THE DIGITALIS DERIVATIVES (R_f)[4]

Compound[a]	Mobile phase			
	1	2	3	4
1c	0.15	0.28	0.45	0.65
2a	0.15	0.28	0.46	0.63
2b	0.15	0.28	0.46	0.62
2c	0.18, 0.24	0.32, 0.35	0.70	0.86
2d	0.19, 0.26	0.33, 0.36	0.72	0.86
2e	0.19	0.21	0.47	0.61
2f	0.20	0.24	0.48	0.65
2g	0.25, 0.34	0.43, 0.53	0.74	0.89
2h	0.25, 0.36	0.48, 0.59	0.83	0.91
3a	0.38	0.48	0.76	0.87
3b	0.39	0.49	0.77	0.87
4	0.35	0.50	0.70	0.81

Table TLC 4 (continued)
SEPARATION OF THE DIGITALIS
DERIVATIVES (R$_f$)[4]

Layer: Silica Gel G (Merck)
Mobile phase: 1. Methylene chloride-ethyl acetate-meth-
               anol (40:2:1)
               2. Methylene chloride-ethyl acetate-meth-
               anol (20:2:1)
               3. Methylene chloride-ethyl acetate-meth-
               anol (10:3:1)
               4. Methylene chloride-ethyl acetate-meth-
               anol (10:5:1)

[a] See Figure 1 (Chapter 8, Section I) for structures.

Table TLC 5
RELATIVE MOBILITIES OF CARDENOLIDES[5]

| Cardenolides | Positions of active hydroxyl groups | | R$_f$ | |
	Sugar chain	Genin	Without boric acid	With boric acid
Primary glycosides				
1. Lanatoside A	22′,24′	—	0.311	0.252
2. Lanatoside B	22′,24′	14β,16β	0.221	0.286
3. Lanatoside C	22′,24′	—	0.172	0.126
4. Purpureaglycoside A	22′,24′	—	0.107	0.076
Secondary glycosides				
5. Digitoxin	15′,16′		0.967	0.874
6. Digoxin	15′,16′	—	0.672	0.563
7. Gitoxin	15′,16′	14β,16β	0.754	0.983
8. 7β-Hydroxydigoxin	15′,16′	7β,14β	0.590	0.672
9. Acetyldigitoxin	—	—	1.311	1.353
10. Acetyldigoxin	—	—	1.025	1.042
Genins				
11. Digitoxigenin	—	—	1.147	1.146
12. Digoxigenin	—	—	0.721	0.706
13. Gitoxigenin	—	14β,16β	0.902	1.294
14. 7β-Hydroxydigoxigenin	—	7β,14β	0.492	0.706
15. 14-Anhydrodigitoxigenin	—	—	1.467	1.546
16. 14-Anhydrodigoxigenin	—	—	1.066	1.092
17. 14,16-Dianhydrogitoxigenin	—	—	1.443	1.513
18. Gitoxigenin 16-acetate	—	—	1.180	1.210
19. Gitoxigenin 3,16-diacetate	—	—	1.606	1.664
20. Gitoxigenin 14,16-phenylboronate	—	—	1.434	1.513
21. 7β-Hydroxydigoxigenin 7,14-phenyl-boronate	—	—	0.975	1.017
22. 7β-Hydroxydigoxigenin 7,14-phenyl-boronate 3,12-diacetate	—	—	1.787	1.781
23. 7β-Hydroxydigoxigenin 3,12-diacetate	—	7β,14β	1.508	1.723

Layer: Silica Gel 60 and same impregnated with boric acid
Mobile phase: Chloroform-methanol-acetic acid (90:10:1)

Table TLC 6
THIN LAYER CHROMATOGRAPHY OF
CARDENOLIDES[6]

Cardenolide	Mobile phase		Fluorescence color (254 nm)
	1	2	
Digitoxigenin	0.74	0.84	Yellow
Gitoxigenin	0.61	0.76	Blue
Digoxigenin	0.50	0.71	Blue
Diginatigenin	0.41	0.61	Blue
Gitaloxigenin	0.68	0.81	Blue
Dtg-Dgx	0.54	0.80	Yellow
Dtg-Dgx-Dgx	0.50	0.77	Yellow
Dtg-Dgx-Dgx-Dgx	0.41	0.76	Yellow
Purpureaglykosid A	0.01	0.46	Yellow
Purpureaglykosid B	0.00	0.37	Blue
Lanatoside A	0.07	0.58	Yellow
Lanatoside B	0.04	0.54	Blue
Lanatoside C	0.01	0.50	Blue

Layer: DC-Alufolien KG 60 (Merck)
Mobile phase: 1. Cyclohexane-acetone-acetic acid (49:49:2)
                2. Chloroform-methanol (80:20)
Detection: Spray with trichloroacetic acid and heat

Dig Digitoxigenin
Dgx Digitoxose

REFERENCES (TLC)

1. **Hunter, I. R., Walden, M. K., Wagner, J. R., and Heftmann, E.**, *J. Chromatogr.*, 118, 259, 1976.
2. **Zullich, G., Braun, W., and Lisboa, B. P.**, *J. Chromatogr.*, 103, 396, 1975.
3. **Starstein, L.**, *J. Chromatogr.*, 117, 87, 1976.
4. **Yoshioka, K., Fullerton, D. S., and Rohrer, D. C.**, *Steroids*, 32, 511, 1978.
5. **Szeleczky, Z.**, *J. Chromatogr.*, 178, 53, 1979.
6. **Karting, T.**, *J. Chromatogr.*, 234, 437, 1982.

Chapter 9

ECDYSTEROIDS

SECTION I. PREPARATION

The separation and quantitation of the natural juvenile hormones pose difficulties because of their instability and their close similarity to other compounds found in the source from which the substances are isolated. The extremely low concentration in natural sources has also intensified the difficulty. Silk moths show a concentration of the hormones of approximately 300 ppb. Other species may have as little as 0.1 to 5 ppb. GC has advanced to the stage where these determinations can be carried out. PLC has been used to separate the products from in vitro organ cultures. GC methodology requires the preparation of derivatives. Since the procedures involved are specialized for ecdysteroids, the pertinent ones will be described in this section rather than in the general section on preparation of derivatives.

Structures and nomenclature for the common ecdysteroids are given in Figure 1. As in the case of natural products, the success of the separation of this class of steroids depends on how the sample is prepared. Thus this section also begins with methods for preparation of the sample. Because of the low concentration as well as the small size of the sample, preparation for chromatography is of even more importance.

Extraction of phytoecdysones (method of Hikins et al.[1]) — The dried underground part (10.0 kg) and dried aerial part (6.5 kg) of *Diplazium donianum*, collected at Yang-ming-shan, Taipei, Taiwan, were separately extracted with refluxing ethanol (30:1 and 20:1, respectively, for 5 hr each, 3 times). Both of the ethanol extracts were diluted with water and continuously extracted with ethyl acetate to give extracts of 15 and 62 g, respectively, which showed insect moulting hormone activity by means of bioassay, the *Sarcophaga* test. Both of the ethyl acetate extracts were combined and subjected to column chromatography using silica gel (846 g). Elution with benzene-ethyl acetate (3:1) gave fraction A (10 mg) and fraction B (10 mg).

Zooecdysones (method of Mijazaki et al.[2]) — 10 pupae were freeze-dried with 20 mℓ acetone and dry ice under reduced pressure and temperature under 50°C; 5 μg cyasterone and 15 g sea sand were added to the residue and ground to a fine powder. The powder was filtered in a filter paper thimble and extracted with 100 mℓ tetrahydrofuran (THF) in a modified Soxhlet apparatus for 24 hr. This apparatus was designed in such a way that a glass container for the filter paper thimble was headed by the vapor of the solvent. The THF extract was concentrated to half the volume, and 3 g Carplex No. 80 (silica acid) were added. The THF was completely eliminated by rotary evaporation. The residue was put into a filter paper thimble and extracted in the Soxhlet apparatus successively for 1 hr with 50 mℓ of each of the following solvents: *n*-hexane, benzene, ether, and THF. The THF extract was passed through a 3 g silica gel (Merck Silica gel 60) column prepared in a glass filter (15 AG-3). The column was washed with THF, and the eluate and washings were combined and the THF was evaporated in vacuum. The residue was purified by preparative TLC (silica gel plate, 20 × 20 cm, 500 μm-thick) using chloroform-ethanol (2:1) as the mobile phase. A wide band containing compounds with R_f values from 0.45 to 0.90 was extracted with THF using an ultrasonic generator (over all yield of recovery was 75%), the THF was evaporated, the residue was dissolved in 100 $\mu\ell$ TSIM, and the solution was heated at 100°C for 30 min. For GLC analysis using electron capture detectors, derivatization with 100 $\mu\ell$ heptafluorobutyrylimadazole and 10 $\mu\ell$ heptafluorobutyric anhydride were added to the trimethylsilylimidazole solution and the reaction mixture was heated at 50°C for 2 hr. Benzene (0.2 mℓ) was added, and the mixture was poured into ice water. The organic layer was

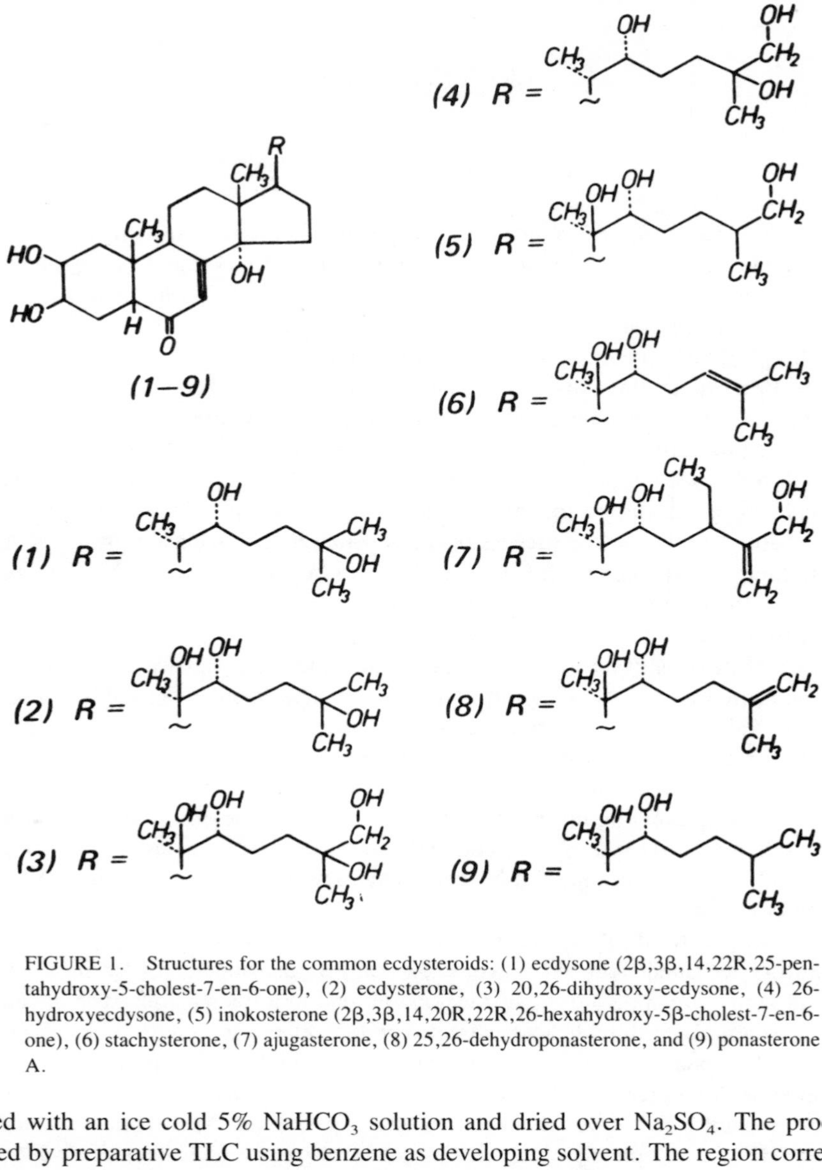

FIGURE 1.   Structures for the common ecdysteroids: (1) ecdysone (2β,3β,14,22R,25-pentahydroxy-5-cholest-7-en-6-one), (2) ecdysterone, (3) 20,26-dihydroxy-ecdysone, (4) 26-hydroxyecdysone, (5) inokosterone (2β,3β,14,20R,22R,26-hexahydroxy-5β-cholest-7-en-6-one), (6) stachysterone, (7) ajugasterone, (8) 25,26-dehydroponasterone, and (9) ponasterone A.

washed with an ice cold 5% $NaHCO_3$ solution and dried over Na_2SO_4. The produce was purified by preparative TLC using benzene as developing solvent. The region corresponding to ecdysone was extracted with benzene and analyzed by GLC.

Extraction of ecdysteroids from frass of the tobacco hornworm, *Manduca sexta* (method of Kaplanis et al.[4]) — The Frass (361 g) was homogenized for 5 min with 3 mℓ methanol per gram (wet weight). The homogenate was transferred to 250 mℓ glass cups and spun at 2500 rpm for 10 min. The supernatants were decanted, and the residues were pooled and rehomogenized twice with 75% methanol (1.5 mℓ/g). The extracted residues in the tubes were pooled and radioassayed (50 g, 1.5×10^5 dpm total) and discarded. The supernatants were combined and dried in vacuum (33 g), and the residue was partitioned between 100 mℓ water and 50 mℓ butanol. After centrifugation, the butanol phase was removed and the aqueous phase extracted twice with 25 mℓ butanol. The aqueous phase (27 g, 3.3×10^5 dpm) was discarded. The butanol extracts were combined, washed once with 2% sodium carbonate solution and then rinsed with water until neutral. The pooled butanol extracts were taken to dryness (1.7 g, 6.0×10^6 dpm total). The residue was partitioned between 100

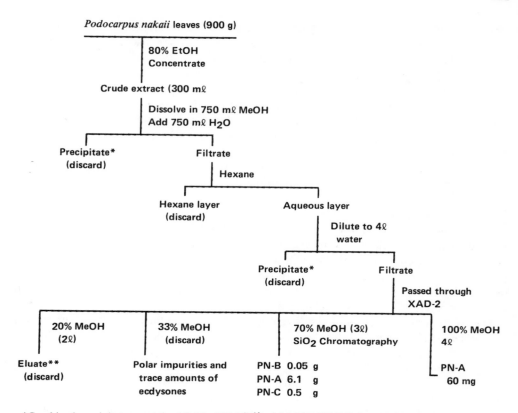

Podocarpus nakaii leaves (900 g)

80% EtOH
Concentrate

Crude extract (300 mℓ

Dissolve in 750 mℓ MeOH
Add 750 mℓ H$_2$O

Precipitate*
(discard)

Filtrate

Hexane

Hexane layer
(discard)

Aqueous layer

Dilute to 4ℓ
water

Precipitate*
(discard)

Filtrate

Passed through
XAD-2

| 20% MeOH (2ℓ) | 33% MeOH (discard) | 70% MeOH (3ℓ) SiO$_2$ Chromatography | 100% MeOH 4ℓ |

Eluate**
(discard)

Polar impurities and
trace amounts of
ecdysones

PN-B 0.05 g
PN-A 6.1 g
PN-C 0.5 g

PN-A
60 mg

*Combined precipitates contained 0.55 g PN-A/B[10] and 0.08 g PN-C (LC analysis).
**Contained 2 mg PN-A/B[10]; PN-C was nondetectable.

MeOH = methanol, EtOH = ethanol, PN = ponasterone.

FIGURE 2. Scheme for extraction of phytoecdysones.

mℓ each of 70% methanol and hexane, the phases were separated, and each phase was re-extracted with 100 mℓ of the appropriate upper or lower phase. The hexane fractions (0.6 g, 8.8 × 10^4 dpm total) were discarded. The 70% methanol fractions were combined and dried (1.1 g, 6.4 × 10^6 dpm total), and the residue was dissolved in 2.5 mℓ methanol and adjusted with benzene to give a final concentration of benzene-methanol (95:5).

Extraction of ecdysteroids from the barnacle, *Balanus balanoides* (method of Wilson et al.[5]) — A sample of *Balanus balanoides* (100 g) was ground in methanol and the methanol-soluble portion was partitioned between light petroleum (bp 40 to 60°C) and aqueous methanol. The aqueous extract was partitioned between butanol and water, and the butanol extract was finally partitioned between ethyl acetate and water. The residue from the aqueous portion was taken up in methanol (2 mℓ), and ecdysone (10 μg) was added and the sample was silylated, cleaned-up by TLC, and subjected to GC. To the residue from a similar sample in methanol (2 mℓ) was added ecdysone (13.2 μg), and 10 μℓ of the mixture was injected onto a C$_{18}$ HPLC column.

Ecdysteroids from *Manduca sexta* (method of Judy et al.[6]) — Culture media were extracted 3 times by vigorous shaking with equal volumes of highly purified ethyl acetate. The pooled organic phases were reduced to a small volume under N$_2$. The extract was then applied to thin layer plates (Brinkmann GF-254 0.25-mm thick) which were washed with hexane-ethyl acetate (1:1). The mobile phase also used hexane-ethyl acetate (1:1). The chromatograms were scanned in a Packard model 7201 radiochromatogram scanner. Appropriate zones were eluted with pure ethyl acetate and stored in that solvent at $-20°C$.

Extraction of ecdysterone and inokosterone from *Achyranthis radix* (method of Ogawa et al.[7]) — Methyl ethyl ketone extract: to 10 g finely powdered *Achyranthis radix* (dried over P_2O_5 in vacuum), 100 mℓ methyl ethyl ketone (MEK) were added and the mixture was refluxed for 1 hr. The MEK layer was transferred and the residue was extracted 3 times with 100 mℓ MEK. The MEK extracts were combined and evaporated to dryness under reduced pressure. Finely powdered dried *Achyranthis radix* (10 g) was extracted with 200 mℓ MEK for 5 hr in a Soxhlet extractor. The MEK extract was evaporated to dryness under reduced pressure. Methanol extract: two kinds of extracts (the MeOH extracts mentioned above) were prepared under the same procedure described above.

Derivative Formation

Preparation of acetates of ecdysteroids (method of Koolman et al.[8]) — Zones of TLC were scraped from the plate. The labeled compounds were eluted with ethanol, and acetylation was performed at room temperature in 50 $\mu\ell$ pyridine to which 50 $\mu\ell$ acetic anhydride were added. In order to observe the kinetics of the acetylation of the different hydroxyl groups of the acdysteroids, samples were taken from the reaction mixture at 0, 2, 10, 30, 60, 120, and 140 min and analyzed by TLC directly.

Preparation of acetonides of ecdysteroids (method of Koolman et al[8]) — The labeled compounds were scraped and eluted as described above. Acetonide formation was performed in 100 $\mu\ell$ 2,2-dimethoxypropane-acetone (1:1) with 20 $\mu\ell$ 10 μM p-toluenesulfonic acid in acetone as catalyst. The reaction at room temperature was stopped after 2 hr by the addition of solid sodium hydrogen carbonate. After evaporation of the solvent under a stream of nitrogen, methanol was used to transfer the reaction products to a thin layer plate.

Formation of TMS ethers of ecdysones (method of Morgan and Poole[9]) — Ecdysone was treated at room temperature with TMS. The unhindered hydroxyl groups (at 2, 3, 22, and 25) were all rapidly silylated. In order to silylate the fifth hydroxyl (14), heating at 140°C overnight was required. The pentakis (trimethylsiloxyl) ether was obtained quantitatively by heating 20-hydroxyecdysone at 100°C for 4 hr. The hexakis-ether was prepared by heating at 140°C for 20 hr.

Heptafluorobutyryl and trimethylsilyl derivatization (method of Mijazaki et al.[10]) — Ecdysterone (0.5 mg) was dissolved in 20 $\mu\ell$ trimethylsilylimidazole in a small Teflon® capped tube, and the solution was heated at 100°C for 1 hr. To this solution 20 $\mu\ell$ heptafluorobutyrylimidazole and 2 $\mu\ell$ heptafluorobutyric anhydride were added, and the mixture was heated at 50°C for 2 hr. Benzene (0.1 mℓ) was added. This was washed with ice-cold 5% $NaHCO_3$ and then with water. After drying over Na_2SO_4, the solvent was evaporated. For ecdysterone, the nonheptafluorobutyl-penta-trimethyl-silyl derivative is formed.

REFERENCES

1. **Hikins, H., Mohri, K., Okujama, T., and Takemoto, T.,** *Steroids,* 28, 649, 1976.
2. **Mijazaki, H., Ishibashi, M., and Mori, C.,** *Anal. Chem.,* 45, 1164, 1973.
3. **Schooley, D. A., Weiss, G., and Nakanishi, K.,** *Steroids,* 20, 377, 1972.
4. **Kaplanis, J. N., Thompson, M. J., Dutky, S. R., and Robbins, W. E.,** *Steroids,* 34, 333, 1979.
5. **Wilson, J. D., Scalia, I. D., and Morgan, E. D.,** *J. Chromatogr.,* 212, 211, 1981.
6. **Judy, K. J., Schooley, D. A., Dunham, L. L., Hall, M. S., Bergot, B. J., and Siddall, J. B.,** *Proc. Natl. Acad. Sci. U.S.A.,* 70, 1509, 1973.

7. **Ogawa, S., Yoshida, A., and Kato, R.,** *Chem. Pharm. Bull.,* 25, 704, 1977.
8. **Koolman, I., Reum, L., and Karlson, P.,** *Z. Physiol. Chem.,* 1355, 1979.
9. **Morgan, E. D. and Poole, C. F.,** *J. Chromatogr.,* 116, 333, 1976.
10. **Mijazaki, H., Ishibashi, M., and Mori, C.,** *Anal. Chem.,* 45, 1164, 1973.

SECTION II. GAS CHROMATOGRAPHY

Because of their high molecular weight along with large polarities, the ecdysteroids generally must be derivatized before GC. The silanizing reagents described in Section III. can be used successfully for these compounds. The specific methods discussed earlier in this section were developed for this class of compounds. The examples given reflect this use of derivatization. However, GC is little used in analysis of the ecdysteroids.

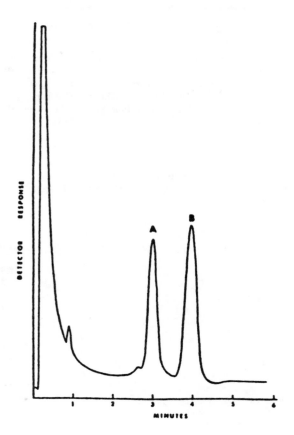

FIGURE GC 1.   Separation of ecdysones by GC. Instrument: Varian 920, flame ionization detector; column: 1.5% OV-101 on Chromosorb W (80—100 mesh); temperature: column 290°C, detector 310°C, injector 310—325°C; nitrogen flow: 45 mℓ/min. (A) α-Ecdysone-TMS (1 ng), (B) β-ecdysone (1 ng). (From Borst, D. W. and O'Connor, J. D., *Steroids,* 24, 837, 1974. With permission.)

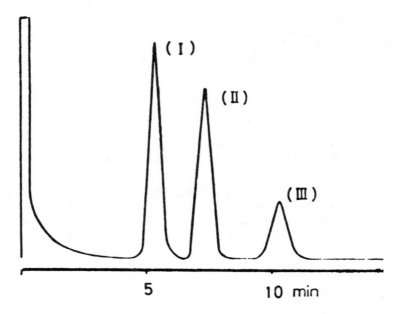

FIGURE GC 2. Separation of ecdysone derivatives. Instrument: Shimadzu Model 5-AP, flame ionization detector; column: 1.5% OV-101 on Chromosorb W-HP (100 cm × 4 mm); temperature: column 230°C, detector 280°C, injector 280°C; nitrogen flow: 60mℓ/min. (I) Di-HFB derivative, (II) Mono-HFB-monoTMS derivative, (III) Di-TMS derivative. (Reprinted with permission from Miyazaki, H., Ishibashi, M., and Mori, C., *Anal. Chem.*, 45, 1164, 1973. Copyright 1973 American Chemical Society.)

Table GC 1
ECDYSTEROID TRIMETHYLSILYL ETHERS[3]

Parent ecdysteroid	No. of TMS groups	Retention time (min)
Ecdysone	4	1.85
	5	1.65
20-Hydroxyecdysone	4	2.25
	5	2.45
	6	1.90
Inokosterone	4	2.55
	5	2.85
	6	2.20
2-Deoxy-20-hydroxyecdysone	4	3.9
	5	2.4
Poststerone	2	0.6 (1.3)
	3	0.9
Cyasterone		8.0
3-Dehydroecdysone	3	2.15
3-Dehydro-20-hydroxyecdysone	4	2.70

Instrument: Pye 104, flame ionization
Column: 1.5% OV-101 on Chromosorb Q (100—120 mesh)
Temperatures: Column 270—280°C, detector 300°C
Flow: Nitrogen 50—60 mℓ/min

REFERENCES (GC)

1. **Borst, D. W. and O'Connor, J. D.**, *Steroids*, 24, 837, 1974.
2. **Miyakazi, H., Ishibashi, M., and Mori, C.**, *Anal. Chem.*, 45, 1164, 1973.
3. **Brelby, C. R., Gaube, A. R., Morgan, E. D., and Wilson, I. D.**, *J. Chromatgr.*, 164, 43, 1980.

SECTION III. HIGH PERFORMANCE LIQUID CHROMATOGRAPHY

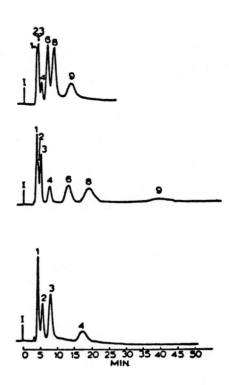

FIGURE HPLC 1. Separation of ecdysteroids with three different mobile phases. Instrument: Dupont 830; column: Corasil II (37—50 μm) (3—4 cm × 2 mm, 3 columns coupled in series); mobile phase: (top) chloroform-ethanol (4:1), (middle) chloroform-ethanol (9:7), (bottom) chloroform-ethanol (14:1); detector: 245 nm. (1) 2β,14α-Dihydroxy-5β-cholest-7-en-3,6-dione, (2) 2β,β,3β,14α-trihydroxy-5α-cholest-7-en-6-one, (3) 2β,3β,14α-trihydroxy-5β-cholest-7-en-6-one, (4) 22-deoxyecdysone, (5) 5α,α-ecdysone, (6) α-ecdysone, (7) 20-hydroxyecdysone.

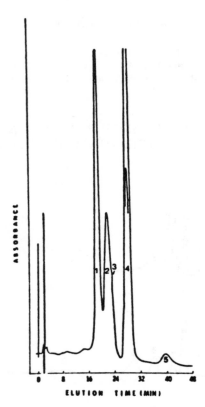

FIGURE HPLC 2.   Separation of ecdysteroids. Instrument: Milton Roy pump with pulse damper and gradient former; column: μBondapak, phenyl-corasil (10 cm × 3 mm); mobile phase: gradient of water to 20% ethanol in water; flow: 1 mℓ/min. (1) Ecdysterone, (2) inokosterone, (3) makisterone, (4) α-ecdysterone, (5) ponasterone. (From Ogarva, S., Yoshida, A., and Kato, R., *Chem. Pharm. Bull.*, 25, 904, 1977. With permission.)

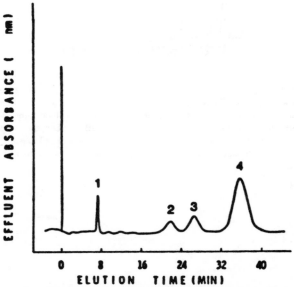

FIGURE HPLC 3.   Separation of ecdysteroids as acetates. Instrument: Milton Roy pump with LDC damper and gradient former; column: Corasil II (1000 × 3 mm); mobile phase: chloroform-ethanol (8:1); flow: 1.1 mℓ/min; detector: 254 nm; temperature: 20°C. (1) Ecdysterone triacetate, (2) α-ecdysone, (3) makisterone A, (4) ecdysterone + inokosterone. (From Gilgan, M. W., *J. Chromatogr.*, 129, 447, 1976. With permission.)

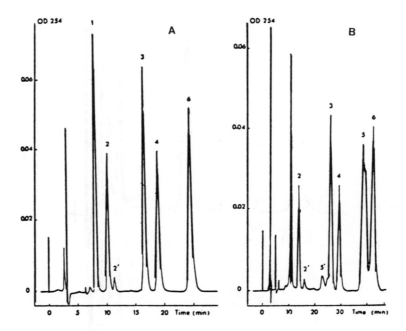

FIGURE HPLC 4. Separation of ecdysteroids. Instrument: Dupont 848 pump and Model 8303 gradient former; column: (A) Zorbax ODS, 25 cm × 4.6 mm, (B) Zorbax ODS, 25 cm × 4.6 mm; mobile phase: (A) dichloromethane-ethanol-water (840:145:15), (B) dichloromethane-isopropanol-water (125:25:2); flow rate: 1 mℓ/min; detector: UV 254 nm; peak identification: (1) poststerone, (2) cyasterone, (3) ecdysone, (4) makisterone, (5) inokosterone, (6) ecdysterone, (2′) cyasterone impurities, (5′) inokosterone impurities. (From Lafont, R., Martin-Somme, S., and Chambet, J. C., *J. Chromatogr.*, 170, 185, 1979. With permission.)

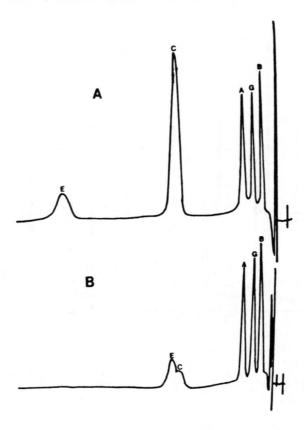

FIGURE HPLC 5.   Separation of ponasterone and 2-deoxyec-dysone. Instrument: LDC Constantimetric III pump and Pye LL3-UV detector; column: Spherisorb ODS, 5 μm (150 × 3 mm); flow: 1 mℓ/min; mobile phase: (A) acetonitrile-water (15:85) (B) tetrahydrofuran-water (10:90). (A and B) E = ponasterone and C = 2-deoxyeadysone. (From Mayer, R. T., Durrant, J. L., Holman, G. M., Weirich, G. F., and Svoboda, J. A., *Steroids*, 34, 555, 1979. With permission.)

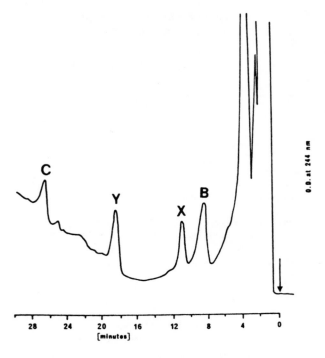

FIGURE HPLC 6. Separation of conjugated ecdysteroids. Instrument: Pye pump LL3 × P and Pye Unicam LC-UV detector; column: Spherisorb ODS, 5 μm (10 cm × 5 mm); mobile phase: (linear gradient) 4%/min from 25 to 40% methanol in 0.4 *M* ammonium acetate at pH 7 (0.8 mg/ℓ of tetrabutylammonium hydroxide added for ion pairing); flow: 1 mℓ/min; detection: 244 nm. (B) Conjugated 20-hydroxyecdysone, (X) conjugated ecdysone, (C) conjugated 2-deoxyecdysone, (Y) unknown compound, not ecdysteroidal. (From Wilson, I. D., Brelby, C. R., and Morgan, E. D., *J. Chromatogr.*, 238, 457, 1982. With permission.)

Table HPLC 1
ECDYSTEROIDS[7]

Using Partisil-10 ODS

	Retention volume (mℓ)			
	Mobile phase			
Ecdysteroid	1	2	3	4
Ecdysone	75	38	14	35
20-Hydroxyecdysone	31			22
3-Epiecdysone	75			35
3-Epi-20-hydroxyecdysone	31			22
3-Dehydroecdysone	75			35
3-Dehydro-20-hydroxyecdysone	31			22
26-Hydroxyecdysone	45			28
20,26-Dihydroxyecdysone	20			16
2-Deoxyecdysone	≥120	136	32	54
Ecdysone-2,3-acetonide			75	

Table HPLC 1 (continued)
ECDYSTEROIDS[7]

Using APS Hypersil

| | Retention volume (mℓ) | | | | |
| | Mobile phase | | | | |
Ecdysteroid	5	6	7	8	9
Ecdysone	61	44	36	28	
20-Hydroxyecdysone	97	65	54	42	
3-Epiecdysone	107	68			
3-Epi-20-hydroxyecdysone	184	110		52	
3-Dehydroecdysone	18	12			
3-Dehydro-20-hydroxyecdysone	27				
26-Hydroxyecdysone	$\gg$		124	103	24
20,26-Dihydroxyecdysone	$\gg$	323		166	35
5-Ecdysone	40				
2-Deoxyecdysone	20		14		
3-Epi-2-deoxyecdysone	27				

Instrument: Waters ALC/100
Columns: APS-Hypersil, 5 μm and Partisil 10 ODS, 7 μm (65 cm × 4.6 mm)
Mobile phase: Methanol-water: 1. 1:3 2 mℓ/min
2. 3:7 4 mℓ/min
3. 2:3 4 mℓ/min
4. Linear gradient 1:3 to 3:2 at 2 mℓ/min, 0—40 min
Dichloromethane-methanol-isopropanol:
5. 95:1:4
6. 95:2:3
7. 95:3:2
8. 95:4:1
9. 92:8 (Dichloromethane-methanol)
Detection: 242 nm

REFERENCES (HPLC)

1. **Nigg, H. N., Thompson, M. J., Rapianis, J. N., Svoloda, J. A., and Robbins, W. E.**, *Steroids*, 22, 507, 1974.
2. **Ogarva, S., Yoshida, A., and Kato, R.**, *Chem. Pharm. Bull.*, 25, 904, 1977.
3. **Gilgan, M. W.**, *J. Chromatogr.*, 129, 447, 1976.
4. **Lafont, R., Martin-Somme, S., and Chambet, J. C.**, *J. Chromatogr.*, 170, 185, 1979.
5. **Mayer, R. T., Durrant, J. L., Holman, G. M., Weirich, G. F., and Svoboda, J. A.**, *Steroids*, 34, 555, 1979.
6. **Wilson, I. D., Brelby, C. R., and Morgan, E. D.**, *J. Chromatogr.*, 238, 457, 1982.
7. **Scalia, S. and Morgan, E. D.**, *J. Chromatogr.*, 238, 457, 1982.

SECTION IV: THIN LAYER CHROMATOGRAPHY

Table TLC 1
SENSITIVITY OF VARIOUS DETECTION METHODS FOR α- AND β-ECDYSONE ON THIN LAYERS[1]

Detection method	Amount spotted (μg)			
	50	10	1	0.5
Fluorescence quenching	4+	4+	4+	3+
Vanillin-sulfuric acid spray	4+	3+	−	−
Anisaldehyde spray	4+	4+	+	−

(−) Negative, (1+, 2+, 3+, 4+) positive.

Table TLC 2
R_f VALUES OBTAINED FOR α- AND β-ECDYSONE USING VARIOUS MOBILE PHASES[1]

Mobile phase	R_f	
	α-Ecdysone	β-Ecdysone
Chloroform-methanol (3:2)	No separation	
Chloroform-methanol (4:1)	0.68	0.60
Chloroform-methanol (9:2)	0.29	0.22
Chloroform-methanol-water (75:24:4)	0.78	0.73
Chloroform-ethyl acetate (1:1)	No separation	

Layer: Whatman LK5F silica gel
Detection: Quench

Table TLC 3
R_f VALUES OF METABOLITES OF ECDYSONE IN THE BLOWFLY[2]

Compound	R_f	Acetates R_f	Acetonides R_f
Ecdysone	0.27	0.71	0.55
Ecdysterone	0.21	0.73	0.72
Inokosterone	0.19	0.71—0.78	0.78
26-Hydroxyecdysone	0.15	0.97	0.73
20,26-Dihydroxyecdysone	0.11	0.73	0.78

Layer: Silica Gel G (Merck)
Mobile phase: Chloroform-methanol (80:20)

Table TLC 4
SEPARATION OF PHYTOECDYSONES AS
ACETATES FROM DISPLAZIUM DONIANUM[3]

Steroid	Mobile phase	R_f	Color
Makisterone A triacetate	Benzene-ethyl Acetate (1:2)	0.35	Brown with H_2SO_4 + heat
Makisterone A	Chloroform-methanol (4:1)	0.60	
Makisterone A	Ethyl acetate-methanol (20:1)	0.21	
Makisterone B letra acetate	Benzene-ethyl acetate (1:2)	0.70	Pink with H_2SO_4 + heat
Makisterone B	Chloroform-methanol (4:1)	0.60	
Makisterone B	Ethyl acetate-methanol (20:1)	0.29	
Makisterone D	Chloroform-methanol (4:1)	0.53	
	Ethyl acetate-methanol (20:1)	0.34	

Layers: Silica Gel G
Acetylation: Acetic anhydride in pyridine

Table TLC 5
SEPARATION OF ECDYSTEROIDS AND
ACETONIDES[4]

Ecdysteroid	R_f free	Reaction with acetone	R_f (acetonide)
Ecdysone	0.47	No reaction	—
Epi-ecdysone	0.52	Diacetonide	0.62
Hydroxyecdysone	0.40	Monoacetonide	0.11
Epi-20-hydroxy-ecdysone	0.44	Triacetonide	0.77
Epi-29,26-dihy-droxyecdsone	0.27	Diacetonide	0.30

Layer: Silica Gel 60F-254 (Merck)
Mobile phase: Chloroform-ethanol (65:35)

Table TLC 6
VARIATION OF R_f VALUES OF ECDYSONE (α) AND 20-
HYDROXYECDYSONE (β) ON REVERSED PHASE LAYERS[5]

Solvent	Proportions	C_2		C_8		C_{18}	
		α	β	α	β	α	β
Methanol-water	50:50	0.38	0.53	0.28	0.44	0	0
	80:20	0.88	0.88	0.75	0.82	0.76	0.82
Ethanol-water	40:60	0.38	0.53	0.34	0.52	0.41	0.57
	70:30	0.89	0.89	0.88	0.88	0.88	0.88

Table TLC 6 (continued)
VARIATION OF R_f VALUES OF ECDYSONE (α) AND 20-HYDROXYECDYSONE (β) ON REVERSED PHASE LAYERS[5]

		C_2		C_8		C_{18}	
Solvent	Proportions	α	β	α	β	α	β
2-Propanol-water	20:80	0.18	0.46	0.19	0.46	0.24	0.46
	60:40	0.87	0.87	0.83	0.83	0.81	0.81
Acetonitrile-water	30:70	0.44	0.60	0.33	0.53	0.35	0.50
	60:40	0.86	0.86	0.90	0.90	0.90	0.90
Acetone-water	30:70	0.35	0.58	0.16	0.53	0.26	0.45
	70:30	0.79	0.87	0.86	0.86	0.84	0.91

Layer: C_2, C_8, and C_{18} reversed phase, EM Science
Mobile phase: As in table

Table TLC 7
SEPARATION OF
ECDYSTEROIDS[6]

Compound	R_f
3-Dehydroecdysone	0.88
3-Dehydro-20-hydroxyecdysone	0.77
3-Epiecdysone	0.53
Ecdysone	0.45
3-Epi-20-hydroxyecdysone	0.34
20-Hydroxyecdysone	0.28

Layer: Silica Gel GF 254, EM Science
Mobile phase: Chloroform-ethanol (9:2); develop continuously for 2 hr
Detection: Quench

Table TLC 8
R$_f$ VALUES OF VARIOUS ECDYSTEROIDS ON THREE TYPES OF RP-TLC PLATES[7]

Compound	C$_2$			C$_8$			C$_{18}$		
	30% PrOH	60% MeOH	40% ACN	30% PrOH	60% MeOH	40% ACN	30% PrOH	60% MeOH	40% ACN
Ecdysone	0.38	0.52	0.66	0.39	0.38	0.61	0.43	0.31	0.64
20-Hydroxy-ecdysone	0.53	0.65	0.75	0.58	0.53	0.72	0.59	0.47	0.75
2-Deoxy-20-hydroxy-ecdysone	0.36	0.53	0.62	0.42	0.32	0.6	0.39	0.33	0.57
Inokosterone	0.49	0.67	0.71	0.53	0.53	0.72	0.55	0.50	0.73
Muristerone	0.27	0.57	0.56	0.28	0.38	0.57	0.29	0.30	0.55
Makisterone A	0.45	0.63	0.67	0.5	0.47	0.72	0.51	0.39	0.71
Cyasterone	0.45	0.66	0.60	0.53	0.53	0.65	0.57	0.47	0.64
Poststerone	0.36	0.63	0.57	0.47	0.47	0.61	0.49	0.40	0.59
Ponasterone A	0.14	0.44	0.41	0.13	0.21	0.41	0.14	0.16	0.39
Polypodine B	0.52	0.71	0.70	0.63	0.55	0.78	0.61	0.48	0.77
Ajugasterone C	0.2	0.60	0.60	0.26	0.38	0.57	0.26	0.31	0.59
Carpesterol	0.0	0.0	0.0	0.0	0.0	0.0	0.0	0.0	0.0

Layer: C$_2$,C$_8$,C$_{18}$ RPTLC (Merck)
Mobile phase: PrOH = 2-propanol; MeOH = methanol; ACN = acetonitrile; in each case the other component is water

REFERENCES (TLC)

1. **Ruh, M. F. and Black, C.,** *J. Chromatogr.,* 116, 486, 1976.
2. **Koolman, J., Reum, L., and Karlson, P.,** *Z. Physiol. Chem.,* 360, 1355, 1979.
3. **Hikino, H., Mohri, K., Okujama, T., and Takemoto, T.,** *Steroids,* 28, 649, 1979.
4. **Kaplanis, J. N., Thompson, M. J., Dutky, S. K., and Robbins, W. E.,** *Steroids,* 34, 333, 1979.
5. **McCarthy, J. F.,** *Steroids,* 34, 799, 1979.
6. **Dinan, L. N., Donnaley, P. L., Rees, H. H., and Goodwin, T. W.,** *J. Chromatogr.,* 205, 189, 1981.
7. **Wilson, I. D., Scalia, I. D., and Morgan, E. D.,** *J. Chromatogr.,* 212, 211, 1981.

Chapter 10

ESTROGENS

SECTION I. PREPARATIONS

Chromatographic techniques such as thin layer chromatography (TLC), gas chromatography (GC), and liquid chromatography (LC) have played important roles in the elucidation of the mechanism of action of the estrogenic steroids and their determination in a wide variety of analytical matrices. This section will be concerned with analyses of estrogens, estrogen conjugates, and synthetic estrogens as well as estrogens in pharmaceutical preparations.

The estrogenic steroids derive their chemical structure from the parent hydrocarbon estrane. The natural estrogens exhibit similar characteristics, which include an aromatic A ring, the hydroxyl at the C-3 position.

Figure 1 lists the chemical structures of a number of estrogens. Estrone, estriol, and estradiol 17β constitute the three major estrogens in the human. Equilin, equilenin, and estrone are found in horses.

Table 1 gives the nomenclature of the estrogen conjugates which have been found in biological fluids.

Figure 2 gives a schematic for the extraction of 2-hydroxyestrogens from urine.

Stabilizing the Estrogens

Since the estrogenic steroids, particularly the 2-hydroxyderivatives, are somewhat labile, it is appropriate to describe methods for their protection during chromatography. In a method described by Gelbke and Kruppen,[2] thin layers of silica gel or alumina were impregnated with ascorbic acid (see Figure 2). Column materials were prepared with slurries containing ascorbic acid. Stock solutions were made with added ascorbic acid; the desired steroid was dissolved in 50 mℓ methanol and 0.3 g ascorbic acid was added. The solution used for impregnation of paper or thin layer chromatograms was 400 mℓ methanol, 15 g ascorbic acid, and 4 mℓ glacial acetic acid.

Thin layer plates were immersed in the standard solution for impregnation, blotted with filter-paper, and dried in a horizontal position at room temperature for at least 30 min. A typical mobile phase was acetic acid-chloroform-methyl cyclohexane (1:2.2) saturated with ascorbic acid. The ascorbic acid does not interfere with the chromatography.

Alumina plates were impregnated with standard ascorbic acid solution. The solvent system was 2-butanone-acetic acid (1:1), saturated with ascorbic acid.

Column Chromatography Methods

Silica gel columns—Silica gel (E. Merck, 0.05 to 0.2 mm) was impregnated with ascorbic acid by stirring 30 g in a solution of 7.5 g ascorbic acid in 200 mℓ methanol. After 30 min, the silica gel was collected and dried overnight at room temperature. The ascorbic acid-impregnated silica gel thus obtained showed a yellowish color, which became more intense after several days. Therefore, the impregnated silica gel was used within the following 2 or 3 days. The columns (0.9 × 8 cm) were prepared by suspending the impregnated silica gel in the solvent system acetic-chloroform-cyclohexane (1:2:2), which was saturated with ascorbic acid.

Alumina columns — Protection of the steroid by ascorbic acid during chromatography was achieved as follows: columns (0.9 × 6 cm) were prepared with a slurry of 5 g alumina (Aluminum Oxide Woelm acid, activity grade I; Woelm, Eschwege, West Germany) in

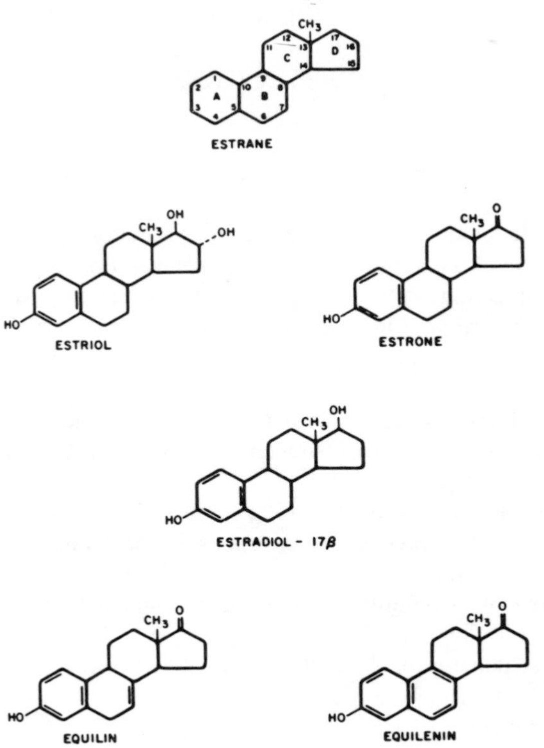

ESTRANE

ESTRIOL

ESTRONE

ESTRADIOL - 17β

EQUILIN

EQUILENIN

FIGURE 1.   Chemical structures of estrogens. (Mono-ols): 17-Deoxyestrone, estrone, 6-dehydroestrone, equilin, equilenin. (Diols): Estradiol-17α, estradiol-17β, 2-hydroxyestrone, 6-dehydroestradiol, 17α-dihydroequilin, 3,16α-estradiol, 17α-dihydroequilenin, 16-ketoestradiol, 16α-hydroxyestrone, 6-ketoestradiol. (Triols): 16-Epiestriol, 17-epiestriol, 2-hydroxyestradiol, estriol, 6α-hydroxyestradiol, 16,17-epiestriol, 6-ketoestriol. (Tetraols): 2-Hydroxyestriol, 6α-hydroxyestriol.

Table 1
ESTROGEN NOMENCLATURE

Trivial	Symbol	IUPAC
Estrone	E_1	3-Hydroxy-1,3,5(10)-estratrien-17-one
Estradiol	E_2	1,3,5(10)-Estratrien-3,17β-diol
Estriol	E_3	1,3,5(10)-Estratrien-3,16α,17β-triol
17α-Dihydroequilenin	17α-Eqe	d-1,3,5(10),6,8-Estrapentaen-3,17β-diol
Equilenin	Eqe	1,3,5(10),6,8-Estrapentaen-3-Ol-17-one
Estrone-3-β-D-glucuronide	E_1-3G	3-Hydroxy-1,3,5(10)-estratrien-17-one-3-(0-1β)-D-glucopyranosid-uronic acid

Table 1 (continued)
ESTROGEN NOMENCLATURE

Trivial	Symbol	IUPAC
17β-Estradiol 17-β-D-glucuronide	E_2-17G	1,3,5(10)-Estratrien-3,17β-diol-17-(0-1β)-D-glyucopyranosiduronic acid
Estriol-3-β-D-glucuronide	E_3-3G	1,3,5(10)-Estratrien-3,16α,17β-triol-3-(0-1β)-D-glucopyranosiduronic acid
Estriol-17-β-D-glucuronide	E_3-17G	1,3,5(10)-Estratrien-3,16α,17β-triol-17-(0-1β)-D-glucopyranosiduronic acid
Estrone-3-sulfate	E_1-3S	3-Hydroxy-1,3,5(10)-estratrien-17-one-3-sulfate
17β-Estradiol-3-sulfate	17β-E_2-3S	1,3,5(10)-Estratrien-3,17β-diol-3-sulfate
17β-Estradiol 17-Sulfate	17β-E_2-17S	1,3,5(10)-Estratrien-3,17β-diol-17-Sulfate
Estriol-3-sulfate	E_3-3S	1,3,5(10)-Estratrien-3,16α,17β-triol-3-sulfate

Hot acid hydrolysis of the 24 hr-pregnancy urine in the presence of KI
↓
Addition of (4-^{14}C)-2-OHE$_3$ saturation with NaCl
↓
Extraction with benzene (organic layer discarded)
↓
Extraction with ethyl acetate (aqueous phase discarded)
↓
Extraction with aqueous ascorbic acid solution of pH 10.5, saturated with NaCl (aqueous phase discarded)
Recovery: 81%
↓
First column chromatography on silica gel
(chloroform acetic acid (7:5) ascorbic acid)
Recovery: 62%
↓
First paper chromatography
(formamide-ascorbic acid-chloroform-ethyl acetate (5:1))
Recovery: 47%
↓
Second paper chromatography
(water-acetic acid-1,2-dichloroethane (3:7:10)-ascorbic acid)
Recovery: 36%
↓
Second column chromatography on silica gel
(n-hexane-ethyl acetate-methanol (10:10:1)-ascorbic acid)
Recovery: 28%
↓
by gas chromatography

FIGURE 2. Separation of 2-hydroxy estrogens from urine. (From Gelbke, H. P. and Kruppen, R., *J. Chromatogr.*, 71, 465, 1972. With permission.)

methanol containing 1.9% (w/v) ascorbic acid. After passage of a further 20 mℓ of the same solution estrogen was applied to the column, elution of the steroid being carried out with another 15 mℓ 1.9% (w/v) ascorbic acid in methanol.

Preparation of TLC plates with ascorbic acid (method of Morreal and Dao[2]) — A solution of 500 mg ascorbic acid in 80 mℓ glass distilled water was extracted with 2 × 50 mℓ ether to remove nonpolar organic impurities. The aqueous solution was then used to form a slurry with silica gel H (35 g) which was used to prepare glass TLC plates (5 × 20 cm) of 250 μm thickness. These plates were activated by heating at 110° for 30 min.

Between 20 and 200 ng of either estrone, estradiol, or estriol and a corresponding radio-active tracer to facilitate recovery calculations were applied on each plate under a stream of nitrogen. The plates were developed in ethanol-benzene (5:95) for estrone and estradiol and methyl acetate for estriol. The plates to which estrogens had been applied were placed in a holding tank flushed with nitrogen while the plates on which standard material was applied for reference were sprayed with a solution of 30% phosphoric acid in ethanol and heated at 100°C for color development.

Extraction Methods

Estrogens are present both free and as conjugates with glucuronic and sulfuric acid. There have been also determinations of the "double conjugate", the sulfoglucuronide. There are two methods for hydrolyzing these conjugated steroids: acid or enzymatic means. In the preferred method of hydrolysis, a buffered solution of the enzyme is incubated with an aliquot of a 24-hr urine specimen. Gotelli et al.[3] used the equivalent of 1800 units of enzyme for hydrolyzing a 5-mℓ volume of urine at an incubation temperature of 60°C for 30 min. Acidic hydrolysis was accomplished by heating the sample with HCl at 110°C for 30 min. After hydrolysis, the free estrogens are usually extracted with an appropriate organic solvent.

Enzymatic hydrolysis displays some significant advantages over acidic hydrolysis procedures, depending upon the extent of interference caused by other substances that may be present in the sample. For example, estimates of estriol concentration in urine samples containing glucose, phenolphthalein, and methenamine mandelate may be low when acid hydrolysis is used. Wantanabe et al.[4] showed that hydrochlorothiazide behaves similarly. They suggested that this may be due to the formation of copolymers of estriol and formaldehyde, the formaldehyde being generated from hydrochlorothiazide during acid hydrolysis.

Urines containing salicylate may likewise give low estriol concentrations with enzymatic hydrolysis. This is due to the excretion of salicylate as the glucuronide, which competes with the estrogen conjugates for available enzymes.[5] Increasing the amount of enzyme used will solve this problem.

Graef and co-workers[6] studied the enzymic activity of β-glucuronidase on estrogen glucuronides in the presence of urinary inhibitors. They found that the use of enzyme obtained from different sources, specifically bovine liver, *Helix pomatia*, and *Escherichia coli*, affected the efficiency of hydrolysis. For fast hydrolysis, they found β-glucuronidase preparations from *E. coli* to be preferable.

More recent methods involving HPLC have focused on separating the conjugates after removal of the free compounds or allowing the system to separate the free compounds from the conjugates on the column.

Extraction of estriol from pregnancy urine (method of Fantl et al.[7]) — 2 mℓ urine were hydrolyzed by 50 μℓ potent β-glucuronidase preparation from *E. coli* (Boehringer Mannheim) at 45°C for 10 min. A single ether extraction (10 mℓ), removal of the aqueous layer, drying with Na$_2$SO$_4$, and evaporating the ether to dryness produced an extract suitable for chromatography. The residue was dissolved in isopropanol (200 μℓ), and 10 μℓ was injected onto a μBondapak column. A single major peak corresponding to estriol was obtained. A polar (3:1) heptane-isopropanol combination enabled estriol to be eluted within about 10 min. A similar chromatogram with a large peak corresponding to estriol was obtained on a Porasil column. The minor impurities eluted near the estriol peak can be eliminated if the ether extract is washed with carbonate buffer before being dried.

Using the same urine sample, hydrolysis with Ketodase, a β-glucuronidase preparation from beef liver, yielded a similar concentration of estriol, but hydrolysis had to be performed overnight.

Preparation of extracts of urine for estriol determinations by TLC (method of Wortmann et al.[8]) — Urine (5 mℓ) was diluted to 20 mℓ with water. Labeled estriol (5000 cpm), which served as an internal standard to check the level of recovery through the extraction, was added, followed by 3 mℓ concentrated HCl. Reflux was carried out for 15 min, timing from the start of boiling. The solution was cooled in an ice bath and extracted successively with two 40-mℓ portions of diethyl ether. The aqueous phase was discarded, and the ether extracts combined and washed with 10 mℓ NaHCO$_3$ solution (80 g/ℓ). The ether was extracted 3 times with 15 mℓ portions of 1 N NaOH and discarded. The combined aqueous extracts were adjusted to pH 2 to 4 by adding 10 mℓ 5 N HCl and extracted twice with 40 mℓ diethyl ether. The aqueous portion was discarded and the combined ether extracts washed once with 10 mℓ NaHCO$_3$ solution and once with 10 mℓ water. The ether was evaporated to reduce the volume, transferred to a 1.5 mℓ conical centrifuge tube, and dried in a stream of nitrogen. The residue was dissolved in 20 μℓ acetone; 5 μℓ were applied to the layer with alternate lanes reserved as reference lanes for double beam scanning. A standard amount of estriol was spotted on the same plate as an external standard. An aliquot of the sample was taken for the counting of radioactivity for recovery.

Extraction of estriol of pregnancy for GC (method of Mitchell[9]) — (1) 20 mℓ urine were adjusted to pH 5.0 with glacial acetic acid in a 50-mℓ Teflon®-lined screw capped centrifuge tube. To this was added 2 mℓ of 0.2 N acetate buffer and 0.5 mℓ enzyme hydrolyzer. The solution was incubated in a heating block at 50°C overnight (17 hr). (17 hr).

(2) After cooling, the hydrolyzed urine was extracted 3 times with 30 mℓ ether. The combined ether extracts were pooled in a separatory funnel and washed once with 20 mℓ water, followed by a wash of 20 mℓ 9% sodium hydrogen carbonate, and then once again with 20 mℓ water. The ether layer containing the estrogens was then extracted 3 times with 20 mℓ of 1 N sodium hydroxide, each sodium hydroxide extract being added in turn to a beaker containing 6 mℓ concentrated HCl. The pooled sodium hydroxide extracts were adjusted to pH 3.0 with 1 N hydrochloric acid, cooled, and re-extracted 3 times with 25 mℓ ether. The ether was dried by filtering through anhydrous sodium sulfate with ether washings and evaporated to dryness with a rotary evaporator *in vacuo* at 50°C.

(3) The residue was quantitatively transferred with acetone to a 20 mℓ Teflon®-lined screw capped tube and evaporated to dryness under vacuum. At this stage, the internal standard (cholesterol) was added to the residue. (The exact amount added depended on the level of estriol that could be expected at that particular stage of pregnancy. For the majority of cases with normal 24-hr urine volumes after 32 weeks, the amount was 400 μg, or 2 mℓ of the 200 μg/mℓ cholesterol standard). The solution was evaporated to dryness. No trace of moisture should remain after this stage. The residue was dissolved in an appropriate volume of solvent for the preparation of the solution to be injected.

Estrogen extraction for LC (method of Gotelli et al.[10]) — The hydrolysis step was performed as previously described. Add 1 mℓ buffer enzyme to 5 mℓ urine. Add 2 drops of chloroform, gently mix, and incubate at 60°C for 30 min. A 5-mℓ aliquot of the working estriol standard is processed in the same way as the unknown, except that the hydrolysis step is eliminated. After hydrolysis, add 1 mℓ internal standard solution to the cooled hydrolysate and the working estriol standard, mix, and pour the mixture into a 50-mℓ glass centrifuge tube. Add 1 mℓ saturated sodium carbonate and at least 5 g sodium chloride.

Mix, add 20 mℓ diethyl ether-petroleum ether (60:40), and shake the mixture for 5 min on a mechanical shaker. Centrifuge until the layers separate, then plunge an aspirator pipette below the ether layer into the aqueous layer and aspirate the aqueous layer. Pour the remaining ether layer into a tube for drying, and evaporate it at 60°C. Reconstitute the residue in 200 μℓ methanol and inject 20 μℓ onto the column.

Extraction of pregnanediol and estriol (method of Bottema and Belderok[11]) — Dilute a specimen of centrifuged 24-hr urine with water (3 mℓ to 50 mℓ if the volume is less than 1 ℓ; 5 mℓ to 50 mℓ if the volume is greater than 1 ℓ). Pipette into each of four 25-mℓ glass-stoppered centrifuge tubes 1 mℓ of the diluted urine. Add 0.1 mℓ acetate buffer and 0.05 mℓ Suc d'*Helix pomatia*. Incubate at 55°C. Alternatively 0.1 mℓ phosphate buffer and 0.02 mℓ glucuronidase (Boehringer) is added, followed by incubation for 20 min at 46°C. After cooling, extract the mixture with 10 mℓ ethyl acetate, shake mechanically for 5 min, and then centrifuge. Pipette into 2 centrifuge tubes 1 mℓ internal standard A and into 2 others 1 mℓ internal standard B, then add 8 mℓ ethyl acetate extract to each tube. Evaporate the ethyl acetate in a water bath at 55 to 60°C in a nitrogen atmosphere and dry for a short time in an oven at 100°C. After cooling, add 2 mℓ acetyl chloride and place in a boiling water bath until dry. Dry in an oven at 100°C for a short time. After cooling, dissolve the residue in 0.25 mℓ absolute ethanol. Of this solution, 5 μℓ can be subjected to chromatography.

Extraction of estrogens from urine (method of Ronco et al.[12]) — 5 mℓ of a 24 hr (or 12 hr) filtered urine collection were brought to pH 6.3 (± 0.2) by addition of 30% acetic acid and incubated with 15 IU of β-glucuronidase for 30 min at 50°C. Meanwhile, powdered magnesium silicate was suspended in a Bio-Rad column by adding double-distilled water. Water was then allowed to pass completely through the column.

After cooling, the hydrolyzate, brought to pH 1.0 with 6 *M* HCl, was applied to the column. The eluate was discarded. Elution of the estrogens was carried out with two 5-mℓ portions of 1 *M* potassium hydroxide. Two drops of phenol red aqueous solution were added to the eluate. After neutralization with 30% sulfuric acid, the eluate was brought to basic pH (red color) with 90 g/ℓ sodium carbonate solution. Free TE extraction was carried out with two 10-mℓ portions of diethyl ether. After pooling and the addition of IS (0.5 mℓ 100 μg/mℓ epicoprostanol ethanolic solution), the ether phase was evaporated to dryness under a stream of nitrogen. After silylation with 100 μℓ BSA-TMCS-chloroform solution (5:1:9) 1 μℓ mixture was injected into the gas chromatograph.

Extraction of estrogen in urine (method of Dolphin[13]) — A 50-mℓ aliquot of urine was heated at reflux, after addition of 7.5 mℓ concentrated HCl for 30 min. This hydrolysis was designed to break down the estrogen conjugates yielding free estrone, estradiol, and estriol. The hydrolyzed sample was extracted 3 times with 50 mℓ diethyl ether, and the ethereal extracts were combined. Unwanted acidic components were removed from the diethyl ether by shaking with 20 mℓ sodium carbonate solution at pH 10.5, and then with 4 mℓ 0.1 *M* sodium hydroxide, adjusting the pH to 10.0 with 0.05 *M* sodium hydrogen carbonate. Finally, the ethereal phase was washed with 4 mℓ 8% (w/w) solution of sodium hydrogen carbonate, followed by 3 mℓ deionized water. The final extracts were transferred to 1 mℓ graduated Reacti-vials (Pierce Products) fitted with PTFE-lined septa and stored under refrigeration so as to avoid further evaporation of the solvent.

In nonpregnancy urine, the estrogens would be present mainly as free compounds and as such should be removed, using the simple ether extraction. Pregnancy urine, on the other hand, contains principally the glucuronide and sulfate conjugates of the estrogens, in which case the full extraction would be more effective.

Urine of normal women (method of Joe et al.[14]) — A minimum aliquot of 50 mℓ from a 24-hr pool was centrifuged for 20 min at 3200 rpm in a Model PR-6 International Centrifuge (International Instruments, Needham Heights, Mass.). The supernatant was hydrolyzed at 37°C for 48 hr using 0.5 mℓ Glusulase. The total hydrolysis volume was adjusted to pH

5.2 with 20% acetic acid and sodium acetate buffer. Prior to incubation, 2000 units of penicillin per milliliter were added to inhibit bacterial growth. The hydrolyzate was extracted 4 times with 50 mℓ portions of ether-ethyl acetate (2:1). The extracts were combined and washed twice with 5 mℓ portions of a 9% sodium bicarbonate solution. The organic phase was maintained, and the bicarbonate wash discarded. The resultant emulsion was subsequently eliminated by washing with 7 consecutive 50 mℓ portions of distilled water. The water washes were discarded, and the washed extract was concentrated to dryness. Alternatively, the sample could be washed twice with 5 mℓ portions of bicarbonate and once with 10 mℓ distilled water. The organic phase was evaporated to dryness by rotary evaporation.

Preparation of urine for estrogen determination by HPLC (method of Schmidt et al.[15]) — 1 mℓ urine and 0.3 mℓ concentrated HCl were transferred to a 16 × 100-mm screw capped tube, mixed well for 15 sec, and heated at 100°C for 30 min. The tube was cooled and 5 mℓ ether were added. The tube was placed on a rotary rack for 5 min and centrifuged for 2 min at 2000 rpm (710 g). The ether layer (upper layer) was transferred to another 16 × 100-mm tube and evaporated at 40°C with air. The sample was derivatized according to the reaction procedure below.

Less than 100 μg estrogen was transferred to a 16 × 100-mm screw capped tube, and the solvent was evaporated at 40°C with air; 100 μℓ pH 10.5 buffer was added followed by 200 μℓ dansyl chloride working solution and mixed vigorously for 15 sec. The tube was capped and heated at 100°C for 15 min. The tube was cooled, and a 5 μℓ aliquot was injected into a liquid chromatograph.

Estrogens in meat (method of Wortberg et al.[16]) — A 50-g amount of meat was minced, mixed with 100 mℓ ethanol and extracted by shaking for 1 hr. After centrifugation at 2500 g, the supernatant was decanted, and the extraction procedure was repeated twice by stirring the residue with 100 mℓ ethanol.

The combined ethanolic extracts were evaporated on a rotary vacuum evaporator at 40°C. The residue was treated with 10 mℓ dichloromethane and, after addition of 10 mℓ water, the organic phase was separated. Two further extractions of free and conjugated steroids from tissues, an enzymatic treatment of the aqueous portion with β-glucuronidase/arylsulfatase followed by solvolysis according to general methods, should be added. The combined dichloromethane extracts were washed 3 times with 10 mℓ 10% sodium carbonate solution (pH 10.5) and evaporated under vacuum. The residue, dissolved in 5 mℓ tetrahydrofuran was separated from lipids by gel chromatography on Fractogel 6000 PVA with tetrahydrofuran. The fraction eluted with 125-250 mℓ (containing the steroids and zeranol) was collected and concentrated under vacuum. The residue was dissolved in 30 mℓ benzene and washed twice with 10 mℓ saturated sodium chloride solution. Steroidal estrogens and zeranol were extracted from benzene solution with 30 mℓ of 1% sodium hydroxide solution. The aqueous phase, collected with 20 mℓ water in a separating funnel, was acidified with 2 mℓ concentrated HCl and re-extracted with benzene (3 × 15 mℓ). After washing the benzene phase with water to neutrality, it was dried with anhydrous sodium sulfate and evaporated to dryness on a rotary evaporator. The residue was dissolved in 1 mℓ ethanol, re-evaporated, and the final residue dissolved in 0.1 mℓ ethanol for derivatization.

Estrogen conjugates in cyst fluid or serum from postmenopausal women (method of Raja et al. [17]) — To 4 mℓ cyst fluid, approximately 2000 cpm (in 0.2 mℓ water) of each estriol conjugate was added. The fluid was extracted 3 times with equal volumes of ether. About 500 cpm of ³H-estriol (internal standard) were added to the combined ether extracts. The ether was evaporated under nitrogen and 0.2 mℓ benzene-methanol (85:15) were added. The solution was stored at −15°C until the conjugates were processed to this stage. The aqueous part was extracted for the conjugates.

The four conjugates were separated as follows: to the aqueous phase from the above, 4 vol absolute ethanol were added to precipitate the proteins. After centrifugation, the supernatant was evaporated under vacuum, and the conjugates were converted to their triethyl-ammonium salts and applied to a Sephadex® LH-20 column. Estriol conjugates were eluted in the order E_3-3S-16G, E_3-3S, and E_3-16G by adding 15 mℓ each of 5, 7.5, 10, 15, and 20% n-butanol in ethylene dichloride containing 0.2% triethylamine; 3 mℓ fractions were collected. Finally, E_3-3G was eluted with 20 mℓ methanol. In the course of study it was found that considerable heterogeneity existed between batches of Sephadex® LH-20. Occasional batches afforded poor separations of estriol conjugates, whereas other batches permitted the successive elution of E_3-3S-16G, E_3-3S, and E_3-16G with 15 mℓ each of 5 and 7.5%, followed by 60 mℓ 10% n-butanol in ethylene dichloride containing 0.2% triethylamine, eliminating the 15 and 20% eluants. Then E_3-3G was eluted with methanol in the usual way. Each estriol conjugate was hydrolyzed with Glusulase. The estriol was partitioned between benzene-hexane (1:1) and water. The water phase was extracted twice with ether, and the ether was evaporated under nitrogen. The estriol was chromatographed on a Sephadex® LH-20 column using benzene methanol (85:15) as mobile phase.

Serum was analyzed in the same way, except that 5 mℓ were used and the assay was repeated on a subsequent day. With each batch of cyst fluid or serum, distilled water or male plasma, respectively, was run in parallel to serve as a blank for the assay.

Estrogen conjugates from amniotic fluids (method of Raja et al.[17]) — To 4 mℓ fluid from which the "free" steroids were removed by ether extraction, 4 vol absolute ethanol were added. After centrifugation, the supernatant was evaporated under vacuum. The conjugates were then converted to the triethylammonium salts and separated on a Sephadex® LH-20 column. The solution was applied to the top of the column and the conjugates eluted using 15 mℓ vol solvent as noted below, collecting 3 mℓ fractions.

%Butanol in ethylene dichloride containing 0.2% triethylamine	Conjugate eluted
5.0	E_3-3S-16G
7.5	E_3-3S
10.0	E_3-16G
15.0	
20.0	
Methanol only	E_3-3G

REFERENCES

1. **Gelbke, H. P. and Kruppen, R.**, *J. Chromatogr.*, 71, 465, 1972.
2. **Morreal, C. E. and Dao, T. L.**, *Steroids.*, 27, 421, 1976.
3. **Gotelli, G. R., Kabra, P. M., and Marton, C. J.**, *Clin. Chem.*, 23, 165, 1977.
4. **Watanabe, F., Nakahara, M., Tsubota, N., Tsukida, K., Saiki, K., and Ito, M.**, *Clin. Chem. Acta*, 88, 21, 1978.
5. **Stempfl, R. S., Jr., Sidbury, S. B., Jr., and Migeon, C. J.**, *J. Clin. Endocrinol. Metab.*, 20, 817, 1960.
6. **Graef, V., Furuya, E., and Ishikaze, O.**, *Clin. Chem.*, 23, 532, 1977.
7. **Fantl, V., Lim, C. K., and Gray, C. H.**, in *High Pressure Liquid Chromatography in Clinical Chemistry*, Dixon, P. F., Gray, C. H., Lim, C. K., and Stroll, M. S., Eds., Academic Press, New York, 1976, 55.
8. **Wortmann, W., Wortmann, B., Schnabel, C., and Touchstone, J. C.**, *J. Chromatogr. Sci.*, 12, 377, 1974.
9. **Mitchell, M. A.**, *Can. J. Med. Tech.*, 106, 1972.
10. **Gotelli, G. R., Wall, J. H., Kabra, P. M., Martozy, L. J.**, *Clin. Chem.*, 24, 2132, 1978.
11. **Bottema, J. K. and Belderok, W. J.**, *Clin. Chem. Acta*, 68, 199, 1976.

12. **Ronco, G., Desmet, G., and Bezoc, J. F.,** *Clin. Chem. Acta,* 81, 119, 1977.
13. **Dolphin, R. J.,** *J. Chromatogr.,* 83, 421, 1973.
14. **Joe, H. T., Leach, C. S., and Lassiter, C. B.,** *Chromatographia,* 11, 671, 1978.
15. **Schmidt, G. J., Van de Mark, F. L., and Slavin, W.,** *Anal. Biochem.,* 91, 636, 1978.
16. **Wortberg, B., Woller, R., and Chalamorkot, T.,** *J. Chromatogr.,* 156, 205, 1978.
17. **Raja, M., Ganguly, M., and Levitz, M.,** *J. Clin. Endocrinol. Metab.,* 45, 429, 1977.

SECTION II. GAS CHROMATOGRAPHY

The estrogens have been separated both as free and derivatized compounds. They are relatively stable under many GC conditions.

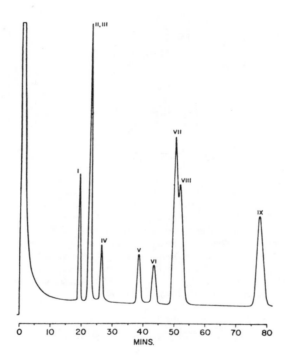

FIGURE GC 1. Separation of methoxime-trimethylsilylated equine estrogens. Instrument: Bendix 2500 with flame ionization detector; column: 3% OV-225 (12 ft × 1/4 in. U shape). (I) Estradiol-17α, (II) estradiol-17β, (III) dihydroequilin-17α, (IV) dihydroequilin-17β, (V) dihydroequilenin-17α, (VI) dihydroequilenin-17β, (VII) estrone, (VIII) equilin, (IX) equilenin (as methoxime-trimethylsilyl ethers). (From McErlane, K. M., *J. Chromatogr. Sci.,* 12, 97, 1974. With permission.)

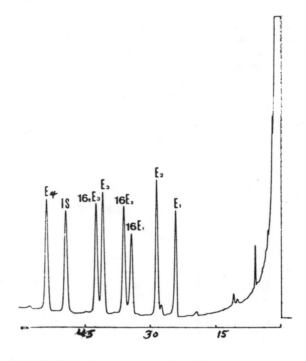

FIGURE GC 2. Gas chromatography of silylated estrogens. Instrument: Perkin-Elmer 30 with flame ionization detector; column: 1% OV-100 on Chromosorb WHP (100—120 mesh); temperature: Column — start at 170°C, raised to 270°C at 1.5°/min. (1) Estrone, 16-hydroxyestrone, (2) estradiol, 16-hydroxyestradiol, (3) estriol, 16-epiestriol, (4) estratetrol (all as trimethylsilyl ethers). IS = Epicoprostanol. (From Adessi, G., Eichenberger, D., Whuan, T. Q., and Joyle, M. G., *Clin. Chim. Acta*, 55, 323, 1974. With permission.)

Table GC 1
SEPARATION OF ESTROGEN HEPTAFLUOROBUTYRATES[3]

	Millimeters from injection peak		
Steroid	**Free**	**Free + HFBA[a]**	**Sulfate + HFBA[b]**
Estrone	110	105	105
Estriol	260	116	116
DHEA	110	104	104
16α-Hydroxy-DHEA	360	142	—[c]

Instrument:	Glowall 310 with 12.5 μCi radium foil detector
Column:	5% OV-210 and 2.5% OV-17 mixed phase (6 ft [1.8 m] × 3.4 mm)
Temperature:	Column 225°C, detector 240°C, injector 270°C

[a] Free steroid reacted with heptafluorobutyric anhydride (HFBA).
[b] Steroid sulfate reacted with heptafluorobutyric anhydride.
[c] Steroid sulfate standard not available.

Table GC 2
RETENTION TIMES OF MPPH AND MAJOR ESTROGENS AS MPPH DERIVATIVES[4]

Compound	Retention time (min)
MPPH[a]	3.4
Estradiol	4.3, 6.2
Estrone	—
16-Epiestriol	8.5, 10
16-Ketoestradiol	8.5
16-Hydroxyestrone	8.8, 10, 12
Estriol	6.6

Instrument:	Hewlett-Packard 5731A with flame ionization detector
Column:	3% SP2250 on Supelcopak (100—120 mesh) (6 ft [1.8 m] × 2 mm)
Temperatures:	Column 230°C, detector 250°C, injector 350°C

[a] MPPH: 5-(4-methylphenyl)-5 phenyl hydantoin.

On column derivatization with methanolic tetramethylammonium hydroxide was used.

Table GC 3
ELUTION SEQUENCE AND SEPARATION OF ESTROGEN STEROIDS[5]

Component[a]	RRT vs. 6-dehydroestrone
16-Estratetraene	0.094
17α-Estradiol	0.130
17β-Estradiol	0.154
Estradiol benzoate	0.160
α-Dihydroequilin	0.166
β-Dihydroequilin	0.195
Estriol	0.197
17α-Dihydroequilenin	0.342
Methylestradiol methyl ether	0.378
17β-Dihydroequilenin	0.389
17-Desoxyestrone	0.425
Estra-4-ene-3-o1-17-one	0.491
$\Delta^{5,7,9}$-Estratrien-3-o1-17-one	0.624
16-Hydroxyestrone	0.646
17α-Estradiol–17-acetate	0.657
Estrone	0.781
16-Oxo-17β-estradiol	0.782
Oxoestradiol	0.805
Equilin	0.875
9-Dehydroestrone	0.914
8-Dehydroestrone	0.953
6-Dehydroestrone	1.000
14-Isoequilenin	1.300

Table GC 3 (continued)
ELUTION SEQUENCE AND SEPARATION OF ESTROGEN STEROIDS[5]

Component[a]	RRT vs. 6-dehydroestrone
2-Methoxyestrone	1.460
1-Methylestrone	1.580
Equilenin	1.650

Instrument:	F & M 400 with flame ionization detector
Column:	1.7% Diethylene glycol succinate on Gas Chrom Q (100—120 mesh)
Temperatures:	Column 195°C, detector 240°C, injector 225°C
Flow:	Helium 60—80 mℓ/mm

[a]   All components are eluted as the trimethylsilyl ethers.

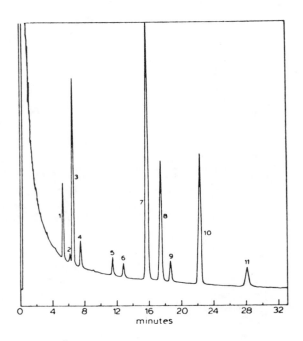

FIGURE GC 3.   Separation of estrogens as their trimethysilyl ethers. Instrument: Hewlett Packard 421 with flame ionization detector; column: fused silica (10 cm × 0.24 mm) coated with SP-1000; temperature: column 225°C, detector and injector 280°C; flow: 1 mℓ/min helium. (1) α-estradiol, (2) β-estradiol, (3) α-dihydroequilin, (4) β-dihydroequilin, (5) α-dihydroequilenin, (6) β-dihydroequilenin, (7) estrone, (8) equilin, (9) $\Delta^{8,9}$-dehydroestrone, (10) 1-methylestrone, (11) equilenin (all estrogens present as trimethylsilyl ethers). (From Lyman, G. W. and Johnson, R. N., *J. Chromatogr.*, 234, 234, 1982. With permission.)

REFERENCES (GC)

1. **McErlane, K. M.**, *J. Chromatogr. Sci.,* 12, 97, 1974.
2. **Adessi, G., Eichenberger, D., Whuan, T. Q., and Joyle, M. G.**, *Clin. Chim. Acta,* 55, 323, 1974.
3. **Touchstone, J. C. and Dobbins, M. F.**, *J. Steroid Biochem.,* 6, 1389, 1975.
4. **Milner, S. N. and Taber, C. A.**, *Clin. Chim. Acta,* 71, 67, 1976.
5. **Johnson, R. N., Masserans, R. P., Kho, B. T., and Adams, W. P.**, *Pharmacol. Sci.,* 67, 1218, 1978.
6. **Lyman, G. W. and Johnson, R. N.**, *J. Chromatogr.,* 234, 234, 1982.

SECTION III. HIGH PERFORMANCE LIQUID CHROMATOGRAPHY

Table HPLC 1
SEPARATION OF
ESTROGENS AS
DANSYL DERIVATIVES[1]

Compound	RT[a]
Dansyl estradiol cypionate	—[b]
Dansyl estradiol valerate	—[b]
Dansyl estrone	160
Testosterone enanthate	1.78
Dansyl ethinyl estradiol	2.26
Norethindrone acetate	2.28
Estradiol cypionate	2.35
Estradiol valerate	2.70
Dansyl α-estradiol	3.00
Estradiol benzoate	3.07
Dansyl estradiol	3.92
Estrone	4.24
Norethindrone	6.08
Ethinyl estradiol	6.71
α-Estradiol	8.71
Estradiol	9.48
Chloroform	1.00

Instrument:	Dupont 841 with detector below
Column:	LiChrosorb Si-60 (25 cm × 3.2 mm)
Mobile phase:	Heptane-chloroform (20:80)
Flow:	1 mℓ/min
Detection:	Fluorescence

[a]   Relative to chloroform as 1.
[b]   Less than 1.05.

Table HPLC 2
SEPARATION OF SYNTHETIC ESTROGENS[2]

	RRT		
	Column A		Column B
Synthetic estrogens	15%	30%	20%
Methallenestril	2.53		1.14
Dienestrol	3.03		5.10
Diethylstilbestrol	3.62		5.42
Hexestrol	3.78		5.19
Benzestrol	9.20		7.28
Promethestrol	14.3		9.10
Dienestrol diacetate	20.2	1.70	5.32
Chlorotrianisene		4.30	20.8
Diethylstilbestrol dipropionate		5.22	13.8
Promethestrol dipropionate		6.64	16.7
Methanol emergence (min)	1.34	1.85	1.73
Pressure at inlet (psig)	1000	1000	1000
Column temperature	50°C	50°C	70°C

Instrument: Dupont 820 with UV detector
Column: A. Permaphase ODS (1 m × 2.1 mm)
                B. Permaphase ETH
Mobile phase: 15, 30, and 20% 2-propanol in water
Detection: 254 nm

Note: Retention times expressed relative to methanol as 1.00.

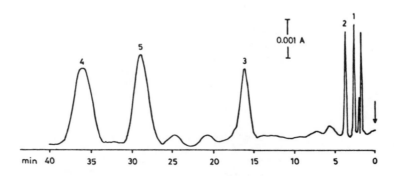

FIGURE HPLC 1.   Separation of estrogen conjugates. Instrument: Waters 6000; column: LiChrospher Si-100 (10 μm) (tetrapropyl ammonium 0.1 *M* added to column (pH 7.4)); mobile phase: chloroform-1-butanol (9:1); flow: 1.5 mℓ/sec; peaks: (1) 3-equilin sulfate, (2) 17α-dihydro-3-equilin sulfate, (3) 3-estradiol sulfate, (4) 17-estradiol sulfate, (5) 17-estriol sulfate, (6) 3-estriol sulfate. (From Fransson, B., Wahlund, I. M., Johansson, I. M., and Schill, G. I., *J. Chromatogr.*, 125, 327, 1976. With permission.)

Table HPLC 3
SEPARATION OF FREE
ESTROGENS[4]

Relative Retention Times and Elution Volumes of Estrogens

Steroid	RRT	Elution vol (mℓ)
Estriol	0.32	4.0—6.0
Estradiol	1.00	13.0—16.0
Estrone	1.47	20.0—23.0

Instrument:	Waters 6000
Column:	Waters μBondapak C_{18} (30 cm × 4 mm)
Mobile phase:	Acetonitrile-water (40:60)
Detection:	280 nm
Flow:	7 mℓ/min

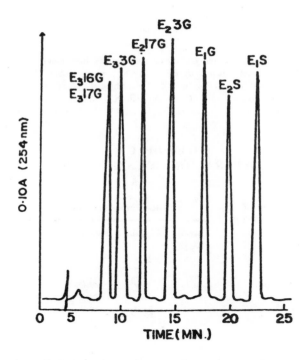

FIGURE HPLC 2.   Separation of estrogen conjugates. Instrument: Milton Roy pump, Valco 6 injector; column: μPartisil SAX (25 cm × 4.0 mm), two columns in series; mobile phase: 0.1 *M* NaCl pH 4.8; flow 0.8 mℓ/min. (E_1) estrone, (E_2) estradiol, (E_3) estriol, (G) glucuronide, (S) sulfate. (From Musey, P. I., Collins, D. C., and Preedy, J. K. R., *Steroids*, 31, 2271, 1978. With permission.)

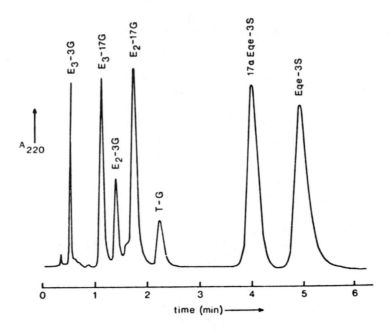

FIGURE HPLC 3.   Separation of estrogen conjugates. Instrument: Hewlett-Packard 1010A; column: LiChrosorb-RP2 (150 × 3 mm); mobile phase: 0.05 *M* phosphate pH 8.0-butanol (93:7), 0.05 *M* phosphate-acetonitrile (85:15); temperature: 70°C for column oven. (G) Glucuronide, (S) sulfate. (From Van der Wal, S. and Huber, J. F. K., *J. Chromatogr.*, 149, 431, 1978. With permission.)

Table HPLC 4
SEPARATION OF ESTROGENS[7]

Grams of AgNO$_3$/500 mℓ of mobile phase	Retention time (min)			
	Estriol	Equilin	Estrone	Estradiol
0	11.8	27.8	30.7	34.1
2.48	11.7	25.7	30.0	33.7
5.46	11.6	23.9	29.6	33.2
7.60	11.7	22.7	29.3	32.9
10.12	11.7	21.7	29.7	33.3

Instrument: Milton Roy 396/2396 pump and loop injector
Column: μBondapak C$_{18}$
Mobile phase: Methanol-water (60:40) with amounts of silver nitrate as in table
Detection: 280 nm

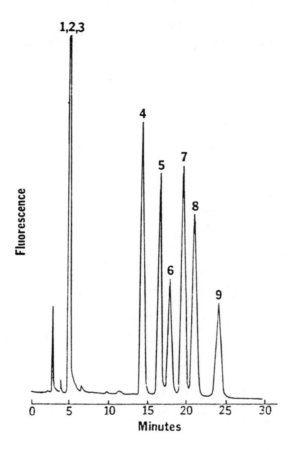

FIGURE HPLC 4.   Separation of dansyl derivatives of estrogens. Instrument: Dupont 841 with Model 836 fluorescence detector; column: LiChrosorb Si-60 (250 × 3.2 mm); mobile phase: chloroform-n-heptane (1:1); flow: 0.98 mℓ/min; detection: fluorescence; key (dansyl derivatives): peaks (1) estrone, (2) equilin, (3) equilenin, (4) α-estradiol, (5) α-dihydroequilin; (6) α-dihydroequilenin, (7) β-estradiol, (8) β-dihydroequilin, (9) β-dihydroequilenin. (From Roos, R. W. and Medwick, T., *J. Chromatogr. Sci.*, 18, 626, 1980. With permission.)

Table HPLC 5
RETENTION TIMES FOR NATURAL ESTROGENS[9]

Compounds	Retention time (min)			
	1	2	3	4
Mono-ols				
17-Deoxoestrone		3.25	>75	
Estrone		9.5	32	
6-Dehydroestrone		9.5	28.5	
Equilin		9.5	28.5	
Equilenin		10.5	25.5	
Diols				
Estradiol-17α		16.25	25.5	
Estradiol-17β	4.75	17.75	23	

Table HPLC 5 (continued)
RETENTION TIMES FOR NATURAL ESTROGENS[9]

	Retention time (min)			
Compounds	1	2	3	4
2-Hydroxyestrone		19.5[a]	16	
6-Dehydroestradiol		20.25	18.5	
17α-Dihydroequilin		22.5	18.5	
3,16α-Estradiol		22.5	21	
17α-Dihydroequlenin		24.25	16	
16-Ketoestradiol		27.5	6.75	22.5
16α-Hydroxyestrone		30.75	7.25	25.5
6-Ketoestradiol	6.75	32	7.25	25.5
Triols				
16-Epiestriol	8.25		9.25	
17-Epiestriol	8.25		13.5[a]	
2-Hydroxyestradiol	8.25[a]		13.5	
Estriol	11		4.25	13
6α-Hydroxyestradiol	11		4.25	10
16,17-Epiestriol	12.75		4.25	13
6-Ketoestriol	16.75(15)[b]		2.5	6.25
Tetraols				
2-Hydroxyestriol	19.25[a]		3.5	7.75
6α-Hydroxyestriol	28		2.5	4

Instrument: Modular assembly
Column: Zorbax BP-Sil (7—8 μm) (250 × 4.6 mm) or Zorbax-ODS (250 × 4.6 mm)
Mobile phases: 1. BP-Sil column, hexane-ethanol (9:1) 2 mℓ/min, pressure 200 psi
2. Same as 1, but hexane-ethanol (97:3)
3. Zorbar-ODS column, acetonitrile-water (35:65) pressure 700 psi
4. Zorbax BP-ODS, acetonitrile-water (25:75) pressure 700 psi
Detection: 280 nm

[a] Broad peaks.
[b] Samples also gave a minor peak at 15 min.

Table HPLC 6
SEPARATION OF ESTRADIOL (E₂) DERIVATIVES[10]

	Retention time[a]			
Mobile phase	4-OHE₂	2-OHE₂	E₂ (min)	2-Methyl-E₂
A	0.40	0.50	1.00 (24.05)	—
B	0.39	0.50	1.00 (21.36)	—
C	0.41	0.51	1.00 (19.31)	1.32
D	0.44	0.53	1.00 (12.11)	—
E	—	0.69	1.00 (8.65)	1.15

Table HPLC 6 (continued)
SEPARATION OF ESTRADIOL (E$_2$) DERIVATIVES[10]

Instrument: Hewlett-Packard 1081 B and electrochemical detector
Column: Hewlett-Packard C$_{18}$ (250 × 4.6 mm)
Mobile phases: A = Water-methanol (65:35)
             B = Water-methanol (62:38)
             C = Water-methanol (60:40)
             D = Water-methanol (55:45)
             E = Water-methanol (36:65)
             (All made acid with 0.2 N acetic acid)

[a]   Retention time relative to estradiol (E$_2$).

Table HPLC 7
SEPARATION OF HYDROXY-ESTROGENS[11]

Compound	μPorasil	μBondapak C$_{18}$	
	1	2	3
2-OHE$_1$	2.67	0.54	0.59
2-OHE$_2$	8.27	0.46	0.76
4-OHE$_1$	1.75	0.62	0.62
4-OHE$_2$	3.08	0.49	0.68
2-MeOE$_1$	1.00	1.13	1.82
2-MeOE$_2$	2.25	0.87	2.27
3-MeOE$_1$	1.00	1.18	1.65
E$_2$	1.83	0.77	1.21
E$_3$	>10	0.23	0.38
E$_1$	1.00 (4.8 min)	1.00 (15.0 min)	1.00 (11.2 min)

Instrument: Waters ALC/GPC 202, UV detector
Column: 1. μ Porasil
             2. 3. μBondapak C$_{18}$
Mobile phase: 1. Cyclohexane-ethyl acetate-acetic acid (175:25:3)
             2. Acetonitrile — 0.5% ammonium dihydrogen phosphate (pH 3) (1:2)
             3. Methanol — 0.5% ammonium dihydrogen phosphate (pH 3) (70:60)
Detection: 280 nm
Flow: 2 mℓ/min

Table HPLC 8
RETENTION TIME (IN MIN) OF
ESTROGENS[12]

Estrogen	Mobile phase					
	1	2	3	4	5	6
E$_3$-3G	14	11	8	9	2	
E$_2$-3SO$_4$,17G	15	12	14	7	2	
E$_2$-3,17SO$_4$	16	13	9	11	2	

Table HPLC 8 (continued)
RETENTION TIME (IN MIN) OF ESTROGENS[12]

Estrogen	1	2	3	4	5	6
E_3-3SO_4	16	14	9	15	2	
E_3-16G	19	16	16	17	4	
E_3-17G	19	16	16	17	4	
E_3-17SO_4	19	17	16	20	5	
E_1-3G	21	18	25	22	4	
E_2-3G	21	18	28	26	4	
E_2-17G	22	19	42	32	6	
15α-OH-E_3	24	20			12	90
2-OH-E_3	24	20				97
E_2-17SO_4		20	26		9	
E_1-3SO_4	24	21	24		7	
E_2-3SO_4	24	21	26		9	
6α-OH-E_2	25	22				76
E_3	27	24				71
16-Epi-E_3	31	28				61
αE_2-17Glyc		28				120
2-OH-E_2	34	30				71
2-OH-E_1	34	30				54
E_1	36	34				34
E_2	37	34				46
2-MeO-E_2	39	38				37
2-MeO-E_1		38				21
E_1-3-Met						14
E_2-3-Met						20
E_3-3-Met						39

Instrument: Waters 6000 A pump, 660 gradient programmer and U6K injector

Columns:
1. LiChrosorb RP-18 (5μm) (250 × 10 mm)
2. Same as 1 but with guard column of μBondapak C_{18} Corasil (25 × 6 mm)
3. μBondapak C_{18} (10 μm) (300 × 3.9 mm) 2 in series
4. Same as 3
5. LiChrosorb C_2 (10 μm) (250 × 3.2 mm)
6. Chromegabond Diol (10 μm) (300 × 4.6 mm)

Mobile phase:
1. For 1 and 2 columns; 50 min convex gradient, 10% methanol in 0.01 *M* ammonium acetate buffer (pH 6.9) to 100% methanol, 2 mℓ/min
2. 45% methanol in 0.01 *M* ammonium acetate buffer (pH 3.97) (acetic acid), 1 mℓ/min
3. Same as 2 except 35% methanol adjusted to pH 7.74 with ammonium hydroxide
4. 25% Methanol in 0.01 *M* ammonium acetate pH 7.5 with ammonium hydroxide 1 mℓ/min
5. 100% Hexane program to hexane-isopropanol (80:20) 1.5 mℓ/min

SO_4 = sulfate, MeO = methoxy, G = glucoronide, Met = methyl ether, gly = glucoside.

Note: See Table HPLC 1 for nomenclature.

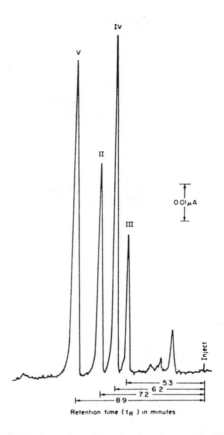

FIGURE HPLC 5.   Separation of estramustines. Instrument: Waters 6000 pump, UK 6 injector, and F5-970 fluorescence detector; column: Partisil PX 5/25 (250 × 4.6 mm); mobile phase: hexane-ethanol (95.5:7.5); detector: fluorescence. (II) Estramustine, (III) 80 ng estrone-3N-bis(2-chloroethyl)carbamate, (IV) 110 ng estrone, (V) 20 ng estradiol. (From Brooks, M. A. and Dixon, R., *J. Chromatogr.*, 182, 387, 1980. With permission.)

REFERENCES (HPLC)

1. **Roos, R. W.,** *J. Pharm. Sci.,* 63, 598, 1974.
2. **Roos, R. W.,** *J. Chromatogr. Sci.,* 14, 14, 1976.
3. **Fransson, B., Wahlund, I. M., Johansson, I. M., and Schill, G. I.,** *J. Chromatogr.,* 125, 327, 1976.
4. **Satyaswaroop, P. G., de la Osa, E. L., and Guipide, E.,** *Steroids,* 30, 139, 1977.
5. **Musey, P. I., Collins, D. C., and Preedy, J. K. R.,** *Steroids,* 31, 2271, 1978.
6. **Van der Wal, S. and Huber, J. F. K.,** *J. Chromatogr.,* 149, 431, 1978.
7. **Tscherne, R. and Capitano, G.,** *J. Chromatogr.,* 136, 337, 1977.
8. **Roos, R. W. and Medwick, T.,** *J. Chromatogr. Sci.,* 18, 626, 1980.
9. **Lin, J. and Heftman, E.,** *J. Chromatogr.,* 212, 239, 1981.
10. **Prescott, W. R., Boyet, B. K., and Seaton, J. F.,** *J. Chromatogr.,* 234, 513, 1982.
11. **Shimada, K., Tanaka, T., and Nambara, T.,** *J. Chromatogr.,* 178, 350, 1979.
12. **Seikker, W., Jr., Lipe, G. W., and Newport, G. D.,** *J. Chromatogr.,* 224, 205, 1981.
13. **Brooks, M. A. and Dixon, R.,** *J. Chromatogr.,* 182, 387, 1980.

SECTION IV. THIN LAYER CHROMATOGRAPHY

TLC has been widely used for separation of estrogens, and a large bibliography exists on these substances. Both free and conjugated estrogens are amenable to separation. With densitometry becoming more common, it would be well to investigate separation by TLC followed by densitometry since micromole and picomole amounts have been quantitated in the methods described here and those of the more recent literature. The earlier literature is covered in the first volume in this series.

Table TLC 1
MOBILITIES OF ESTROGENS AND THEIR DERIVATIVES[1]

	R_f value	
Estrogen	**1**	**2**
Estrone	0.30	0.34
Estrone-3-methyl ether	0.41	0.50
Estrone-3-acetate	0.49	0.51
Estrone-3-monochloroacetate	0.48	0.50
Estrone-3-*p*-bromobenzoate	0.53	0.53
Estrone-3-bromomethyldimethylsilyl ether	—	0.51
Esterone-17-pentafluorophenylhydrazone	0.49	0.52
17β-Estradiol	0.16	0.16
Estradiol-3,17β-diacetate	0.56	0.56
Estradiol-3,17β-dimonochloroacetate	0.66	0.67
Estradiol-3,17β-di-*p*-bromobenzoate	0.61	0.60
Estradiol-3,17β-dibromomethylsilyl ether	0.62	0.52

Layer: Silica Gel GF 254 (Merck)
Mobile phase: 1. Benzene-ethyl acetate (60:10)
                 2. Benzene-ethyl acetate (50:10)
Detection: Heat at 70°C for 30 min after spray with aqueous 50% sulfuric acid

Table TLC 2
MOBILITIES OF ESTRONE AND ESTRADIOL-17β IN VARIOUS MOBILE PHASES[1]

Mobile phase	Estrone	17β-Estradiol
Benzene-ethyl acetate (2:1)	0.55	0.35
Benzene-ethyl acetate (3:1)	0.45	0.26
Benzene-ethyl acetate (5:1)	0.37	0.19
Chloroform-diethyl ether (5:2)	0.64	0.36
Hexane-chloroform (4:1)	0.00	0.00
Hexane-dichloromethane (1:2)	0.16	0.05
Hexane-methanol (4:1)	0.02	0.06
Hexane-ethyl acetate (5:2)	0.28	0.15
Hexane-ethyl acetate (5:2)[a]	0.17	0.09
Hexane-ethyl acetate (5:2) + 0.007 vol 15 *N* NH₄OH	0.13	0.06
Hexane-ethyl acetate (1:1)	0.52	0.38
Hexane-ethyl acetate (1:1)[a]	0.54	0.43

Table TLC 2 (continued)
MOBILITIES OF ESTRONE AND ESTRADIOL-17β IN VARIOUS MOBILE PHASES[1]

Mobile phase	Estrone	17β-Estradiol
Hexane-ethyl acetate (1:1)[b]	0.41	0.30

Layer: Silica Gel G (Merck)

[a] Silica gel prepared with 0.1 M sodium bicarbonate in the slurry.
[b] Silica gel prepared with 0.2 M sodium bicarbonate in the slurry.

Table TLC 3
R_f OF ESTROGENS AND DANSYL DERIVATIVES[2]

	R_f values					
	Estrone		Estradiol-17β		Estriol	
Mobile phase	Free	Dansyl	Free	Dansyl	Free	Dansyl
A	0.15		0.07		0.00	
B	0.55	0.70	0.29	0.48	0.02	0.04
C	0.60	0.73	0.48	0.58	0.16	0.14
D	0.14		0.07		0.00	
E	0.44	0.16	0.17	0.076	0.02	0.00
F	0.48	0.65	0.34	0.50	0.05	0.11
G	0.55		0.46		0.13	

Layer: Silica Gel G (Merck)
Mobile phases: A. Chloroform-benzene (8:2)
        B. Chloroform-dioxane (94:6)
        C. Chloroform-dioxane (8:2)
        D. Carbon tetrachloride-dioxane (9:1)
        E. Cyclohexane-ethyl acetate (3:1)
        F. Chloroform-benzene-ethanol (18:2:1)
        G. Chloroform-dioxane-carbon tetrachloride (3:2:5)
Detection: Fluorescence

Table TLC 4
SEPARATION OF ESTROGENS AND DERIVATIVES[3]

	Estrogen		Derivative	
Mobile phase	Estrone	Estradiol	Estradiol benzoate	Estradiol diundecylate
1	0.12	0.05	—	—
2	0.55	0.33	0.44	0.40
3	0.55	0.33	0.43	0.40
4	0.65	0.45	0.55	0.57
5	0.68	0.41	0.88	0.65
6	0.66	0.40	0.80	0.77
7	0.69	0.76	0.80	0.80

Table TLC 4 (continued)
SEPARATION OF ESTROGENS AND DERIVATIVES[3]

| | Estrogen | | Derivative | |
| | | | Estradiol | Estradiol |
Mobile phase	Estrone	Estradiol	benzoate	diundecylate
8	0.70	0.76	0.80	0.80
9	0.70	0.60	—	—
10	0.60	0.55	—	—

Layer: Silica Gel G (Merck)
Mobile phases: 1. Chloroform
2. Chloroform-ethyl acetate (80:20)
3. Chloroform-acetone (90:10)
4. Chloroform-acetone (80:20)
5. Cyclohexane-chloroform-acetic acid (70:20:10)
6. Methylene chloride-acetone (80:20)
7. Chloroform-acetic acid (90:10)
8. Methylene chloride-acetone-acetic acid (25:75:1)
9. Methylene chloride-acetone (75:25)
10. Cyclohexane-ethyl acetate (25:75)

Table TLC 5
SEPARATION OF ESTRONE AND DERIVATIVES[4]

| | Mobile phase ($R_f \times 100$) | | | | | |
Estrogen	1	2	3	4	5	Color
17β-Hydroxy-16-oximinoestrone-3-methyl ether	4	10	13	29	16	Yellow
16-Oximinoestrone 3-methyl ether	26	38	57	37	57	Violet
16,17-*seco*-17-Oxoestrono-16-nitrile-3-methyl ether	74	83	94	47	86	Lemon yellow
Estrone-3-methyl ether	83	91	98	78	93	Orange
Estrone	46	52	80	40	75	Orange-yellow

Layer: Silica Gel G (Merck)
Mobile phases: 1. Benzene-ethyl acetate (7:1)
2. Benzene-acetone (8:1)
3. Chloroform-ethyl acetate (4:1)
4. *n*-Hexane-acetone (2:1)
5. Carbon tetrachloride-ethyl acetate (2:1)
Detection: Color was developed by spraying with 50% sulfuric acid in methanol followed by heating in
an oven at 100—110°C for 10—15 min

Table TLC 6
R_f VALUES OF ESTRADIOL, DIETHYLSTILBESTROL, AND THEIR IODINATED DERIVATIVES[5]

| | Mobile phase | | | | | | |
Compound	A[a]	A	B[a]	B	C[a]	C	D[a]
17β-Estradiol	0.59	0.71	0.29	0.10	0.32	0.29	0.72
4-Iodoestradiol	0.71	0.79	0.47	0.24	0.58	0.49	0.83
2-Iodoestradiol	0.76	0.79	0.56	0.36	0.74	0.58	0.83

Table TLC 6 (continued)
R_f VALUES OF ESTRADIOL, DIETHYLSTILBESTROL, AND THEIR IODINATED DERIVATIVES[5]

	Mobile phase						
Compound	A[a]	A	B[a]	B	C[a]	C	D[a]
2,4-Diiodoestradiol	0.81	0.84	0.73	0.49	0.80	0.68	0.85
cis-Diethylstilbesterol	0.32		0.04				
trans-Diethylstilbesterol	0.55		0.04				
Iododiethylstilbesterol	0.75		0.30				
Diiododiethylstilbesterol	0.82		0.30				
Triiododiethylstilbesterol	0.85		0.30				

Layer: Silica Gel (Mylar backed) Eastman (with fluorescence indicator)
Mobile phase: A. Ethyl acetate-benzene (1:1)
                B. Chloroform
                C. Methylene chloride
                D. Benzene-methanol (1:1)
Detection: Fluorescence quench

[a]   Plate washed with absolute ethanol before chromatography.

Table TLC 7
R_f VALUES OF ESTROGENS AND EQUILINS[6]

		Colors	
Estrogen	R_f	A	B
Estrone	0.88	Brown	Yellow
Equilin	0.83	Brown	Purple
Equilenin	0.66	Blue	Blue
17α-Estradiol	0.57	Brown	Yellow
17β-Estradiol	0.48	Brown	Yellow
17α-Dihydroequilin	0.44	Brown	Purple
17β-Dihydroequilin	0.36	Brown	Purple
17α-Dihydroequilenin	0.24	Blue	Blue
17β-Dihydroequilenin	0.18	Blue	Blue

Layer: Silica Gel G impregnated by dipping in
                25% formamide in acetone for 5 min
Mobile phase: Chloroform-trichloroethylene (80:20)

(A)   After spraying with Fast Blue B Salt (0.5% aqueous),
(B)   after heating 20 min at 105°C after spraying with A.

Table TLC 8
R_f VALUES AND SPOT COLORS FOR ESTROGENS[7]

Estrogen	Relative R_f	Color
17α-Estradiol	1.16	Yellow
17β-Estradiol	1.00	Yellow
17α-Dihydroequilin	0.99	Purple

Table TLC 8 (continued)
R$_f$ VALUES AND SPOT COLORS FOR
ESTROGENS[7]

Estrogen	Relative R$_f$	Color
17β-Dihydroequilin	0.79	Purple
17α-Dihydroequilenin	0.65	Blue
17β-Dihydroequilenin	0.50	Blue

Layer:	Silica Gel G impregnated with formamide by dipping in 25% formamide in acetone for 5 min
Mobile phase:	Xylene-diethylamine (200:1)
Detection:	Expose to UV light

Table TLC 9
SEPARATION OF ESTROGEN CONJUGATES[8]

Compound	R^f 1	2	3	4
Estriol	0.59	0.74	0.58	0.77
Methyl 17βE$_2$-17-glucosiduronate 3-methyl ether		0.67		0.75
Methyl 17βE$_2$-17-glucosiduronate		0.65		
Methyl 17βE$_2$-3-glucosiduronate	0.40	0.58		0.72
17αE$_2$-17-glucopyranoside 3-methyl ether	0.27	0.49		
17αE$_2$-17-glucopyranoside	0.19	0.45		
17βE$_2$-17-glucopyranoside	0.18	0.44		
17αE$_2$-17-galactopyranoside	0.12	0.39		
17αE$_2$-17-N-acetylglucosaminide	0.07			
17βE$_2$-3-glucosiduronate	0	0	0	0.11

Layer:	Reversed phase C$_{18}$ (Merck)	
Mobile phases:	1.	Ethyl acetate-hexane-ethanol (16:3:1)
	2.	Chloroform-ethanol (3:2)
	3.	Chloroform-ethanol (4:2)
	4.	Ethyl acetate-ethanol-acetic acid (9:10:1)

Table TLC 10
R_f VALUES AND COLORS OF ESTROGEN DERIVATIVES[9]

Estrogen	1	2	3	4	5	Colors
Estrone	56	46	89	56	18	Orange-yellow
3-Methoxyestra-1,3,5(10)-trien-17-one	80	72	95	90	54	Orange
3-Methoxy-16-oximino-estra-1,3,5(10)-trien-17-one	45	30	73	48	11	Violet
3-Methoxy-16-oximino-estra-1,3,5(10)-trien-17β-ol	10	5	18	33	4	Yellow
3-Methoxy-17-oxo-16,17-seco-estra-1,3,5(10)-triene-16-nitrile	74	65	95	65	25	Lemon yellow
3-Methoxy-17-hydroxy-16,17-seco-estra-1,3,5(10)-triene-16-nitrile	33	18	55	45	9	Lemon yellow
3-Methoxy-17-hydroxy-16,17-seco-estra-1,3,5(10)-trien-16-amine hydrochloride	0	0	0	6	0	Grey
3-Methoxy-17-hydroxy-16,17-seco-estra-1,3,5(10)-trien-16-amine	0	0	2	11	0	Grey
3-Methoxy-17-oxa-D-homo-estra-1,3,5(10)-trien-16-one	62	42	95	63	22	Lemon yellow
3-Methoxy-17-oxa-D-homo-estra-1,3,5(10)-trien-16-ol	36	24	63	61	20	Dark red
3-Methoxy-16,17-seco-estra-1,3,5(10)-trien-16,17-diol	16	7	32	62	9	Orange
3-Methoxy-17-oxa-D-homo-estra-1,3,5(10)-triene	81	73	95	90	69	Red
3-Methoxy-17-aza-D-homo-estra-1,3,5(10)-triene-16,17a-dione	40	25	78	41	11	Yellow

Layer: Silica Gel G (Merck)

Mobile phases:
1. Benzene-acetone (8:1)
2. Benzene-ethyl acetate (7:1)
3. Chloroform-ethyl acetate (4:1)
4. Hexane-acetone (2:1)
5. Cyclohexane-acetone (1:1)

Detection: Spray with 50% sulfuric acid in methanol and heat at 100—110°C for 10—15 min

Table TLC 11
SEPARATION OF ESTROGEN DERIVATIVES[10]

	Mobile phase						
Estrogen	1	2	3	4	5	6	7
E	0.72	0.63	0.67	0.84	0.61	0.56	0.53
E[16]	0.65	0.48	0.63	0.78	0.56	0.51	0.44
16β,17β-ox-E	0.69	0.56	0.66	0.80	0.59	0.54	0.47
16α,17α-ox-E	0.69	0.54	0.66	0.79	0.60	0.51	0.47
E-17-one	0.69	0.53	0.65	0.80	0.59	0.51	0.46
E$^{9(11)}$-17-one	0.65	0.50	0.64	0.79	0.56	0.52	0.45
E$^{6.8}$-17-one	0.63	0.47	0.63	0.77	0.58	0.50	0.42
E-7,17-one	0.61	0.36	0.55	0.72	0.55	0.48	0.44
7α-ol-E-17-one	0.42	0.12	0.40	0.60	0.47	0.30	0.17
11β-ol-E-17-one	0.56	0.27	0.50	0.72	0.51	0.42	0.26
15α-ol-E-17-one	0.48	0.18	0.46	0.66	0.45	0.30	0.21
4-ol,3-MeO-E-17-one	0.66	0.53	0.76	0.79	0.67	0.63	0.69
17β-ol-E	0.61	0.40	0.52	0.74	0.53	0.42	0.30
17α-ol-E	0.61	0.43	0.55	0.75	0.52	0.45	0.34
17β-ol-E[6]	0.57	0.36	0.52	0.73	0.49	0.41	0.29
17β-ol-E[7]	0.57	0.36	0.53	0.74	0.50	0.41	0.30
E-7α,17β-ol	0.40	0.09	0.22	0.59	0.37	0.24	0.08
E-11β,17β-ol	0.34	0.08	0.29	0.54	0.39	0.20	0.09
E-16α,17β-ol	0.29	0.06	0.21	0.48	0.38	0.17	0.07
2-MeO-E-17β-ol	0.58	0.36	0.58	0.72	0.53	0.45	0.41

Layer: Silica Gel G (Merck)
Mobile phase: 1. Cyclohexane-ethyl acetate-ethanol (45:45:10)
                2. Cyclohexane-ethyl acetate (50:50)
                3. Chloroform-ethanol (90:10)
                4. Ethyl acetate-*n*-hexane-acetate acid ethanol (72:13.5:10:4.5)
                5. Benzene-ethanol (80:20)
                6. Benzene-ethanol (90:10)
                7. Chloroform-ethanol (95:5)

E = estra-1,3,5 (10)-trien-3-ol. Double bonds are indicated by superscripts. Ketone, hydroxyl, aldehyde, methoxyl, and epoxy functions are indicated by -one, -ol, -al, -MeO, and -ox, respectively.

REFERENCES (TLC)

1. **Rojkowski, K. M. and Broadheat, G. D.,** *J. Chromatogr.,* 69, 373, 1972.
2. **Penyes, L. P. and Oertel, G. W.,** *J. Chromatogr.,* 74, 359, 1972.
3. **Barr, P. and Stancy T.,** *Pharmacia,* 26, 155, 1978.
4. **Petrovic, J. A. and Petrovic, S. M.,** *J. Chromatogr.,* 119, 625, 1976.
5. **Mansinger, D., Marcus, C. S., and Wolf, W.,** *J. Chromatogr.,* 130, 26, 1977.
6. **Ruh, T. S.,** *J. Chromatogr.,* 121, 82, 1976.
7. **Crocker, L. E. and Lodge, B. A.,** *J. Chromatogr.,* 691, 419, 1972.
8. **Rao, P. N., Purdy, R. H., Williams, M. C., Moore, P. H., Jr., Goldzreher, J. W., and Layne, D. S.,** *J. Steroid Biochem.,* 10, 179, 1978.
9. **Petrovic, S. M., Traljic, E., and Petrovic, J. A.,** *J. Chromatogr.,* 205, 223, 1981.
10. **Mattox, V. R., Litviller, R. D., and Carpenter, P. C.,** *J. Chromatogr.,* 175, 243, 1979.

Chapter 11

PREGNANES AND CORTICOIDS (C_{21})

Corticoids is a term applied to hormones or steroids of the adrenal cortex or to other natural or synthetic compounds having similar activity. These are all C_{21} substances differing only in the number and location of the oxygen functions. Most of the literature on chromatography is devoted to analysis with a small portion devoted to the important preparation of the sample. Considering the cost of columns in LC and GC, it is important that attention be paid to this since the success is related to the ability to remove the extraneous material prior to subjecting the sample to chromatography. Figure 1 shows some of the common corticoids with which one will have to contend. The figure does not include the metabolites of the various compounds. Table 1 gives the names of metabolites and parent corticoids.

SECTION I. PREPARATIONS

Extraction of Corticoids (C_{21}) from Blood

Extraction of cortisol and corticosterone from plasma (method of Von Hesse et al.[1]) — To 1 mℓ serum in a 50 mℓ Erlenmeyer flask was added 0.5 mℓ of a solution of 2 μg/ℓ predisone, 0.1 mℓ 2 M sodium hydroxide and 10 mℓ dichloromethane; after 15 min on a magnetic stirrer the organic layer was separated. The aqueous phase was extracted twice with 10 mℓ dichloromethane. The combined extracts were taken to dryness. The residue was dissolved in 40 $\mu\ell$ dichloromethane and in an aliquot injected into a high pressure liquid chromatograph.

One milliliter 0.1 N HCl was added to 1 mℓ serum, and the mixture was extracted wtih 5 mℓ distilled diethyl ether by shaking on a mechanical shaker for 10 min at room temperature. After centrifuging at 2000 g, the organic phase was transferred to another tube, 0.2 mℓ 0.1 N sodium hydroxide solution was added, and the extraction and centrifuging were repeated as above, followed by transfer of the organic phase into a 5 mℓ ampoule where it was evaporated to dryness under a stream of nitrogen at 50°C. Special care was taken to ensure that the volumes of organic phase transferred did not vary from sample to sample. This extract was subjected to TLC.

Extraction of spirolactone (method of Van der Merwe et al.[2]) — To 1 mℓ serum, 1 mℓ 0.1 N HCl was added and the mixture was extracted with 5 mℓ distilled diethyl ether by shaking on a mechanical shaker for 10 min at room temperature. After centrifuging at 2000 g the organic phase was transferred to another tube, 0.2 mℓ 0.1 N sodium hydroxide solution was added, and the extraction and centrifuging were repeated as above, followed by transfer of the organic phase into a 5 mℓ ampoule where it was evaporated to dryness under a stream of nitrogen at 50°C. Special care was taken to ensure that the volumes of organic phase transferred did not vary from sample to sample. This extract was subjected to TLC.

Extraction of serum cortisol (method of Kabra et al.[3]) — 1 mℓ serum or plasma containing 1 mℓ working internal standard (2 μg equilenin) was extracted with 8 mℓ methylene dichloride on a mechanical shaker for 15 min. The solution was centrifuged at 2500 rpm (210 $\times$ g) for 10 min, and the organic phase was removed into a labeled 13 $\times$ 100 mm glass tube. The solution was evaporated with a nitrogen stream of gas or filtered air at 37°C. The residue was dissolved in 100 $\mu\ell$ methanol, and all of the solution was injected into the chromatograph.

Extraction of plasma for HPLC (method of Reardon et al.[4]) — Blood was sampled by venipuncture; the serum was separated by centrifugation within 1 hr after the blood was drawn and stored at -20°C until assay. Methylene chloride (10 mℓ) was added to 1 mℓ

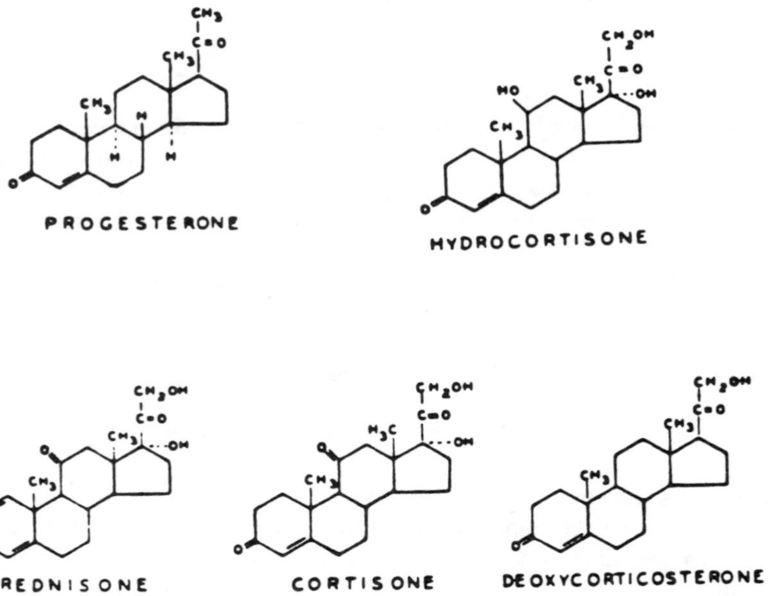

FIGURE 1.   Structure of some common corticoids.

Table 1
SOME COMMON CORTICOIDS AND THEIR METABOLITES

Trivial name	Systematic name
Pregnanediol (PD)	5β-P-3α,20α-diol
Pregnanetriol (PT)	5β-P-3α,17α,20α-triol
Tetrahydro substance S (THS)	3α,17α,21-Trihydroxy-5β-P-20-one
Tetrahydrodeoxycorticosterone (THDOC)	3α,21-Dihydroxy-5β--P-20-one
allo-Tetrahydro substance S (a-THS)	3α,17α,21-Trihydroxy-5α-P-20-one
Pregnanetriolone (11 KPT)	3α,17α,20α-Trihydroxy-5β-P-11-one
5-Pregnenetriol	5-P'-3α,17α,20α-triol
Tetrahydrocortisone (THE)	3α,17α,21-Trihydroxy-5β-P-11,20-dione
Tetrahydrocorticosterone (THB)	3α,11β,21-Trihydroxy-5β-P-20-one
Tetrahydrocortisol (THF)	3α,11β,17α,21-Tetrahydroxy-5β-P-20-one
allo-Tetrahydrocortisol (a-THF)	3α,11β,17α,21-Tetrahydroxy-5α-P-20-one
Cortolone (HHE)	3α,17α,20α,21-Tetrahydroxy-5β-P-11-one
β-Cortolone (β-HHE)	3α,17α,20β,21-Tetrahydroxy-5β-P-11-one
Cortol (HHF)	5β-P-3α,11β,17α,20α,21-pentol
β-Cortol (β-HHF)	5β-P-3α,11β,17α,20β,21-pentol
Substance S (S)	17α,21-Dihydroxy-4-P'-3,20-dione
Cortisone (E)	17α,21-Dihydroxy-4-P'-3,11,20-trione
Cortisol (F)	11β,17α,21-Trihydroxy-4-P'-3,20-dione

P = Pregnane, P' = 4-pregnene, D = diol, T = triol in PT and in THE = tetrahydro.

serum, mixed, and centrifuged at 3500 rpm for 15 min. The aqueous layer was removed and the organic layer was washed with 1 mℓ NaOH (0.1 mol/ℓ) and 1 mℓ water, then dried under a nitrogen stream at 40°C. The extract was dissolved in methanol-water (60:40) before analysis. To determine analytical recovery by the method, 100 ng cortisol and 100 ng 11-deoxycortisol were added to the serum samples before the extraction procedure.

Extraction of aldosterone from urine (method of Herkner et al.[5]) — To 0.5 mℓ of a

24 hr urine was added 15 mℓ 6 *N* HCl and 1000 cpm 1,2,³H-androsterone as a control. This was made to pH 1 with concentrated sulfuric acid and left at room temperature for 20 hr. The androsterone was extracted with 7.5 mℓ dichloromethane. The organic phase was removed and washed once with 0.5 mℓ ice-cold 0.1 *N* NaOh, then twice with water. The extract was evaporated under nitrogen and taken up in 50 μℓ dichloromethane. This was subjected to TLC.

Urine samples (24 hr) were collected in tared bottles (51, ACI) containing 15 g boric acid as preservative. After weighing and thoroughly mixing the sample, a 10 mℓ portion was diluted with distilled water (total volume 15 mℓ). The solution was transferred to a column of wet XAD-2 resin (1 × 10 cm) and washed with distilled water (15 mℓ). The absorbed steroids and steroid conjugates were eluted with methanol (45 mℓ) and water (60 mℓ). The dried residue containing the steroids was suspended in acetate buffer (5 mℓ, 0.2 mol, pH 4.4) and the *Helix pomatia* extract (0.4 mℓ) added for hydrolysis. The mixture was incubated overnight (approximately 16 hr) at 49°C. The mixture was extracted with ethyl acetate (2 × 15 mℓ), and the organic extract was washed with sodium bicarbonate solution (15 mℓ 10% aqueous) and dried over anhydrous sodium sulfate. Prior to evaporation of the solvent, *n*-tetracosane and *n*-dotriacontane were added as internal standards. After derivatization this extract was suitable for GC.

Extraction of urinary neutral steroids (method of Phillipou et al.[6]) — Urine samples (24 hr) were collected into tared bottles (51, ACI) containing 15 g boric acid as preservative. After weighing and thoroughly mixing the sample, a 10 mℓ portion was diluted with distilled water (total volume 15 mℓ). The solution was transferred to column of wet XAD-2 (1 × 10 cm) and washed with distilled water (15 mℓ). The adsorbed steroids and steroid conjugates were then eluted with methanol (45 mℓ) and the solvent removed on a rotary evaporator at 40°C. Regeneration of the resin was achieved by alternate washes with methanol (40 mℓ) and water (60 mℓ).

The dried residue containing the steroids was suspended in acetate buffer (5 mℓ, 0.2 mol, pH 4.4), the *Helix pomatia* extract (0.4 mℓ) added and the mixture incubated overnight (approximately 16 hr) at 49°C. The mixture was then extracted with ethyl acetate (2 × 15 mℓ), and the organic extract washed with sodium bicarbonate solution (15 mℓ 10% aqueous) and dried over anhydrous sodium sulfate. Prior to evaporation of the solvent, *n*-tetracosane and *n*-dotracontane were added as internal standards. After derivatization this extract is suitable for GC.

Urinary steroids (method of Trocha and D'Amato[7]) — To 50 mℓ screw top test tubes was added 7 mℓ urine sample, 2 mℓ concentrated HCl, and 7 mℓ 1,2-dichloroethane containing 30 nmol (13 μg) of the internal standard, 8,24(5a)-cholestandien-4,4,14a-trimethyl-3β-ol (65% pure). The tubes were tightly capped and placed in a 75 ± 2°C water bath for 10 min. After cooling the tubes and agitating their contents with a vortex-type mixer for 15 sec, centrifuging for 2 min at 2000 × *g*, the aqueous (upper) layer aspirated. The dichloroethane layer was extracted with NaOH (2*M*) by agitating the tubes for at least 15 sec on a vortex-type mixer, centrifuged, and the aqueous (upper) layer aspirated. The organic phase was washed with 5 mℓ water and the water layer discarded. The organic extract was evaporated under reduced pressure at 66°C and the residue dissolved in 0.20 mℓ equivolume mixture of ethyl ether and ethanol, transferred to vials, and re-evaporated. This extract, after treatment with trimethylultrifluoro-acetamide was suitable for GC.

Extraction of Urine of Newborns to Recover Corticords[8]

Separation of unconjugated and glucuronic acid conjugated steroids — After mixing each urine well, a 20-mℓ aliquot was chromatographed on a 60 × 1 cm DEAE-Sephadex® A-25 column. The column was eluted with NaCl gradient in water, linearly increasing from zero concentration and reaching a concentration of 0.3*M* after 400 mℓ eluent had gone through the column. The flow rate was maintained at 1.05 mℓ/min (peristaltic pump), and 10.5 mℓ fractions were collected.

The unconjugated steroids were collected in fractions 2 to 7 and the glucuronides in fractions 13 to 19. To both pools, 9.5 μg 6β-hydrocortisone (6β-OHATHE) were added as a recovery standard. In a pilot study this compound was shown to be not present in neonatal urine.

Amberlite® XAD-2 chromatography — The unconjugated and glucuronic acid conjugated steroids were extracted from the DEAE-Sephadex® eluates by Amberlite® XAD-2 resin chromatography. In order to obtain a satisfactory recovery of the urinary cortisol metabolites in the ethanol eluates, the XAD-2 columns were not washed with water following the application of the urinary sample. The ethanol fractions were collected from the precise moment that the ethanol reached the lower end of the column.

Ketodase hydrolysis — The ethanol eluate containing the glucuronides was filtered and evaporated *in vacuo*. To the dry residue 10 mℓ 0.2 *M* acetate buffer (pH 5.0) and 1 mℓ ketodase (5000 Fishman units of β-glucuronidase) were added. After 65 hr of incubation at 37°C, this solution was subjected again to Amberlite ® XAD-2 resin chromatography.

Prepurification — The material isolated from unconjugated and glucuronide fractions was dissolved in 1.5 mℓ methanol-methylene chloride (3:7) and filtered through 40 × 0.4 cm silica gel columns. The columns were rinsed with 5 mℓ of the same solvent, and eluates were taken to dryness. The residues were applied to 0.25 mm silica gel TLC plate, and the plate was developed in the system water-methanol-chloroform (10:125:865) with THE as marker. Fractions were located by spraying blue tetrazolium reagent on the marker lane and subjecting recovered compounds to HPLC.

General Extraction Procedure[9]

Samples of saliva, plasma, or urine (1 mℓ) were added to 10 mℓ methylene chloride. The internal standard, dexamethasone (125 ng), was added, and the glass culture tubes were shaken for 20 min. The tubes were centrifuged and the aqueous layer and creamy interface aspirated. The organic phase was washed with 1 mℓ 0.1 *N* sodium hydroxide and subsequently with 1 mℓ water. After aspirating the aqueous phase, 1 g of anhydrous sodium sulfate was added to dry the organic phase. The latter was evaporated at 45°C under a nitrogen gas stream.

Extraction of 6β-Hydroxycortisol in Urine[10]

A 1 to 5% aliquot of the total daily urine of humans was usually adequate. To remove lipophilic compounds the urine was extracted for 10 sec with 1/10 of its own volume of chloroform, which was discarded. After adding and dissolving 20% (w/v) sodium sulfate (anhydrous, analytical grade) at 35°C, the urine was extracted once with 3 times its own volume of ethyl acetate (analytical grade) for 3 min at room temperature using a shaking device. The aqueous phase was removed, and the ethyl acetate phase was washed twice with 1/20 vol 0.25 *N* sodium hydroxide saturated with sodium sulfate. Cortisol (200 μg) in alcoholic solution was added to the ethyl acetate phase before evaporation to dryness in a rotary vacuum evaporator. The residue was dissolved in ethanol (6 mℓ) (analytical grade), transferred to a conical evaporation tube, and again completely evaporated. The addition of ethanol (100 μℓ) gave an extract ready for injection into a high performance liquid chromatograph.

Extraction of Mitochondrial Incubations[11]

Aliquots of the incubation (0.8 mℓ) were removed at 0, 1, 3, and 5 min of incubation and placed directly into 10 mℓ methylene chloride containing 10 μg 11-deoxycortisol as an internal standard to correct for procedural losses. Each aliquot was then extracted, the organic phases were separated, and the methylene chloride extract was washed with 1.0 mℓ water and then evaporated to dryness under nitrogen. The dried extract was solubilized with 50 $\mu\ell$ ethanol, and a 2- to 10-$\mu\ell$ sample was injected into the chromatograph. The conversion of DOC to β(11β-hydroxylation) and 18-OH-DOC (18-hydroxylation) was quantitated by comparing peak areas of the sample with peak areas of authentic standards and by correcting for recovery based on the peak area of the internal standard (11-deoxycortisol). The rate of formation of 18-OH-DOC was estimated by subtracting the 1-min values from the 3-min values and dividing by two of the protein concentrations.

Extraction of Rat Adrenal To Recover Natural Corticosteroids[12]

Male albino rats of Wistar strain (mean weight 150 to 350 g) were used; adrenal glands were pooled in groups of 8 or 16, depending on the mean weight of the glands in order to obtain about 250 mg tissue.

The tissue was tritiated with fine quartz sand and extracted with chloroform-methanol (2:1) (25 to 30 mℓ/250 mg) in a graduated cylinder with occasional shaking for about 1 hr. The extracts were filtered and washed with water. The washed extracts were evaporated to dryness under reduced pressure and the residue partitioned between 25 mℓ purified ethanol, diluted to 90%, and three 5-mℓ vol n-heptane in a 50-mℓ separating funnel. The n-heptane layers were discarded, the ethanolic layer was dried under reduced pressure, adding small amounts of purified ethanol in the last steps; the residue was dissolved in this solvent, calculating 0.1 mℓ extract as corresponding to 15 mℓ fresh tissue. This solution (E_s), stored in a refrigerator, was used for the chromatographic analysis. A 150 $\mu\ell$ vol solution E_s was transferred into the injector vial of the HPLC instrument, evaporated to dryness with nitrogen, and the residue dissolved in 150 $\mu\ell$ water-saturated chloroform, just before the analysis; the injection volume was 100 $\mu\ell$. The external standardization injections of the sample were alternated with injections of a standard mixture, without modifying the volume setting of the injector. The standard mixture was prepared by dissolving in 1 mℓ purified ethanol, 15 μg corticosterone, 5.22 μg 18-OH-DOC, and 5 μg aldosterone. For the injections, a volume of 250 or 500 $\mu\ell$ was transferred into the injection vial, evaporated to dryness, and the residue dissolved in 500 $\mu\ell$ water-saturated chloroform: the injection was 100 $\mu\ell$.

REFERENCES

1. **Von Hesse, C., Pritrzik, K., and Hotzel, D.,** *J. Clin. Chem. Clin. Biochem.,* 12, 193, 1974.
2. **Van der Merwe, P. J., Muller, D. G., and Clark, E. C.,** *J. Chromatogr.,* 171, 579, 1979.
3. **Kabra, P. M., Tsai, L. J., and Marton, L. N.,** *Clin. Chem.,* 25, 1239, 1979.
4. **Reardon, D. E., Cadarella, A. M., and Canalis, E.,** *Clin. Chem.,* 25, 122, 1979.
5. **Herkner, K., Nowotny, P., and Waldhansl, W.,** *J. Chromatogr.,* 146, 273, 1978.
6. **Phillipou, G., Scamark, R. F., and Cox, L. W.,** *Aust. N. Z. J. Med.,* 8, 63, 1978.
7. **Trocha, P. and D'Amato, N. A.,** *Clin. CHem.,* 24, 193, 1978.
8. **Derks, H. J. G. M. and Drayer, N. M.,** *Steroids,* 31, 289, 1978.
9. **Rose, J. Q. and Jusko, W. J.,** *J. Chromatogr.,* 167, 273, 1979.
10. **Roots, J., Molbe, R., Hovermann, W., Migam, S., Heinemeyer, G., and Hildebrandt, A. G.,** *Eur. J. Clin. Pharmacol.,* 16, 63, 1979.
11. **Gallant, S., Bruckheimer, S. M., and Brownie, A. C.,** *Anal. Biochem.,* 89, 196, 1978.
12. **Cavina, G., Moretti, G., Alimenti, R., and Gallinella, B.,** *J. Chromatogr.,* 175, 125, 1979.

SECTION II. GAS CHROMATOGRAPHY

Many of the C_{21} steroids such as pregnanediol are readily amenable to GC; however, the corticoids usually contain three or more oxygen functions, which makes them difficult to separate by GC. Consequently, most of the preparative steps involve cleavages of the C_{17} side chain followed by GC, derivatization followed by GC, or both. The sample preparations given reflect this methodology.

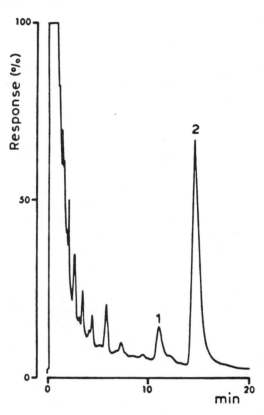

FIGURE GC 1.   Separation of aldosterone gamma-lactone. Instrument: Microtec-220 with flame ionization detector; column: 3% SE-30 on Gas Chrom Q (100—120 mesh) (6ft × 1.8 in.); temperatures: column 250°C, detector 300 °C, and injector 250°C. (From Leung, F. Y. and Griffiths, J., *Clin. Chim. Acta.*, 37, 423, 1972. With permission.)

Table GC 1
SEPARATION OF CORTICOIDS AND METABOLITES AS TRIMETHYL SILYL ETHERS[2]

	Methylene unit values	
	1	2
Tetrahydrodehydrocorticosterone (THA)	31.29	32.36
Tetrahydrocorticosterone (THB)	31.56	32.78
Tetrahydrodeoxycortisol (THS)	29.81	29.92

Table GC 1 (continued)
SEPARATION OF CORTICOIDS AND METABOLITES AS TRIMETHYL SILYL ETHERS[2]

	Methylene unit values	
	1	2
Tetrahydrocortisone (THE)	30.98	31.58
Tetrahydrocortisol (THF)	31.24	32.02
Allo-THF (a-THF)	31.54	32.15
Cortol	32.66	33.91
11-Deoxycortisol (S)	31.78	33.69
Cortisone (E)	32.76	31.06
Cortisol (F)	33.14	35.78
Deoxycorticosterone (DOC)	30.98	33.89
Corticosterone (B)	32.79	36.47

Instrument: Carlo-Erba G1 with flame ionization detector
Column: 1.1% OV-1 or 2% OV-17 on Gas Chrom P (100—120 mesh)
Temperature: Programmed 0 to 190°C at 1.2°C/min

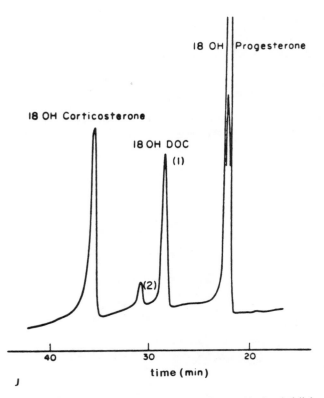

FIGURE GC 2. Separation of 18-hydroxy C_{21} steroid trimethylsilyl ethers. Instrument: Packard 7400 with flame detector; column: 1% OV-1 on Dexsil-300 (100—120 mesh); temperatures: column program from ambient to 180°C, detector 290°C, injector 290°C; flow: 30—40 mℓ/min. (From Prost, M. and Maume, B. F., *J. Steroid Biochem.*, 5, 133, 1974. With permission.)

Table GC 2
RETENTION TIMES OF STEROID
HEPTOFLUOROBUTYRATES[4]

	Retention time relative to 11-oxoprogesterone HFB						
Conditions	1*	2	3	4	5	6	7
1	4.2	3.2	0.53	0.56	0.65	0.44	1.0
2	4.2	2.4	0.52	0.70	0.65	0.33	1.0
3	3.0	2.5	0.78	0.82	0.68	0.36	1.0
4	2.5	1.9	0.82	0.92	0.55	0.42	1.0
5	4.6	2.2	0.76	0.81	0.64	0.44	1.0

Instrument: Pye 104 with ^{63}Ni electron capture detector
Columns and conditions:

	Support	% Phase	Gas flow (mℓ/min)	Temperature (°C)
1.	Chromosorb WHP, 80—100 mesh	1% OV 22	90	225
2.	Chromosorb WHP, 80—100 mesh	3% OV 225	85	245
3.	Supelcoport, 100—120 mesh	1% Dexsil 300 GC	100	245
4.	Supelcoport, 100-120 mesh	3% Dexsil 300 GC	100	250
5.	Supelcoport, 80—100 mesh	3% OV 17	90	245

(1)* Aldosterone lactone, (2) 18-OH-DOC lactone, (3) corticosterone, (4) DOC, (5) androstenetrione, (6) androstenedione, (7) 11-ketoprogesterone.

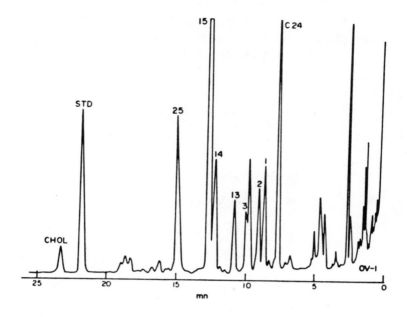

FIGURE GC 3.   Separation of trimethyl silyl ethers of C_{19} and C_{21} steroids. Instrument: Carlo Erba GI with flame ionization detector; column: 1% OV-1 on Gas Chrom P; temperature: column 240°C, separator 250°C; flow: to mass spectrometer. (1) Androsterone, (2) etiocholanolone, (3) dehydroepiandrosterone, (13) 32,17α-dihydroxy-pregnan-20-one, (14) 3α-hydroxy-5α-pregnan-20-one, (15) 3α-hydroxy-5β-pregnan-20-one, (25) 2,3-dihydroxy-5-pregnan-20-one, (CHOL) cholesterol, (STD) 5β-cholestane-3α-ol. (From Begue, R. et al., *J. Steroid Biochem.*, 7, 211, 1976. With permission.)

Table GC 3
RETENTION INDEXES OF PREGNANES AS TRIMETHYL SILYL ETHER METHOXIMES[6]

	Steroid	Retention index
1.	3α-Hydroxy-5α-androstan-17-one	2511
2.	3α-Hydroxy-5β-androstan-17-one	2526
3.	3β-Hydroxy-5-androsten-17-one	2572
4.	3α-Hydroxy-5α-androstane-11,17-dione	2613
5.	3α-Hydroxy-5β-androstane-11,17-dione	2613
6.	3α,11β-Dihydroxy-5α-androstan-17-one	2693
7.	3α,11β-Dihydroxy-5β-androstan-17-one	2710
8.	3α-Hydroxy-5β-pregnan-20-one	2717
9.	3α,17α-Dihydroxy-5β-pregnan-20-one	2693
10.	5β-Pregnane-3α,20α-diol	2767
11.	5β-Pregnane-3α,17α,20α-triol	2800
12.	5-Pregnene-3β,20α-diol	2837
13.	5-Androstene-3β,17β,16α-triol	2844
14.	3α,17α,21-Trihydroxy-5β-pregnan-20-one	2867
15.	3α,21-Dihydroxy-5β-pregnan-20-one	2888
16.	3α,17α,20α,-Trihydroxy-5β-pregnan-11-one	2909
17.	5β-Pregnan-3α,17α,20α-tetrol	2928
18.	5-Pregnene-3β,17α,20α-triol	2957
19.	3α,17α,21-Trihydroxy-5β-pregnane-11,20-dione	2965
20.	3α,11β,21-Trihydroxy-5α-pregnan-20-one	2997

Table GC 3 (continued)
RETENTION INDEXES OF PREGNANES AS TRIMETHYL SILYL ETHER METHOXIMES[6]

	Steroid	Retention index
21.	3α,11β,17α,21-Tetrahydroxy-5β-pregnan,20-one	3025
22.	3α,11β,17α,21-Tetrahydroxy-5α-pregnan-20-one	3035
23.	3α,17α,20α,21-Tetrahydroxy-5β-pregnan-11-one	3052
24.	3α,17α,20β,21-Tetrahydroxy-5β-pregnan-11-one	3083
25.	5-Cholesten-3β-ol	3097
26.	5β-Pregnane-3α,11β,17α,20α,21-Pentol	3122

Instrument: Pye 104A with flame ionization detector
Column: Capillary column coated with SE-30 (20 m × 0.5 mm)
Temperatures: Column 200 to 260°C at 1°C/min, detector and injector 250°C

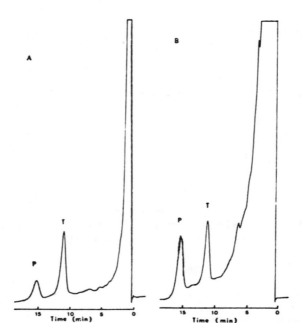

FIGURE GC 4.   Detection at picogram level of pentafluorobutyl derivatives of progesterone and testosterone. Instrument: Packard-Backer 417 with [63]Ni electron capture detector; column: 3% OV-17 Chromosorb GAW (2 m × 2 mm); temperatures: column 285°C, detector 310°C, injector 300°C; flow: argon with 5% methane at 60 mℓ/min; (A) Progesterone di-*O*-PFBO (P) and testosterone *O*-PFBO-PFB (T). (B) Sample of 2.8 mℓ plasma obtained from a pregnant rat (4th day of pregnancy). (From Wehner, R. and Handke, H., *J. Chromatogr.*, 177, 237, 1979. With permission.)

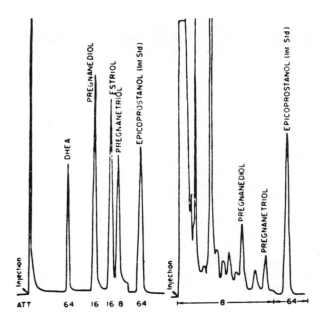

FIGURE GC 5. Separation of pregnanediol and pregnanetriol of urine as trimethylsilyl ethers. Instrument: Shimadzu 4 BM PFE with flame ionization detector; column: 3% OV-1 on Gas Chrom Q (80—100 mesh) (1.5 m × 4 mm); temperatures: column programmed from 255—275°C at 1°C/min, detector 290°C; flow: 55 mℓ/min. (Left) Standards, (right) urinary steroids. Dehydroepiandrosterone (DHEA), 5-androsten-3β-ol-17-one; pregnanediol, 5β-pregnan-3α,20α-diol; estriol, 1,3,5, (10)-estratrien-3,16α,17β-triol; pregnanetriol, 5β-pregnan-3α,17α,20α-triol; epicoprostanol, 5β-cholestan-3α-ol. (From Hammon, J. E. et al., *Ann. Clin. Lab. Sci.*, 9, 416, 1979. With permission.)

Table GC 4
SEPARATION OF CORTICOIDS AND SOME C_{19} STEROIDS METHYLENE UNIT VALUES GIVEN[8]

	Methylene unit values	
Steroid	1	2
3α-Hydroxy-5α-A-17-one	25.02	25.3
3α-Hydroxy-5β-A-17-one	25.22	25.6
3β-Hydroxy-5-A'-17-one	25.68	26.0
3α-Hydroxy-5α-A-11,17-dione	26.05	26.7
3α-Hydroxy-5β-A-11,17-dione	26.05	26.8
3α,11β-Dihydroxy-5α-A-17-one	26.90	27.0
3α,11β-Dihydroxy-5β-A-17-one	27.02	27.30
5β-P-3α,20α-diol	27.52	27.62
5β-P-3α,17α,20α-triol	27.88	27.88
3α,17α,21-Trihydroxy-5β-P-20-one	28.58	28.65
3α,21-Dihydroxy-5β-P-20-one	28.58	28.88
3α,17α,21-Trihydroxy-5α-P-20-one	28.58	29.02
3α,17α,20α-Trihydroxy-5β-P-11-one	29.05	29.02
5-P'-3α,17α,20α-triol	29.45	29.40

Table GC 4 (continued)
SEPARATION OF CORTICOIDS AND SOME C$_{19}$ STEROIDS METHYLENE UNIT VALUES GIVEN[8]

Steroid	Methylene unit values	
	1	2
3α,17α,21-Trihydroxy-5β-P-11,20-dione	29.58	29.82
3α,11β,21-Trihydroxy-5β-P-20-one	29.93	30.05
3α,11β,17α,21-Tetrahydroxy-5β-P-20-one	30.21	30.22
3α,11β,17α,21-Tetrahydroxy-5α-P-20-one	30.32	30.35
3α,17α,20α,21-Tetrahydroxy-5β-P-20-one	30.48	30.65
3α,17α,20β,21-Tetrahydroxy-5β-P-20-one	30.78	30.92
5β-P-3α,11β,17α,20α,21-pentol	31.20	31.15
5β-P-3α,11β,17α,20β,21-pentol	30.78	30.70
5-C′-3β-ol	30.78	30.92
17α,21-Dihydroxy-4-P′-3,20-dione	30.95	31.15
17α,21-Dihydroxy-4-P′-3,11,20-trione	31.98	32.65
11β,17α,21-Trihydroxy-4-P′-3,20-dione	32.65	32.92
5-C′-3β-ol isobutyrate	33.10	33.75

Instrument: Packard-Becker 419 with flame ionization detector
Column: 1. 1% OV-1 on Gas Chrom Q (100—120 mesh) 3.76 m × 2 mm)
    2. Mixed bed: 1% OV-1 and 3% OV-225 (9:1) on Gas Chrom Q (100—120 mesh) (2.7 m × 2mm)
Temperatures: 1. 190°C for 5 min, then to 270°C at 1°C/min
    2. 190°C for 10 min, then to 240°C at 10°C/min
    Detector and injector: 270°C
Flow: Nitrogen (30 mℓ/min)

(A) Androstane; (P) pregnane; (A′) androstene; (P′) pregnene; (C′) cholestene.

Table GC 5
SEPARATION OF PREGNANES AS TRIMETHYLSILYL ETHER METHOXIMES[9]

Steroid	Methylene unit values	
	1	2
3β-Hydroxy-4-pregnen-20-one	27.78	28.20
3β,21-Dihydroxy-4-pregnen-20-one	30.00	30.00
3β-Hydroxy-5α-pregnan-20-one	27.88	28.00
3β-Hydroxy-5β-pregnan-20-one	27.22	27.24
4-Pregnene-3,20-dione	28.35	28.79
4-Pregnene-3,11,20-trione	29.42	29.91
11β-Hydroxy-4-pregnene-3,20-dione	30.44	30.80
		30.85
17α-Hydroxy-4-pregnene-3,20-dione	29.26	29.47
6β-Hydroxy-4-pregnene-3,20-dione	29.07	29.32
2α-Hydroxy-4-pregnene-3,20-dione	29.30	29.60
20α-Hydroxy-4-pregnen-3-one	28.87	29.28
20β-Hydroxy-4-pregnen-3-one	28.62	28.71
17α,20α-Dihydroxy-4-pregnen-3-one	30.49	30.81

Table GC 5 (continued)
SEPARATION OF PREGNANES AS TRIMETHYLSILYL ETHER METHOXIMES[9]

Steroid	Methylene unit values	
	1	2
21-Hydroxy-4-pregnene-3,20-dione (DOC)		31.26
18,21-Dihydroxy-4-pregnen-3,20-dione	32.30	32.30
11β,21-Dihydroxy-4-pregnene-3,20-dione (B)	32.70	32.38
21-Dihydroxyprogesterone = 6β-hydroxy-DOC	30.97	31.09
6β,21-Dihydroxy-4-pregnene-3,20-dione	31.09	31.24
11-Hydroxy-4-pregnene-3,11,20-trione	31.66	31.57
11β,21-Dihydroxy-3,20-dioxo-4-pregnen-18-al:18-hemiacetal form	32.14	31.82
		32.18
		32.25
17β,21-Dihydroxy-4-pregnene-3,20-dione		31.18
5-Cholestene-3β-ol	31.07	31.35
Cholesteryl butyrate (3β-butyryloxy-5-cholestene)	34.00	34.19

Instrument:	Packard-Becker 420 with flame ionization detector
Column:	1. 1% OV-1 on Gas Chrom Q (4 m × 3 mm)
	2. Capillary coated with SE-30 (25 m × 0.21mm)
Temperatures:	1. 200 to 290°C at 1°C/min
	2. 180°C at 1°C/min
Flow:	Nitrogen 1. 30 mℓ/min
	2. 1 mℓ/min

REFERENCES (GC)

1. **Leung, F. Y. and Griffiths, J.,** *Clin. Chim. Acta,* 37, 423, 1972.
2. **Chambaz, E. M., Defaye, G., and Madani, C.,** *Anal. Chem.,* 45, 1090, 1973.
3. **Prost, M. and Maume, B. F.,** *J. Steroid Biochem.,* 5, 133, 1974.
4. **Mason, P. A. and Graser, R.,** *J. Endocrinol.,* 64, 277, 1975.
5. **Begue, R., Desgres, J., Gustaffeson, J., and Padrew, P.,** *J. Steroid Biochem.,* 7, 211, 1976.
6. **Phillipou, G., Seamark, R. F., and Cox, L. W.,** *Aust. N. Z. J. Med.,* 8, 63, 1978.
7. **Hammond, J. E., Phillips, J. C., and Savory, J.,** *Ann. Clin. Lab. Sci.,* 9, 416, 1979.
8. **Tomasova, Z., Gregorova, I., and Hochy, K.,** *J. Chromatogr.,* 200, 221, 1980.
9. **Maume, B. F., Millot, C., Patouraux, D., Doumas, J., and Tomori, E.,** *J. Chromatogr.,* 186, 581, 1979.
10. **Wehner, R. and Handke, H.,** *J. Chromatogr.,* 177, 237, 1979.
11. **Trocha, P. and D'Amato, N. A.,** *Clin. Chem.,* 24, 193, 1978.

SECTION III. HIGH PERFORMANCE LIQUID CHROMATOGRAPHY

Unlike GC, HPLC and TLC are able to separate many corticoids and other C_{21} steroids without derivatization. However, preparation of the sample, particularly if it is of biomedical origin, requires careful consideration if the separation desired is to be achieved. Success in HPLC of corticoids was delayed due to lack of suitable columns, and it was in 1973 that Touchstone and Wortmann[1] showed that reversed phase HPLC may be the path to follow. The results in Table HPLC 1 are from this earlier paper.

Table HPLC 1
SEPARATION OF CORTICOIDS[1]

	Steroid	Retention (cm)	Band width (cm)
E:	Cortisone	1.8	0.9
F:	Cortisol	2.0	1.0
A:	Aldosterone	1.7	1.0
B:	Corticosterone	3.1	1.6
S:	17-Hydroxydeoxycorticosterone	3.5	0.9
Prog:	Progesterone	15.0	5.8

Instrument: Nester Faust LC with UV detector
Column: Nester Faust Sil XRP
Mobile phase: Methanol-water (40:60)
Flow: Variable

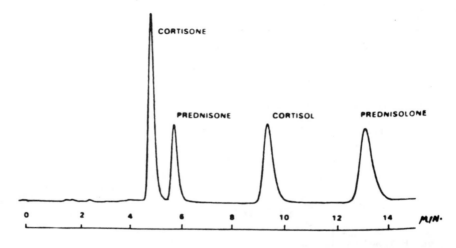

FIGURE HPLC 1.   Separation of natural and synthetic corticoids. Instrument: Dupont 830 with UV detector; column: Zorbax Sil (25 cm × 2.1 mm) with water jacket at 25°C; mobile phase: methylene chloride-methanol-water (960:20:20); flow: 1 mℓ/min; detector: 254 mm. (From Parris, N. A., *J. Chromatogr. Sci.*, 12, 753, 1974. With permission.)

Table HPLC 2
RETENTION TIMES (MIN) FOR CORTICOIDS SEPARATED BY GRADIENT ELUTION[3]

Steroid	Mobile phase 1	2	3
6β-Hydroxycortisol	3	3.5	4
11-Dehydroaldosterone	8	5	15
17-Isoaldosterone	11	6	15
18-Hydroxy-11-dehydrocorticosterone	11.5	6.5	17.5
Prednisone	12	9	23
Aldosterone	13.5	8	19
Estriol	14	8	25
Cortisone	14	9	25.5
18-Hydroxycorticosterone	15.5	7	21

Table HPLC 2 (continued)
RETENTION TIMES (MIN) FOR CORTICOIDS SEPARATED BY GRADIENT ELUTION[3]

Steroid	Mobile phase		
	1	2	3
Prednisolone	15.5	9	26.5
Cortisol	16	9	27
19-Hydroxytestosterone	19	10	24
16α-Hydroxytestosterone	21	11	27
11-Dehydrocorticosterone	21.5	15	29.5
Dexamethasone	24	13	35
11β-Hydroxyandrostenedione	24	17	32
21-Deoxycortisol	26.5	16	34.5
Corticosterone	31	18	36
11β-Hydroxytestosterone	31	15	33.5
18-Hydroxydeoxycorticosterone	31.5	16	34.5
11-Deoxycortisol	33	20	37.5
6α-Hydroxyprogesterone	34	24.5	38.5
16α-Hydroxyprogesterone	35	24	38
11-Ketoprogesterone	37	30	40
11β-Hydroxy-20α-dihydroprogesterone	38	26	39
6β-Hydroxyprogesterone	39	30	41
11-Hydroxyprogesterone	41	33	43
11-Deoxycorticosterone	41	36	43.5
17α-Hydroxyprogesterone	41.5	38	45
17α-Hydroxy-20α-dihydroprogesterone	42.5	32	42
17α-Hydroxy-20β-dihydroprogesterone	43	32.5	43.5
Dehydroepiandrosterone	44	41	44
17α-Hydroxypregnenolone	45.5	39	—
Progesterone	47	45	49
20α-Dihydroprogesterone	47.5	43.5	47.5
20β-Dihydroprogesterone	48.5	46	49.5
Pregnenolone	48.5	47	50
Cholesterol	54	>60	—

Instrument: Dupont 830 with variable wavelength UV detector.
Column and mobile phase gradient:

System	1	2	3
Column	Zorbax-ODS	Zorbax-ODS	Zorbax-ODS
Solvent (start)	40% (v/v) methanol-water	32% (v/v) acetonitrile-water	20% (v/v) dioxane-water
Solvent (finish)	100% methanol	100% acetonitrile	100% dioxane
Gradient	$y = x^3$	$y = x^3$	$y = x^3$
Flow (start), mℓ/min	0.38	0.38	0.38
Flow (finish), mℓ/min	0.67	0.80	0.34

Detection: 240 nm

Table HPLC 3
RETENTION TIMES OF SYNTHETIC STEROIDS[4]

Compound	Mobile phase							
	A				B			
	1	2	3	4	5	6	7	8
Acetone	1.00	1.00	1.00	1.00	1.00	1.00	1.00	1.00
Hydrocortisone hemisuccinate	0.66	0.86	0.82	0.93	0.96	1.20	0.98	1.31
Methylpredisolone sodium succinate	0.69	1.12	0.82	1.07	0.94	1.24	0.98	1.58
Methylparaben	1.15	1.81	1.12	1.58	1.06	1.40	1.06	1.42
Triamcinolone	1.14	1.86	1.00	1.19	0.98	1.56	1.04	1.54
Prednisone	1.22	2.48	1.05	1.44	1.34	2.44	1.06	2.31
Cortisone	1.25	2.57	1.10	1.49	1.13	2.56	1.08	2.38
Fluprednisolone	1.25	2.60	1.00	1.42	1.09	1.80	1.08	1.92
Prednisolone	1.32	3.00	1.10	1.42	1.11	2.08	1.04	2.08
Hydrocortisone	1.32	3.05	1.08	1.46	1.11	2.08	1.09	1.69
Triamcinolone diacetate	1.30	3.38	1.25	2.65	1.21	5.60	1.21	6.31
Propylparaben	1.48	3.95	1.32	2.60	1.13	2.40	1.15	2.69
Betamethasone	1.42	4.10	1.18	1.84	1.13	2.80	1.08	2.88
Dexamethasone	1.42	4.14	1.10	1.88	1.15	2.92	1.13	3.00
Methyl prednisolone	1.50	4.45	1.10	1.67	1.15	2.76	1.11	2.69
Fluocinolone acetonide	1.48	4.48	1.20	2.40	1.15	4.28	1.17	4.31
Prednisolone acetate	1.55	4.86	1.35	2.60	1.28	5.48	1.23	5.38
Triamcinolone acetonide	1.50	4.93	1.22	2.21	1.28	5.40	1.17	4.56
Fluorometholone	1.55	5.00	1.32	2.67	1.23	4.20	1.23	4.81
Flurandrenolide	1.60	5.05	1.28	2.35	1.21	4.40	1.17	3.96
Hydrocortisone acetate	1.52	5.07	1.35	2.63	1.26	5.32	1.21	5.38
Fluorandrenolone	1.59	5.14	1.25	2.37	1.21	4.44	1.17	3.85
Cortisone acetate	1.52	5.14	1.40	2.98	1.42	7.60	1.32	7.62
Betamethasone acetate	1.68	6.60	1.42	3.58	1.32	7.60	1.32	8.08
Paramethasone acetate	1.72	7.33	1.48	3.72	1.30	7.12	1.32	8.23
Methylprednisolone acetate	1.78	—	1.40	3.53	1.34	7.92	1.30	8.15
Desoxycorticosterone acetate	2.54	—	2.28	—	2.79	—	2.08	—
Betamethasone valerate	2.85	—	1.95	8.00	1.96	—	1.70	38.5
Desoxycorticosterone trimethylacetate	5.95	—	4.92	—	6.60	—	4.47	—

Instrument:	Waters 202/401 with UV detector
Column:	A. μBondapak C$_{18}$
	B. μBondapak phenyl/corasil
Mobile phase:	1. Methanol-water (70:30)
	2. Methanol-water (50:50)
	3. Acetonitrile-water (60:40)
	4. Acetonitrile-water (40:60)
	5. Methanol-water (60:40)
	6. Methanol-water (40:60)
	7. Acetonitrile-water (40:60)
	8. Acetonitrile-water (20:80)
Flow:	A. 1.5 mℓ/min
	B. 1.0 mℓ/min
Detector:	254 nm

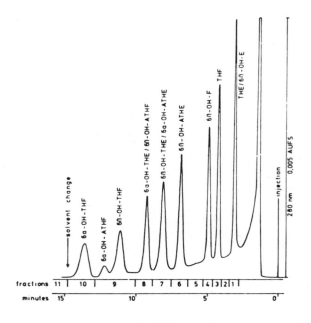

FIGURE HPLC 2.   Separation of corticoid metabolites. Instrument: Waters 6000 pump, U6K injector, and UV detector; column: μPorasil (30 × 0.4 cm); mobile phase: water-methanol-chloroform (7:70:723); flow: 2 mℓ/min; detector: 254 nm. See glossary in Table HPLC 1. (From Derks, J. J. G. M. and Drayer, N. M., *Steroids,* 31, 289, 1978. With permission.)

Table HPLC 4
SEPARATION OF STEROIDAL ACIDS AS *p*-BROMOPHENACYL ESTERS[6]

Steroid acid	Time (min)
C_{21} acids	
11β,17,20α-Trihydroxy-3-oxo-4-pregnen-21-oic acid	18.8
11β,17,20β-Trihydroxy-3-oxo-4-pregnen-21-oic acid	28.6
17,20α-Dihydroxy-3,11-dioxo-4-pregnen-21-oic acid	20.7
17,20β-Dihydroxy-3,11-dioxo-4-pregnen-21-oic acid	38.7
3α,11β,17,20α-Tetrahydroxy-5β-pregnan-21-oic acid	24.0
3α,11β,17,20β-Tetrahydroxy-5β-pregnan-21-oic acid	48.0
3α,17,20α-Trihydroxy-11-oxo-5β-pregnan-21-oic acid	30.9
3α,17,20β-Trihydroxy-11-oxo-5β-pregnan-21-oic acid	51.6
C_{20} acids	
11β,17-Dihydroxy-3-oxo-4-androstene-17β-carboxylic acid	29.0
17-Hydroxy-3,11-dioxo-4 androstene-17β-carboxylic acid	32.4
3α,17,11,β Trihydroxy-5β-pregnane-17β-carboxylic acid	40.3
3α,17-Dihydroxy-11-oxo-5β-pregnane-17β carboxylic acid	60.3

Instrument:	Dupont 830 with UV detector
Column:	Zorbax ODS (4.6 × 250 mm)
Mobile phase:	Methanol-water (60:40)
Flow:	2.2 mℓ/min
Temperature:	40°C
Detection:	254 nm

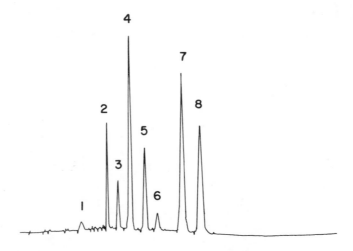

FIGURE HPLC 3.   Separation of corticoids. Instrument: Hewlett-Packard 1084 with UV detector; column: LiChrosorb Si-100 10 μm (25 cm × 4.6 mm); mobile phase: gradient, 10-100% methanol formed over 20 min; flow: 1 mℓ/min; detection: 254 nm; numbers: time in min: (1) deoxycorticosterone, (2) 11-dehydrocorticosterone, (3) 11-deoxycortisol, (4) corticosterone, (5) cortisone, (6) aldosterone, (7) cortisol, (8) prednisolone. (From Cavina, G. et al., *J. Chromatogr.*, 175, 125, 1979. With permission.)

Table HPLC 5
SEPARATION OF SYNTHETIC CORTICOIDS[8]

Steroid	Retention time (min)
Cortisone	4.2
Triamcinolone acetonide	4.3
Prednisone	5.4
Beclomethasone	6.5
Dexamethasone	7.1
Cortisol	8.4
Methylprednisolone	10.5
Prednisolone	11.6
17α,20α,21-Trihydroxy-1,4-pregnadiene-3,11-dione	15.5
17α,20β,21-Trihydroxy-1,4-pregnadiene-3,11-dione	18.0
6-β-Hydroxycortisol	18.4

Instrument:	Waters 6000A solvent delivery with Whatman precolumn of HC Pellosil, U6K injector, and a UV detector
Column:	Zorbax Sil (5—6 μm) (25 cm × 4.6 mm)
Mobile phase:	Methanol-methylene chloride (3:97)
Flow:	2 mℓ/min
Detection:	254 nm

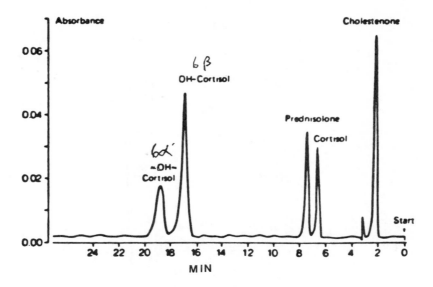

FIGURE HPLC 4.   Separation of 6-hydroxcortisols. Instrument: Siemens S-100 with Zeiss PM-2-DLC detector; column: LiChrosorb Si-100 5 μm (30 cm × 6 mm); mobile phase: methylene chloride-hexane-ethanol-water (40:470:112:15), lower phase; detector: 254 nm. (From Roots, I. et al., *Eur. J. Clin. Pharmacol.*, 16, 63, 1979. With permission.)

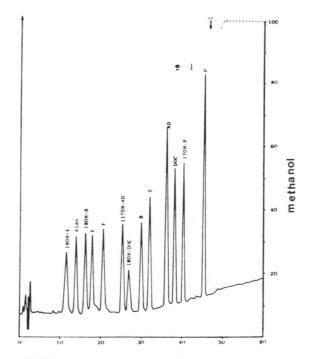

FIGURE HPLC 5.   Separation of 18-hydroxypregnanes. Instrument: Spectra Physics SP-8000 with UV detector; column: Hypersil ODS 10% (5 μm) (150 × 4.6 mm); mobile phase: methanol-water gradient as in figure (40—100% methanol at 1 mℓ/min) detector: 240 nm; flow: 1 mℓ/min. (E) Cortisone, (F) cortisol, (B) corticosterone (S) 11-dioxycortisol. Glossary for letters in Section I: Preparations, Table 1. (From O'Hare, M. J. et al., *J. Chromatogr.*, 198, 23, 1980. With permission.)

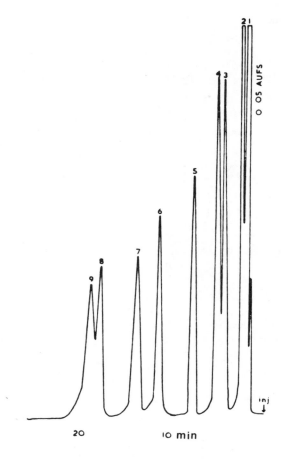

FIGURE HPLC 6.   Separation of pregnanes. Instrument: Applied Chromatography Model 750/03 pump with variable wavelength detector; column: Partisil (5 μm) (250 × 4.5mm); mobile phase: dichloromethane-ethanol (95:4) saturated with water; flow: 1.5 mℓ/min; detection: 240 nm. (1) Progesterone, (2) 17-hydroxyprogesterone, (3) 11-deoxycortisol, (4) corticosterone, (5) prednisone, (6) dexamethasone, (7) cortisol, (8) 6α-methylprednisolone, (9) prednisolone. (From Scott, N. et al., *Anal. Biochem.*, 108, 226, 1980. With permission.)

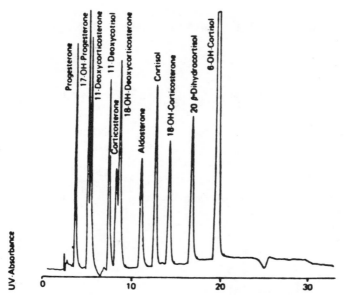

FIGURE HPLC 7.   Separation of corticoids. Instrument: Hewlett-Packard 1084B with UV detector; column: diol volumn (Knauer); mobile phase: Solvent A, *n*-hexane; B, *n*-hexane-isopropanol (75:25), stepwise linear gradient from 50—100% solvent B; flow: 1.3 mℓ/min; temperature: 40°C; detection: 254 nm. (From Shoneshofer, M. et al., *J. Chromatogr.*, 227, 492, 1982. With permission.)

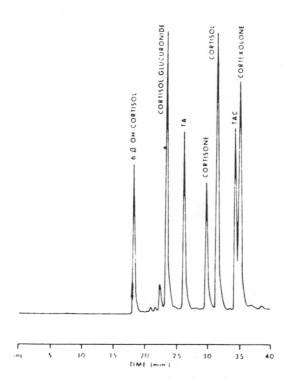

FIGURE HPLC 8.   Separation of corticoid conjugates. Instrument: Waters 6000A pump, U6K injector and UV detector; column: LiChrosorb RP-18 (250 × 10 mm) with a guard column of μBondapak C$_{18}$ Corasil; mobile phase: convex gradient of methanol, 0.1 *M* ammonium acetate (pH 6.9) (10:9) to 100% methanol at 1.5 mℓ/min over 50 min; detection: 254 nm. (From Althus, Z. R. et al., *J. Chromatogr.*, 227, 11, 1982. With permission.)

Table HPLC 6
SEPARATION OF PREGNANE DIOLS AND KETONES[14]

Steroid	Mobile phase Retention time (min)		
	1	**2**	**3**
Diones			
4-Pregnene-3,20-dione (progesterone)	44	—	40
5-Pregnene-3,20-dione	51	—	52
5α-Pregnane-3,20-dione	18	—	62
5β-Pregnane-3,20-dione	22	—	51
Monohydroxymonoketones			
3β-Hydroxy-5-pregnen-20-one (pregnenolone)	37	—	36
20α-Hydroxy-4-pregnen-3-one	—	18	25
20β-Hydroxy-4-pregnen-3-one	—	17	35
3α-Hydroxy-5α-pregnan-20-one	32	—	51
3α-Hydroxy-5β-pregnan-20-one	61	—	41
3β-Hydroxy-5α-pregnan-20-one	39	—	50
3β-Hydroxy-5β-pregnan-20-one	27	—	45
20α-Hydroxy-5α-pregnan-3-one	53	—	43
20α-Hydroxy-5β-pregnan-3-one	100	—	38
20β-Hydroxy-5α-pregnan-3-one	53	—	59
20β-Hydroxy-5β-pregnan-3-one	95	—	56
Diols			
5-Pregnene-3β,20α-diol	—	15	21
5-Pregnene-3β-20β-diol	—	15	27.5
5α-Pregnane-3α,20α-diol	—	15	35
5α-Pregnane-3α,20β-diol	—	16	44
5α-Pregnane-3β-20α-diol	—	15	29
5α-Pregnane-3β,20β-diol	—	13	38
5β-Pregnane-3α,20α-diol	—	26	29
5β-Pregnane-3α,20β-diol	—	22	38
5β-Pregnane-3β,20α-diol	—	14.5	27
5β-Pregnane-3β,20β-diol	—	14.5	38

Instrument: LDC Constametric II pump with Altex 905-23 injector
Column: 1. 2. Partisil 5 μm (600 × 2 min)
                3.   Zorbax PB-005 (500 × 4 min)
Mobile phase: 1. Ethanol-dichloromethane (0.25—99.75)
                2. Hexane-isopropanol (97:3)
                3. Acetonitrile-water (60:40)
Flow: 1 mℓ/min
Detection: Refractive index

REFERENCES (HPLC)

1. **Touchstone, J. C. and Wortman, W.,** *J. Chromatogr.,* 76, 244, 1973.
2. **Parris, N. A.,** *J. Chromatogr. Sci.,* 12, 753, 1974.
3. **O'Hare, M. J., Nice, E. C., Magee, R., and Bullman, H.,** *J. Chromatogr.,* 125, 357, 1976.
4. **Tymes, N. W.,** *J. Chromatogr. Sci.,* 15, 151, 1977.
5. **Derks, J. J. G. M. and Drayer, N. M.,** *Steroids,* 31, 289, 1978.
6. **Farhi, R. L. and Monder, C.,** *Anal. Biochem.,* 90, 58, 1978.
7. **Cavina, G., Morette, G., Alimenti, R., and Gallinella, B.,** *J. Chromatogr.,* 175, 125, 1979.
8. **Rose, J. Q. and Jusko, W. J.,** *J. Chromatogr.,* 162, 273, 1979.

9. **Roots, I., Holbe, R., Hovermann, W., Nigman, S., Heinemeyer, G., and Hildebrandt, A. G.,** *Eur. J. Clin. Pharmacol.,* 16, 63, 1979.
10. **O'Hare, M. J., Nice, E. C., and Capp, M.,** *J. Chromatogr.,* 198, 23, 1980.
11. **Scott, N., Chakroborty, J., and Marks, V.,** *Anal. Biochem.,* 108, 266, 1980.
12. **Shoneshofer, M., Weber, B., and Dulce, H. J.,** *J. Chromatogr.,* 227, 492, 1982.
13. **Althus, Z. R., Rowland, J. M., and Freeman, J. P.,** *J. Chromatogr.,* 227, 11, 1982.
14. **Lin, J. T., Heftmann, E., and Hunter, I. R.,** *J. Chromatogr.,* 190, 169, 1980.

SECTION IV. THIN LAYER CHROMATOGRAPHY

Table TLC 1
SEPARATION OF PREGNANES AND ANDROGENS[1]

Steroids	Mobile phase/R_f 1	2
Androsterone	0.63	0.44
Etiocholanolone	0.63	0.44
Dehydroepiandrosterone	0.61	0.44
11-Hydroxyandrosterone	0.58	0.30
11-Hydroxyetiocholanolone	0.56	0.23
11-Ketoandrosterone	0.58	0.36
11-Ketoetiocholanolone	0.57	0.26
Pregnanediol	0.58	0.27
Pregnanolone	0.68	0.44
Pregnanetriol	0.40	0.05
Pregnanetriol-11-one	0.34	0.03
Tetrahydrocortisol	0.25	0.01
Cortol	0.19	0.01
Sundan	0.77	0.74
Isatin	0.56	0.42

Layer: Silica Gel GF$_{254}$
Mobile phase: 1. Chloroform-methanol (9:1)
2. Ethyl acetate-benzene (1:1)
Detection: UV quench

Table TLC 2
R_f VALUES FOR C_{21} STEROIDS[2]

Steroid	Mobile phase 1	2	3	4	5	6
DOC	66	67	66	82	66	69
THDOC	44	45	50	70	61	51
αTHDOC	50	55	59	75	65	59
S	42	43	52	64	50	55
THS	26	25	28	41	41	32
αTHS	30	33	37	51	44	40
A	55	50	51	72	54	59
THA	28	24	25	45	41	30

Table TLC 2 (continued)
R_f VALUES FOR C_{21} STEROIDS[2]

Steroid	Mobile phase					
	1	2	3	4	5	6
αTHA	33	30	31	53	44	38
THB	26	22	24	41	37	26
B	38	34	40	57	45	43
αTHB	27	25	29	46	40	31
E	36	34	41	50	33	48
THE	21	17	19	25	22	24
F	26	22	27	34	24	33
THF	18	12	14	18	16	18
αTHF	19	15	17	23	19	20
6HB	23	20	21	28	15	30
6HF	14	10	13	13	6	19
Ald	27	23	21	36	30	34

Layer:
Mobile phase: % in dichloroethane
1. 12.5 Ethanol
2. 15 *t*-butanol
3. 33 Acetone
4. 25 Dioxane
5. 22.5 Acetic acid
6. 10 Pyridine

TH = Tetrahydroform, 6H = 6-hydroxy, ald = aldosterone. Glossary in Section I: Preparations, Table 1.

Table TLC 3
SEPARATION OF CORTICOIDS
$(R_f \times 100)^2$

Steroid	Additive (mobile phase)					
	1	2	3	4	5	6
DOC	46	53	55	42	37	58
THDOC	49	61	51	37	48	55
αTHDOC	59	70	61	49	57	62
S	47	62	48	35	35	47
THS	46	64	42	27	42	43
αTHS	53	72	52	38	50	47
A	17	23	33	20	15	36
THA	16	19	20	12	17	30
αTHA	20	29	28	18	20	34
B	29	39	36	22	20	39
THB	27	40	29	18	26	33
αTHB	32	48	38	27	29	39
E	33	44	34	20	18	31
THE	30	43	24	12	21	25
F	34	49	33	17	20	29
THF	30	50	26	14	24	24
αTHF	34	56	31	19	27	26
6HB	21	33	22	14	10	21
6HF	22	35	18	10	9	14
Ald	14	21	20	16	9	31

Table TLC 3 (continued)
SEPARATION OF CORTICOIDS
$$(R_f \times 100)^2$$

Layer: Polygram SiG (Macherey-Nagel)
Mobile phase: % in isopropylether
1. 10 Ethanol
2. 20 Butanol
3. 30 Acetone
4. 20 Dioxane
5. 25 Acetic acid
6. 20 Pyridine

Glossary in Section I: Preparations, Table 1.

Table TLC 4
SEPARATION OF
CORTICOIDS $R_f \times 100^2$

Steroid	Mobile phase			
	1	2	3	4
DOC	53	53	58	60
THDOC	45	42	45	47
αTHDOC	57	59	51	54
S	54	54	48	46
THS	43	40	29	30
αTHS	55	56	41	37
A	32	35	54	49
αTHA	24	24	39	36
B	34	34	43	41
THB	26	25	28	31
αTHB	32	32	30	34
E	41	41	45	36
THE	27	22	27	21
F	39	37	33	31
THF	30	26	20	20
αTHF	36	34	22	24
6HB	26	25	27	27
6HF	26	24	21	19
Ald	18	12	25	29

Layer: Polygram SiG
(Macherey-Nagel)
Mobile phase:
1. Propyl acetate
2. Methylisobutyl ketone
3. Nitromethane-ethanol (7.5:92.5)
4. Anisole-ethanol (10:90)

Glossary in Section I: Preparations, Table 1.

Table TLC 5
SEPARATION OF PREGNANES IN 11 MOBILE PHASES[3]

Compound	Mobile phase/$R_f \times 100$										
	1	2	3	4	5	6	7	8	9	10	11
P-4-ene-3,20-dione	62	59	58	61	63	61	67		44	61	72
17α-Caproyloxy-P-4-ene-3,20-dione	67	63	68	67	44	18	22	0.43	12	33	58
17α-Hydroxy-21-acetoxy-P-4-ene-3,11,20-trione	49	41	23	19	33	17	14	32	11	27	48
11β,17α-Dihydroxy-21-acetoxy-P-4-ene-3,-20-dione	48	40	29	18	17	13	07	19	11	06	19
17α,21-Dihydroxy-P-4-ene-3,11,20-trione	33	22	08	08	15	03	05	16	07	03	17
17α,21-Dihydroxy-P-1,4-diene-3,11,20-trione	35	19	14	05	10	05	03	11	06	03	10
11β,17α,21-Trihydroxy-P-4-ene-3,20,dione	27	18	08	04	09	04	02	11	02	01	09
11β,17α,21-Trihydroxy-6β-Me-P-1,4-diene-3,20-dione	26	17	07	02	09	03	02	11	01	01	09
11β,17α,21-Trihydroxy-P-1,4-diene-3,20-dione	24	15	06	02	70	74	82	69	37	75	79
Dimethylaminoazobenzene	71	66	80	71	45	25	13	68	08	43	
Isatin	55	47	50	40	07	25	07	05		25	
Neutral red	05	05	00	06							

Layer: Wakogel B-5 (1—9) alumina B-10 with 10% gypsum (10—11)
Mobile phases:
1. Hexane-ethyl acetate (2:8)
2. Benzene-ethyl acetate (3:7)
3. Ether
4. Benzene-acetone (4:1)
5. Chloroform-acetone (3:1)
6. Benzene-methanol (97:3)
7. Chloroform-methanol (97:3)
8. Chloroform-methanol (3:1)
9. Petroleum ether-benzene-acetic acid-water (67:33:85:15)
10. Benzene-acetone (4:1)
11. Benzene-methanol (9:1)

P = pregn.

Table TLC 6
SEPARATION OF CORTICOIDS AND SYNTHETIC STEROIDS[4]

Steroid	Mobile phase ($R_f \times 100$)	
	1	2
Prednisone	54	27
Acetate	79	58
Prednisolone	34	13
Acetate	71	45
Dexamethasone	42	17
Acetate	79	52
Hydrocortisone	43	17
Acetate	80	56
Cortisone acetate	83	67
Betamethasone	43	17
Acetate	78	52
Methylprednisolone	36	12
Acetate	77	50
Fluprednisolone	37	14

Table TLC 6 (continued)
SEPARATION OF CORTICOIDS AND SYNTHETIC STEROIDS[4]

Steroid	Mobile phase ($R_f \times 100$)	
	1	2
Fluorocortisone acetate	78	56
Paramethasone acetate	78	55
Triamcinolone	22	07

Layer: Silica Gel 60

Mobile phase: 1. Methylene chloride-dioxane-water (100:50:50)

2. Methylene chloride-dioxane-water (120:30:50)

Table TLC 7
SEPARATION OF CLOSELY RELATED PREGNANES (R_f)[5]

Compound	Mobile phase	
	1	2
5α-Pregnane-3,20-dione		
β-Pregnane-3,20-dione		
3α-Hydroxy-5α-pregnan-20-one	0.67	
3β-Hydroxy-5α-pregnan-20-one	0.52	
20α-Hydroxy-4-pregnen-3-one	0.53	
20β-Hydroxy-4-pregnen-3-one	0.61	
6α-Hydroxy-4-pregnene-3,20-dione	0.24	0.27
6β-Hydroxy-4-pregnene-3,20-dione	0.36	0.42
11α-Hydroxy-4-pregnene-3,20-dione	0.30	0.31
16αa-Hydroxy-4-pregnene-3,20-dione	0.24	0.22
5α-Pregnane-3α,20α-diol	0.32	0.51
5α-Pregnane-3α,20β-diol	0.40	0.65
5α-Pregnane-3β,20α-diol	0.35	0.56
5α-Pregnane-3β,20β-diol	0.37	0.60
5β-Pregnane-3α-20α-diol	0.21	0.44
5β-Pregnane-3α,20β-diol	0.25	0.50
5β-Pregnane-3β,20α-diol	0.45	0.63
5β-Pregnane-3β,20β-diol	0.48	0.67

Layer: 1. Neutral alumina no binder

2. Basic alumina

Mobile phase: 1. Benzene-ethanol (98:2) develop twice

2. Cyclohexane-ethyl acetate-water (125:125:0.4)

Table TLC 8
SEPARATION OF CLOSELY RELATED
PREGNANES (R$_f$)[6]

	Mobile phase	
Compound	**1**	**2**
5α-Pregnane-3,20-dione	0.75	
5β-Pregnane-3,20-dione	0.63	
3α-Hydroxy-5α-pregnan-20-one		
3β-Hydroxy-5α-pregnan-20-one		
6α-Hydroxy-4-pregnene-3,20-dione		0.28
6β-Hydroxy-4-pregnene-3,20-dione		0.44
5α-Pregnane-3α,20α-diol	0.27	0.59
5α-Pregnane-3α,20β-diol	0.39	0.68
5α-Pregnane-3β-20α-diol	0.32	0.62
5α-Pregnane-3β-20β-diol	0.37	0.64
5β-Pregnane-3α,20α-diol	0.16	0.43
5β-Pregnane-3α,20β-diol	0.21	0.51
5β-Pregnane-3β,20α-diol	0.39	0.70
5β-Pregnane-3β,20β-diol	0.43	0.75

Layer: Silica gel GF$_{254}$
Mobile phase: 1. Petroleum ether-ethyl ether (60:40), developed 10 times
2. Petroleum ether-ethyl ether (40:60), developed 5 times

Table TLC 9

SEPARATION OF CORTICOIDS AND SYNTHETIC CORTICORDS[6]

Mobile phases/R_f

Steroids	1	2	3	4	5	6	7	8	9	10	11	12	13	14	15	16	17	18	19	20	21	22
Cortisol	0.62	0.68	0.24	0.55	0.10	0.55	0.66	0.55	0.61	0.11	0.33	0.30	0.81	0.48	0.37	0.82	0.40	0.71	0.70	0.65	0.24	0.11
Corticosterone	0.58	0.66	0.43	0.67	0.20	0.63	0.72	0.48	0.58	0.37	0.67	0.41	0.77	0.74	0.46	0.76	0.57	0.84	0.68	0.59	0.53	0.26
Aldosterone	0.39			0.56			0.65															
11-Desoxy-corticosterone		0.80	0.80		0.53	0.83	0.83	0.70	0.77	1.0	0.78	0.55	0.82	1	0.69	0.83	0.85	1	0.80		0.87	0.62
Progesterone	0.89	0.90	0.90		0.74	0.89	0.88	0.83	.87	1.0	0.82	0.66	0.89	1	0.88	0.89	0.89	1	0.89		0.88	0.78
17α-OH-Progesterone	0.87	0.80	0.80		0.56	0.84	0.85	0.79	0.85	1.0	0.78	0.54	0.88	1	0.63	0.89	0.89	1	0.87		0.78	0.62
Dexamethasone	0.76		0.26		0.08	0.48	0.61	0.67	0.74	0.11	0.32	0.26	0.85	0.47	0.32	0.88		0.86	0.80	0.80	0.38	0.16
Flurcortolone	0.67		0.36		0.14	0.56	0.66	0.50	0.59	0.28	0.58	0.40	0.83	0.73	0.50	0.80	0.57	0.84	0.71	0.61	0.43	0.20
Paramethaone			0.22		0.10	0.54	0.62	0.59	0.73	0.12	0.31	0.35	0.88	0.57	0.40	0.88		0.71	0.77	0.70	0.27	0.14
Prednisolone	0.64		0.20		0.07	0.42	0.55	0.47	0.57	0.08	0.25	0.26	0.80	0.41	0.36	0.81		0.64	0.69	0.63	0.22	0.10
Prednisone	0.67		0.35		0.13	0.55	0.62	0.49	0.62	0.21	0.57	0.44	0.81	0.70	0.39	0.80		0.71	0.70	0.61	0.31	0.12

Layer: Polygram SiG (Macherey-Nagel)

Mobile phases:
1. Dioxane-ethyl acetate (5:95)
2. Dioxane (10:90)
3. Dioxane-chloroform (25:75)
4. Dioxane-chloroform (35:65)
5. Dioxane-benzene (20:80)
6. Dioxane-benzene (40:60)
7. Dioxane-benzene (50:50)
8. Ethylene glycol monomethyl ester-ethyl acetate (2:98)
9. Ethylene glycol monomethyl ester-chloroform (3:97)
10. Ethylene glycol monomethyl ester-chloroform (3:97)
11. Ethylene glycol monomethyl ester-chloroform (7:93)
12. Ethylene glycol monomethyl ester-benzene (10:90)
13. Methanol-ethyl acetate (10:90)
14. Methanol-chloroform (5:95)
16. Propanol-ethyl acetate (10:90)
17. Propanol-chloroform (10:90)
18. Propanol-chloroform (15:85)
19. Acetone-ethyl acetate (10:90)
20. Acetic acid-ethyl acetate (5:95)
21. Acetic acid-chloroform (30:90)
22. Acetic acid-benzene (30:70).

Table TLC 10
MOBILITIES OF CORTICOID ACETATES[7]

Corticosteroid	System A	System B
Betamethasone	0.66	0.26
Cortisone acetate	1.04	1.21
Dexamethasone	0.66	0.24
Fluorocortisone acetate	1.01	0.94
Fluperolone acetate	1.01	1.00
Hydrocortisone	0.62	0.24
Hydrocortisone acetate	0.97	0.85
Paramethasone acetate	1.03	1.12
Prednisolone	0.51	0.19
Prednisolone acetate	0.88	0.72
Prednisone	0.73	0.47
Prednisone acetate	1.01	1.03
Triamcinolone	0.40	0.12
Triamcinolone diacetate	0.88	1.00
Beclomethasone dipropionate		1.62
Beclomethasone 17-propionate		0.82
Beclomethasone 21-propionate		1.16
Betamethasone 17-valerate		0.85
Deoxycortone acetate		1.85
Deoxycortone pivalate		1.94
Methylprednisolone		0.19
Methylprednisolone acetate		0.78
Prednisolone acetate		0.72
Triamcinolone acetonide		0.65

Layer:　　　　　Silica Gel GF$_{254}$
Mobile phases: (A) Methylene chloride-dioxane-
　　　　　　　　　　water (100:50:50), lower layer
　　　　　　　　(B) Ethylene chloride-methanol-
　　　　　　　　　　water (95:5:0.2)

16-Betamethylprednisone acetate R$_f$ = 1.00.

Table TLC 11
SEPARATION OF CORTICOIDS (R$_f$ VALUES)[8]

Steroid	Mobile phase 1	2	3	4
Aldosterone	0.08	0.02	0.07	0.41
Cortisol	0.23	0.06	0.15	0.57
Corticosterone	0.24	0.06	0.24	0.66
Cortisone	0.28	0.075	0.22	0.58
11-Deoxycortisol	0.48	0.22	0.35	0.69
Deoxycorticosterone	0.55	0.28	0.51	0.79
Testosterone	0.61	0.37	0.48	0.76
17α-Hydroxy-pregnenolone	0.75	0.46	0.46	0.75
Androstendione	0.76	0.46	0.63	0.82
Dehydroepiandrosterone	0.77	0.49	0.54	0.77

Table TLC 11 (continued)
SEPARATION OF CORTICOIDS (R_f VALUES)[8]

Steroid	Mobile phase			
	1	2	3	4
17α-Hydroxy-progesterone	0.80	0.46	0.54	0.78
Pregnenolone	0.83	0.59	0.55	0.80
Progesterone	0.84	0.60	0.65	0.89

Layer: Silica Gel G
Mobile phases: 1. Cyclohexane-ethyl acetate (20:80)
                    2. Cyclohexane-ethyl acetate (45:55)
                    3. Benzene-acetone (75:25)
                    4. Benzene-acetone (50:50)

Table TLC 12
SEPARATION OF CORTICOIDS BY TLC[9]

Compound	Mobile phase/R_f						
	1	2	3	4	5	6	7
5αP-3β-ol	0.66	0.48	0.62	0.76	0.64	0.48	0.53
5βP-3α-ol	0.67	0.55	0.65	0.77	0.67	0.52	0.59
5βP-3α,6α-ol	0.31	0.06	0.29	0.48	0.58	0.23	0.12
5αP-3α,20β-ol	0.59	0.34	0.55	0.70	0.56	0.34	0.34
5αP-3β,20α-ol	0.52	0.27	0.50	0.65	0.53	0.30	0.33
5αP-3β,20β-ol	0.57	0.31	0.52	0.70	0.56	0.33	0.35
5βP-3α,20α-ol	0.49	0.22	0.50	0.65	0.52	0.33	0.28
5βP-3α,20β-ol	0.54	0.25	0.52	0.68	0.54	0.36	0.31
5βP-3β,20β-ol	0.59	0.37	0.57	0.70	0.57	0.41	0.40
5βP-3α,17α,20α-ol	0.26	0.05	0.23	0.45	0.39	0.20	0.07
5βP-3β,17α,20α-ol	0.34	0.11	0.35	0.55	0.47	0.26	0.15
5βP-3α,17α,20β-ol	0.32	0.08	0.32	0.51	0.44	0.23	0.12
5βP-3β,17α,20β-ol	0.39	0.16	0.41	0.61	0.49	0.29	0.19
5αP-3β,17α,20α-ol	0.34	0.10	0.33	0.58	0.45	0.25	0.14
20β-ol-5αP-3-one	0.64	0.42	0.71	0.76	0.65	0.51	0.60
3β-ol-5αP-20-one	0.55	0.33	0.62	0.71	0.58	0.42	0.45

Layer: Silica Gel
Mobile phases: 1. Cyclohexane-ethyl acetate-ethanol (45:45:10)
                    2. Cyclohexane-ethyl acetate (50:50)
                    3. Chloroform-ethanol (90:10)
                    4. Ethyl acetate-*n*-hexane-acetic acid-ethanol
                        (72:13.5:10:4.5)
                    5. Benzene-ethanol (80:20)
                    6. Benzene ethanol (90:10)
                    7. Chloroform-ethanol (95:5)

5αP = 5α-pregnan, etc. Ketone, hydroxyl, adlehyde, methoxyl, and epoxy functions are indicated by -one, -ol, -al, -MeO, and -ox, respectively.

REFERENCES (TLC)

1. **Yamaguchi, Y. and Hayashi, C.,** *Jpn. J. Clin. Chem.,* 5, 140, 1976.
2. **Smith, P. and Hall, C. J.,** *J. Chromatogr.,* 101, 202, 1974.
3. **Hara, S. and Mibe, K.,** *Chem. Pharm Bull.,* 23, 2850, 1975.
4. **Byrne, J. A., Brown, I. K., Chaubal, M. G., and Malone, M. H.,** *J. Chromatogr.,* 137, 489, 1977.
5. **Egert, D.,** *J. Chromatogr.,* 135, 481, 1977.
6. **Gleispach, H.,** *Chromatographia,* 10, 40, 1977.
7. **Bailey, K., By, A. W., and Lodge, B. A.,** *J. Chromatogr.,* 166, 299, 1978.
8. **Herkner, K., Nowotny, P., and Waldausel, W.,** *J. Chromatogr.,* 146, 273, 1978.
9. **Mattox, V. P., Litroviller, R. D., and Carpenter, P. C.,** *J. Chromatogr.,* 175, 243, 1979.

Chapter 12

STEROLS AND CHOLESTEROL

Cholesterol and cholesterol esters have an increasing role as indicators of physiological and pathological events. Classical methods such as precipitation are not practical and require large amounts of plasma or other biological samples. The analytical investigation of "fats" of mammalian and vegetable origin has progressed rapidly. Most sterols, because of their high votility or low polarity, are generally amenable to GC. As seen in this section, GC enjoys a strong position in the analysis of these compounds; however, because of this property, it is sometimes difficult to separate members of an individual class. Cholesterol esters are widespread, and these can be saponified or analyzed as such. Figure 1 gives the structures of a number of steroids and cholesterol analogues and acetates. Table 1 gives the names of the compounds listed in Figure 1.

SECTION I. PREPARATIONS

Sample Preparation for Cholesterol

Determination of cholesterol in blood or other biological fluids (method of Dutta et al.[1]) — Blood samples of hospital patients were collected and the serum was separated from the blood as soon as possible after collection. The determination was either carried out immediately or the serum was stored at $-20°C$.

A TLC plate of Silica Gel G was divided lengthwise into 3 lanes that were 5, 5, and 2 cm in width; 0.05 mℓ serum was applied evenly in the form of a band throughout the two 5-cm lanes, and the charged plate was dried in a desiccator over calcium chloride for 10 min. The plate was then subjected to 2 consecutive predevelopments with chloroform-methanol (2:1) in a saturated chamber to a position 1.5 cm above the starting line in each run (the plate was air dried before each run). This was done in order to extract the lipid material from the serum band. After extraction, a mixture of cholesterol and cholesterol palmitate was spotted of the middle of the 2 cm wide lane at a distance of 1.5 cm from the starting line.

The plate was then developed in a saturated chamber to a position 16 cm above starting line with the same solvent. The plate was exposed to iodine vapor and the cholesterol and its ester bands were identified from the position of the reference spots. The cholesterol content of these 2 bands was estimated.

Extraction of free cholesterol in blood (method of Cham et al.[2]) — Blood was taken from fasting normal subjects and collected into tubes containing 100 IU heparin per 10 mℓ blood. Plasma was separated within a $^1/_2$ hr of collection. The lipids were extracted by adding approximately 20 mℓ chloroform-methanol (2:1) to exactly 1 mℓ plasma in a glass-stoppered tube. The mixture was shaken and allowed to stand for 2 hr, after which it was filtered and the precipitate washed 3 times with 5 mℓ chloroform-methanol (2:1). The filtrates were combined and evaporated using a water bath at 70°C. The dried residue was reconstituted in 1 mℓ chloroform-hexane (65:35).

Total cholesterol in serum (method of Duncan et al.[3]) — Serum was saponified as follows: 6 mℓ 33% KOH in H_2O, diluted to 100 mℓ with absolute alcohol was added with a Cornwall syringe to 0.5 mℓ serum in a 20 × 150 mm glass test tube equipped with a Teflon®-lined cap. The tubes were incubated at 45°C for 60 min and then allowed to cool to room temperature. After the addition of 5 mℓ water to each tube, they were cooled to room temperature, 10 mℓ hexane were added to each, and the tubes were shaken mechanically for 10 min. A 4-mℓ aliquot of the hexane layer was pipetted into a 16 × 130 mm glass test tube and evaporated in a 45°C oven at a reduced pressure of 20 to 28 psi. The residue

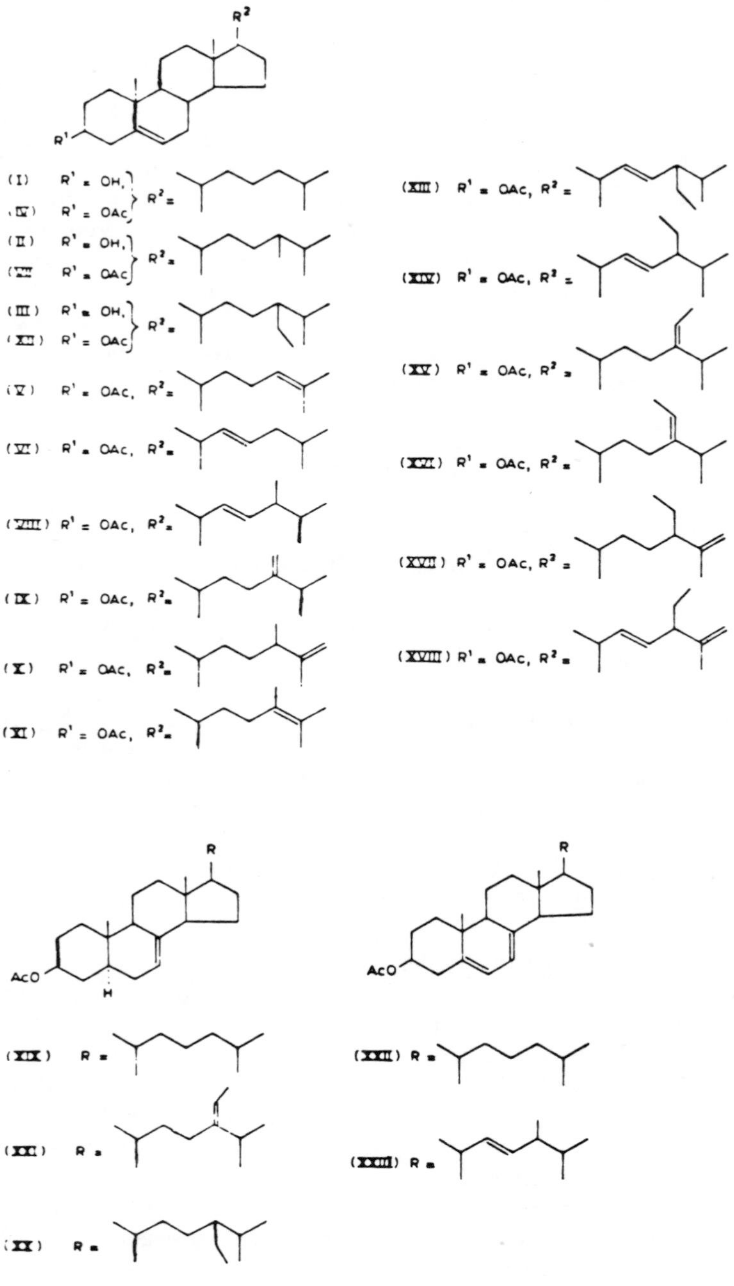

FIGURE 1.   Structures for sterols (see Table 1 for names).

was then dissolved in 800 μℓ isopropanol and the extract was used for injection onto the column of the chromatograph. For the free cholesterol determination, 100 μℓ serum was vortexed with 500 μℓ isopropanol for 2 min, centrifuged, and the supernatant removed. This supernatant was used for the chromatographic analysis of free cholesterol and to obtain a profile of the individual cholesterol esters and other lipids.

Total cholesterol in plasma (method of Fennel et al.[4]) — To each 20 μℓ sample of plasma, 5α-cholestane (20 μg/100 μℓ) as internal standard was added. After saponfication,

Table 1
NOMENCLATURE FOR COMPOUNDS IN
FIGURE 1

I.	Cholesterol (C_{27})
II.	Campesterol (C_{28})
III.	Sitosterol
IV.	Cholesterol acetate
V.	Demesterol
VI.	22-*trans*-Cholestra-5,22-chem-3β-yl acetate
VII.	Campesterol acetate
VIII.	Brassicasteryl acetate
IX.	24-Methylenecholesteryl acetate
X.	24-Methylcholest-5,25-chem-3β-yl acetate
XI.	24-Methylcholest-5,24-dien-3β-yl acetate
XII.	Sitosteryl acetate
XIII.	Stigmasteryl acetate
XIV.	Poriferasteryl acetate
XV.	28-Isofucosteryl acetate
XVI.	Fucosteryl acetate
XVII.	24-Ethyl-cholest-5,25-chem-3β-yl acetate
XVIII.	24-Ethylcholest-5,22,25-trien-3β-yl acetate
XIX.	5α-cholestan-3β-yl-acetate
XX.	5α-stigmast-7-en-3β-yl-acetate
XXI.	(24Z)-5α-stigmasta-7,24(28)-dien-3β-yl-acetate
XXII.	Cholestra-5,7-dien-3β-yl-acetate
XXIII.	Ergosteryl acetate

extraction were performed with 100 μℓ solvent. Since flame ionization detectors are not sensitive to carbon disulfide, it was chosen as the extracting solvent. Phases were separated by centrifugation at 3500 rpm for 10 min. GLC analyses were performed on aliquots of the clear, lower phase. Serum was diluted 10-fold with saline, and 100 μℓ diluted mixture was saponified together with 52 nmol (20 μg) (2,2,3,4-³H) cholesterol in 10μℓ acetone by treatment with 1 mℓ 0.5M potassium hydroxide in ethanol at 70°C for 1.5 hr. In some experiments 1000- or 10,000-fold diluted serum was used. The saponifications were performed in glass-stoppered test tubes. Water (1 mℓ) was added and the cholesterol was extracted with 5 mℓ hexane by vigorous shaking for 1 min. The hexane phase was collected, the solvent was removed under a stream of nitrogen, and the residue was dissolved in 50 μℓ acetone. In the preparation of samples for the standard curve, the serum was substituted with 100 μℓ water and different amounts of cholesterol in acetone.

Cholesterol in microsomal incubates (method of Hansbury and Scallen[5]) — For a total microsomal incubation volume of 3 mℓ, 3 mℓ 15% ethanolic KOH was added to each flask, followed by saponification at 50°C for 2 hr. After cooling, the samples were extracted twice with 6-mℓ portions of petroleum ether, the extracts filtered through a fluoropore filter with prefilter, the petroleum ether removed by a stream of nitrogen, and the samples redissolved in acetone for chromatography.

Sterols of yeast cultures (method of Trocha et al.[6]) — Cells were harvested and saponified with 20% KOH in 50% aqueous methanol (1 g wet weight/5 mℓ) at 85°C for 60 min, and the mixture was extracted 3 times with petroleum ether (bp 30 to 60°C).

The nonsaponifiable fraction remaining after removal of solvent was dissolved in a small volume of CHCl₃, spotted on silica gel thin layer plates, and developed with benzene-ethyl acetate (4:1). Sterol bands, located by spraying a sample plate with Liebermann-Burchard reagent, were scraped off the plates and extracted 3 times with hot CHCl₃, and the solvent was removed at room temperature in a stream of N₂. An aliquot (2 mg) was dissolved in 50 μℓ freshly distilled tetrahydrofuran and chromatographed on 60 cm (two 0.39 × 30 cm

columns) of μBondapak C_{18} with the following solvent systems: (A) tetrahydrofuran-ace-
tonitril-water (5:5:2) at a flow rate of 0.5 mℓ/min; (B) acetonitrile-water (10:1) at a flow
rate of 1.2 mℓ/min. Commercial lanosterol was chromatographed with solvent system A at
a flow rate of 0.5 mℓ/min. Areas under refractive index peaks obtained with 0.1 to 1.2 mg
known sterols were proportional to size of sample.

Sterols of gorgonians (*P. homomalla*) (method of Popov et al.[7]) — Air-dried gorgonians
(*P. homomalla*) were crushed in a mortar and pestle and subsequently ground in a Waring®
blender in pure acetone. The ground material was transferred to a Soxhlet apparatus and
continuously extracted for 36 hr using a tissue to acetone (w/v) ratio of 1 kg/ℓ. The extract
was reduced in volume under reduced pressure at 30 to 40°C. The residue was subjected to
initial SiO_2 column chromatography. A portion of the extracted tissues was then subjected
to base hydrolysis by refluxing in 1 *N* KOH in ethanol for 3 to 4 hr. The hydrolyzate was
extracted with ethyl ether, washed dried (Na_2SO_4), and the sterols isolated by TLC 1 using
a cholesterol marker. A second portion of the extracted gorgonian tissues was subjected to
acidic hydrolysis: refluxing in 5% HCl for 3 to 4 hr followed by ethyl ether extraction
neutralization ($NaHCO_3$), drying (Na_2SO_4) of the extract, and isolation of the sterols present.

*Removal of Sterols in Water by Liquid-Liquid extraction or by Use of Amberlite® XAD-2
Resin*

Detection of fecal pollution of water[8] — Liquid-liquid extraction: The fecal sterols were
extracted by liquid-liquid partitioning. Two milliliters of concentrated HCl and 5 mℓ 20%
(w/v) NaCl were added to each liter of water sample. The sample was extracted by vigorous
mixing with three 100 mℓ portions of hexane for 30 min each. The combined extract was
washed with two 50 mℓ portions of acetonitrite (saturated with hexane), followed by two
50 mℓ portion of 70% ethanol. The hexane was then brought to dryness on a rotary evaporator
under reduced pressure. The sample was redissolved in 100 to 200 μℓ CS_2, and 1 to 5 μℓ
solution was used for GLC analysis. Column extraction: Columns packed with Amberlite®
XAD-2 neutral resin (mesh sizes 40 to 50 or 60 to 120, Brinkman Instruments, Inc.,
Westbury, N.Y.), which were prewashed with distilled water, methanol, acetone, methanol,
and water respectively, were employed.

Water samples (0.25 to 1 ℓ) were passed through the column at a predetermined flow
rate. The column was washed with 20 mℓ distilled water and eluted with 15 to 30 mℓ
acetone. The acetone eluant was evaporated and redissolved as above and 1 to 5 μℓ solution
was used for GLC analysis.

Fecal sterols in water[9] — Chemical work-up: Following the addition of 2ℓ redistilled
hexane, water samples (1 ℓ) were stirred to a deep vortex for 15 min, using a motor-driven
stainless steel shaft and propeller. After separation of the 2 liquid phases, the hexane fraction
was siphoned off, dried over anhydrous sodium sulfate, filtered through a cotton plug, and
the total volume of hexane retrieved noted. The organic extract was evaporated under vacuum
using a Buchi "Rotavapor", and the semi-liquid residue transferred with several washings
to a small capped vial, using acetone. The latter solvent was then removed under a gentle
stream of air, and the product thus obtained reconstituted for GC analysis.

Extraction of tear samples from rabbits and humans for determination of cholesterol[10]
— With the use of micro-capillaries, tear samples were collected from the inferior lacrimal
punctum qand lower cul-de-sac female rabbits (New Zealand White) as well as human
subjects of either sex. A sample of 1 μℓ human tear or 2 μℓ rabbit tear was added to 1.0
mℓ alcoholic potassium hydroxide. After vortex mixing, the saponification was allowed to
proceed for 1 hr at 60°C. The mixture was then allowed to cool and 1.0 mℓ water added.
After mixing, 1.0 mℓ petroleum ether was evaporated off and the silylated derivative of
cholesterol was prepared by adding 25 μℓ BSA, followed by mixing and incubating at 50°C
for 10 min. At the end of the incubation period, the excess silylating agent was evaporated

off (hot water bath under a stream of nitrogen) and was cooled and reconstituted with a 10 μℓ aliquot of a 28 ppm 5α-cholestane solution in carbon tetrachloride for GC.

Cholesterol in amniotic fluid[11] — To 1 mℓ of either amniotic fluid or cholesterol standard was added 5 mℓ ethanolic KOH, the tube contents vortex-mixed for 15 sec, and incubated at 37°C for 90 min. After the mixture had cooled to room temperature, 10 mℓ *n*-hexane was added.

The tubes were vortex-mixed for 1 min and the layers allowed to separate; 7 mℓ hexane (upper) layer was removed and evaporated in a stream of air at room temperature. To the residue were added 0.5 mℓ cholesteryl acetate in pyridine, 0.2 mℓ hexamethyldisilane, and 0.1 mℓ trimethylchlorosilane. The tubes were vortex-mixed for 1 min, and the mixture was allowed to stand for 15 min at room temperature. Aliquots were taken for GC.

Direct GC analysis of cholestanol in plasma[12] — One hundred mℓ plasma were hydrolyzed in a 5 mℓ glass-stoppered centrifuge tube by the addition of 0.5 mℓ 0.5 M tetramethylammonium hydroxide-ethanol (TMH-e) solution containing 5α-cholestane as the internal standard. The TMH-e solution was prepared by diluting 1.0 μg of 5α-cholestane (dissolved in 2 mℓ diethyl ether) and 5 mℓ TMH (2 M, methanolic, Southwestern Analytical Chemicals, Inc.) to 100 mℓ absolute ethanol. The plasma TMH-e mixture was vortex-mixed for 30 sec and was then placed in a heating block at 90°C for 30 min. After cooling, 75 μℓ 1.1 M phosphoric acid and 2 mℓ petroleum ether (40 to 60°C) were added, followed by thorough mixing for 1 to 2 min; 1 mℓ distilled water was then added to induce phase separation. After centrifugation at room temperature for 10 min at 2000 rpm the petroleum ether upper phase was carefully removed with a Pasteur pipette and placed in a Concentratube (LH 101, Laboratory Research Company, Los Angeles). The remaining aqueous phase was re-extracted twice more with 2 mℓ petroleum ether. The petroleum ether extracts were combined and evaporated to dryness under N_2. Prior to analysis, the residue was reconstituted with 10 to 20 μℓ hexane, and 0.5 μℓ were injected into the GLC instrument. Analyses for recovery of cholestanol during this extraction procedure were made by subjecting ³H-cholestanol to the above procedure with subsequent counting of the extract on a Packard 3375 liquid scintillation counter.

Extraction of sterols from food[12] — Grind, homogenize, ball mill, or take measured aliquot or weighed portion of raw sample to obtain uniform, representative sample of product to be extracted. Sample size should contain 1 g fat (analytical balance). (Sample size was readily calculated from label claim for total fat.) The sample was transferred to homogenization vessel (i.e., Virtis, Waring, etc). Sample was homogenized ≥2 min with a volume of chloroform-methanol (2:1) that is 20-fold that of original sample (i.e., 10 g sample uses 200 mℓ reagent). The homogenate was filtered through qualitative (rapid) paper into appropriate size separatory funnel. The blender (bowl, blade, lid, etc.) was rinsed with at least 2 chloroform-methanol rinses (0.2 vol homogenate) and the washes filtered as done previously with homogenate. The residue was washed on the paper with 2 chloroform-methanol rinses (0.2 vol homogenate). All rinses were filtered through paper into the separatory funnel. The extract was washed by thoroughly mixing 0.2 of its volume with water (i.e., 200 mℓ extract uses 40 mℓ water) and allowed to stand until the layers separate. Most of the lower phase (chloroform) was drained into appropriate size round-bottom flasks with 24/40 fitting. The upper phase was washed 3 times with water-methanol (2:1). The volume of these washes was equal to water-methanol volume remaining in separatory funnel. All washes were drained into a round-bottom flask and flash evaporated with 50°C water bath. The dried residue was transferred to a tared receiver with ethyl ether; the solution was evaporated under a stream of nitrogen. The receiver was transferred to the vacuum oven and the extract dried at 60°C and under 26 in. vacuum. When constant weight was achieved, the receiver was transferred to a nitrogen-flushed desiccator, and the sample and receiver allowed to cool to room temperature and weighed. The dried lipid was dissolved in redistilled petroleum ether, quantitatively transferred to a 100 mℓ volumetric flask, and diluted to volume with redistilled petroleum ether.

Occasionally a dried powder will form a suspension in the separatory funnel and a distinct line of demarcation between the water phase and the ether phase never becomes apparent. If this occurs, a new subsample should be placed in a beaker and extracted with petroleum ether, filtered, and the solvent removed. In this case, petroleum ether extraction was repeated 3 times, making a composite solution of combined extracts for injection into GLC.

Extraction of plasma for cholesterol (method of Sheppard et al.[13]) — 0.01% EDTA-plasma (0.2 to 0.5 mℓ) was added to a PTFE-lined screw-cap centrifuge tube (18 mℓ capacity) containing 0.2 to 0.4 mg phospholipase C in 4 mℓ 17.5 mM Tris buffer (pH 7.3), along with 1.3 mℓ 1% CaCl$_2$ and 1 mℓ diethyl ether; the mixture was incubated with shaking for 2 hr at 30°C. The reaction mixture was treated with 5 drops of 0.1 NHCl and extracted once by vigorous shaking with 10 mℓ chloroform-methanol (2:1) containing 150 to 250 μg tridecanoylglycerol as internal standard. The solvent phases were separated by centrifuging for 10 min at 200 g. The clear chloroform phase was removed from the bottom of the tube and was dried by passing through a Pasteur pipette containing 2 g anhydrous sodium sulfate. The solution was evaporated under nitrogen. Cholesterol was derivatized using Trisil-BSA (150 to 250 μℓ) and transferred to a sampling vial.

Total sterols in serum (method of Desager and Harvengt[14]) — In a PTFE-lined, screw-capped, 15 mℓ glass tube were mixed 1 mℓ 0.25 mM solution of epicoprostanol in absolute ethanol, 0.1 mℓ equeous potassium hydroxide (6.0 mM), and 50 μℓ serum. The tube was left for 1 hr at 60°C and, after cooling, 10 mℓ 0.025 mM solution of 5α-cholestane in hexane were added. After mixing, 2 mℓ bidistilled water were added, and the tube was shaken for 1 min on a vortex mixer. A 2 mℓ aliquot of the hexane phase was transferred in a 2 mℓ glass tube and evaporated under nitrogen at 60°C. Silylation: A 0.2 mℓ vol freshly prepared mixture of BSTFA and TMCS (4:1) was added to the dry residue. After 1 hr at 60°C, 0.8 mℓ hexane was added, and 1 μℓ cooled mixture was injected into a gas chromatograph with an all-glass solid injector.

Sterols in feces (method of Roseleur et al.[15]) — In clinical experiments, stools were collected, diluted with an equal amount of distilled water, and homogenized. An aliquot of this homogenate was evaporated to dryness on a boiling water bath by addition of several portions of absolute ethanol. Thereafter, the fecal sample was lyophilized to ensure complete dryness. Extraction of steroids was performed as follows: in a 30 mℓ stoppered tube, 1 mℓ 5α-cholestane solution (4000 mg) and 1 mg 5α-cholanic acid solution (1000 mg) were evaporated to dryness. Approximately 400 mg fecal powder and 5 mℓ glacial acetic acid were added. After heating for 1 hr at 120°C, the tube was cooled and the contents were diluted with 10 mℓ toluene. After shaking and centrifuging at 1600 × g for 5 min, the supernatant was decanted into a second tube.

From the acetic acid-toluene extract (containing sterols and their esters, other lipids and bile, and their conjugates), two 2-mℓ portions were taken and evaporated to dryness on a rotary evaporator in a water bath maintained at 60°C. The residue of each was saponified by adding 0.2 mℓ 10 M sodium hydroxide in water and 1.8 mℓ absolute ethanol and heating them at reflux for 1 hr. After cooling, 5 mℓ petroleum ether (60 to 80°C) and 1.4 mℓ water were added. Tubes were shaken and then centrifuged at 1000 × g for 3 min. This was repeated twice. The organic phases were combined and evaporated to dryness. The sterols were then converted to TMS-ethers.

REFERENCES

1. **Dutta, J., Biswas, A., Saha, S., and Deb, C.,** *J. Chromatogr.,* 124, 29, 1976.
2. **Cham, B. E., Hurwood, J. J., Knowles, B. R., and Powell, L. W.,** *Clin. Chem. Acta,* 49, 109, 1973.
3. **Duncan, I. W., Culbreath, P. H., and Burtis, C. A.,** *J. Chromatogr.,* 162, 281, 1979.
4. **Fenell, W. J., Ripola, D., and Batiskis, J. G.,** *Clin. Biochem.,* 10, 118, 1977.
5. **Hansbury, E. and Scallen, T. J.,** *J. Lipid Res.,* 19, 742, 1978.
6. **Trocha, P. J., Jasne, S. J., and Sprimson, D. B.,** *Biochemistry,* 16, 4721, 1977.
7. **Popov, S., Carlson, R. M. K., Wegmann, A., and Djcrassi, C.,** *Steroids,* 28, 699, 1976.
8. **Wun, C. K., Walter, R. W., and Litsky, W.,** *Water Res.,* 10, 955, 1976.
9. **Dugan, J. and Tan, L.,** *J. Chromatogr.,* 86, 107, 1973.
10. **Ishikawa, T. T., Brazier, J. B., Stewart, L. E., Fallot, R. W., and Gluek, C. J.,** *J. Lab. Chim. Med.,* 87, 345, 1977.
11. **Riuz, R. and Dea, P.,** *J. Chromatogr.,* 146, 321, 1978.
12. **Owen, V. M. J., Ho, F., Mazzuchin, A., Doran, T. A., Liedgren, S., and Porter, C.,** *J. Clin. Chem.,* 22, 224, 1976.
13. **Sheppard, A. J., Newkirk, D. R., Hubbard, W. D., and Osgood, T.,** *J. Assoc. Off. Anal. Chem.,* 60, 1302, 1977.
14. **Desager, J. P. and Harvengt, C. J.,** *High Res. Chromatogr. Chromatogr. Commun.,* 217, 1978.
15. **Roseleur, D. J. and Van Gent, C. M.,** *Clin. Chim. Acta,* 82, 13, 1978.

SECTION II. GAS CHROMATOGRAPHY

Sterols are of the most widely distributed of the steroids and are difficult to analyze. This is due to the close similarity of many of these compounds, their metabolites, and oxidation products. Studies on hyperlipoproteinemias in man have indicated the need for sterol balance determinations. GC has seen considerable success in the determination of this class of steroid.

Table GC 1

RETENTION DATA OF PLANT DESMETHYL STEROL ACETATES RELATIVE TO CHOLESTEROL ACETATE (RELATIVE TIME IN MIN)[1]

Sterol acetate	Abbreviation	OV-101	GE-F-50	Dexsil 300	OV-17	OV-210	SP-2401	OV-25	AN-600	SP-525	OV-225	PMPE	HI-EFF 8BP	CTpA	SP-1000
													Liquid phase		
Cholesterol	FC_{27}	1.00	1.00	1.00	1.00	1.00	1.00	1.00	1.00	1.00	1.00	1.00	1.00	1.00	1.00
Cholestanol	C_{27}	1.02	1.01	1.05	1.00	1.04	1.05	0.99	1.00	1.03	1.03	1.03	1.00	0.96	0.95
Desmosterol	$FC_{27}F$	1.10	1.11	1.12	1.22	1.12	1.24	1.25	1.18	1.28	1.28	1.32	1.27	1.35	1.37
25-Dehydrocholesterol	$FC_{27}F$			1.15			1.28								
Campesterol	FC_{28}	1.31	1.31	1.33	1.34	1.34	1.33	1.33	1.32	1.31	1.31	1.32	1.33	1.30	1.30
Ergostanol	C_{28}	1.33	1.33	1.39	1.38	1.38	1.35	1.31	1.32	1.34	1.34	1.35	1.33	1.24	1.23
Brassicasterol	$FC_{28}F$	1.12	1.11	1.11	1.14	1.09	1.10	1.14	1.08	1.11	1.12	1.08	1.10	1.13	1.12
Ergosterol	$2FC_{28}F$	1.09	1.10	1.10	1.21	1.21	1.44	1.28	1.17	1.19	1.27	1.36	1.35	1.34	1.48
Sitosterol	FC_{29}	1.65	1.65	1.71	1.65	1.64	1.64	1.65	1.65	1.60	1.59	1.57	1.63	1.57	1.60
Stigmasterol	$FC_{29}F$	1.42	1.42	1.41	1.45	1.37	1.38	1.44	1.36	1.36	1.41	1.35	1.37	1.36	1.39
Fucosterol	$FC_{29}F$								1.72	1.79	1.72	1.77	1.77	1.74	
Isofucosterol	$FC_{29}F$	1.66	1.71	1.76	1.84	1.65	1.59	1.75	1.75	1.86	1.80	1.84	1.87	1.86	1.92

Instrument: F & M 5750 with flame detector

Column and temperature (°C):

1. OV-101, 230	9. SP-525, 250
2. GE-F-50, 230	10. OV-225, 235
3. Dexsil-300, 240	11. PMPE, 230
4. OV-17, 240	12. HI-EFF-8BP, 230
5. OV-210, 200	13. CTpA, 230
6. SP-2401, 220	14. SP-1000, 240
7. OV-25, 230	15. Poly-I-110, 240
8. AN-600, 210	

All columns 6 ft × 4 mm U Tube, all packings coated on Gas Chrom Q as 1% loading

270°C

Detector:

Flow: Helium 80 mℓ/min

Table GC 2
RETENTION TIMES FOR
CHOLESTEROLS[2]

Steroid	Mobile phase		
	1	2	3
Cholesterol (cholest-5-en-3β-ol)	1.00	1.00	1.00
Cholesta-2,4,6-triene	0.41	0.41	0.47
Cholesta-3,5-dien-7-one	2.16	2.00	2.14
Cholest-5-ene-3β,7α-diol	2.27	2.16	2.18
Cholest-5-ene-3β,7β-diol	2.46	2.31	2.34
Cholesta-4,6-dien-3-one	3.46	3.31	3.24
β-Hydroxycholest-5-en-7-one	5.01	4.73	4.70

Instrument: Perkin-Elmer 402 with flame ionization detector
Column: 1. 3% OV-210
2. 3% SP-2401
3. 3% QF-1
Temperatures: Column 230°C, detector 250°C, injector 250°C
Flow: 20 mℓ/min

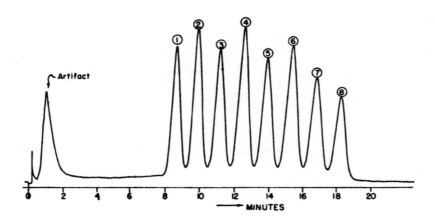

FIGURE GC 1. Separation of sterol esters. Instrument: Barber Colman 5000 with flame ionization detector; column: Polu-S-179 on Gas Chrom Q (100—120 mesh) (212 m × 4 mm); temperatures: detector 250°C, injector 250°C, program 2°C/min from 270—333°C; cholesterol ester: (1) undecanoate, (2) laurate, (3) tridecanoate, (4) myristate, (5) pentadecanoate, (6) palmitate, (7) heptadecanoate, (8) stearate. (From Schwartz, R. D. et al., *J. Chromatogr.*, 112, 111, 1975. With permission.)

Table GC 3
RETENTION TIMES OF
STEROLS AND α-TOCOPHEROL[4]

| Steroid | Stationary phase/time (min) | |
	1	2
Cholestane	3.3	3.3
Cholesterol	8.7	8.7
Cholestanol	8.7	8.7
Desmosterol	10.4	10.4
Campesterol	11.7	11.6
Stigmasterol	12.8	12.6
β-Sitosterol	14.3	14.2
α-Tocopherol	9.4	9.3

Instrument:	Hewlett-Packard 402 B with flame ionization detector
Columns:	1. 3% SP-2550 on Supelcoport (100—120 mesh) (1 m × 2 mm)
	2. 3% OV-17 on Gas Chrom Q (100—120 mesh) (1.8 m × 2 mm)
Temperatures:	Column 260°C, detector 300°C, injector 280°C

Table GC 4
RELATIVE RETENTION TIMES OF STANDARD STEROL TMS DERIVATIVES[5]

No.	Sterol TMS derivatives	Notation	SE-30 "ultra"	OV-1	Dexsil 300 GC	SP1000	STAP	HI-EFF-8BP	Silar 5CP
1	22-trans-240BNorcholesta-5,22-dien-3β-ol	C_{26}:5,22	0.65	0.65	0.63	0.67	0.67	0.63	0.64
2	22-trans-Cholesta-5,22-dien-3β-ol	C_{27}:5,22	0.90	0.90	0.89	0.95	0.95	0.92	0.93
3	Cholestan-3β-ol	C_{27}:0	1.03	1.01	1.04	0.98	0.96	0.99	0.97
4	Cholesterol	C_{27}:5	(15)a	(9)a	(19)a	(10)a	(10)a	(14)a	(16)a
5	Brassicasterol	C_{28}:5,22	1.10	1.11	1.10	1.12	1.11	1.08	1.12
6	Cholesta-5,7-dien-3β-ol	C_{27}:5,7	1.10	1.09	1.13	1.28	1.28	1.32	1.40
7	Desmosterol	C_{27}:5,24	1.10	1.08	1.10	1.28	1.30	1.28	1.34
8	Campesterol	C_{28}:5	1.31	1.30	1.29	1.29	1.29	1.32	1.32
9	Stigmasterol	C_{29}:5,22	1.40	1.39	1.36	1.36	1.35	1.32	1.43
10	24-Methylenecholesterol	C_{28}:5,24(28)	1.23	1.25	1.29	1.41	1.41	1.42	1.49
11	Ergosterol	C_{29}:5,7,22	1.22	1.21	1.36	1.42	1.43	1.42	1.57
12	β-Sitosterol	C_{29}:5	1.64	1.62	1.61	1.57	1.54	1.59	1.61
13	Fucosterol	C_{29}:5,24(28)E	1.60	1.59	1.62	1.73	1.71	1.72	1.82
14	28-Isofucosterol	C_{29}:5,24(28)Z	1.66	1.64	1.69	1.82	1.81	1.84	1.92
	GLC oven temperatures, °C		250	250	260	250	250	245	220

Instrument: Pye Model 104 with flame ionization detector
Column: As in table, 1% on Diatomite CQ (2.5 m × 4 mm)
Temperatures: Column 220—260°C, detector 300°C
Flow: 50 mℓ/min helium

Note: TMS = trimethyl silyl ether.

a Retention time in minutes for cholesteryl TMS ether.

Table GC 5
RETENTION TIMES FOR STEROLS[6]

Steroid	RT		
	1	**2**	**3**
5α-Cholestane	1.00 (6 min)	1.00 (7 min)	1.00 (5 min)
5-Cholestan-3α-ol	5.19	2.45	2.49
5α-Cholestan-3α-ol	5.50	2.44	2.48
5α-Cholestan-3β-ol	6.52	2.68	2.75
5-Cholesten-3β-ol	7.15	2.46	2.73
5.24-Cholestadien-3β-ol	7.89	2.92	3.19
5-Cholesten-24-methyl-38β-ol	9.11	3.17	3.48
5,22-Cholestadien-24-ethyl-3β-ol	9.70	3.33	3.80
5-Cholesten-24-ethyl-3β-ol	11.56	3.68	4.35

Instrument: Perkin-Elmer 57 or Hewlett Packard 402B with flame ionization detector
Column:
    1. 1% SP-1000 on Supelcoport (100—120 mesh)
    2. 3% SP-2401 on Supelcoport (100—120 mesh)
    3. OV-17 on Gas Chrom Q (100-120 mesh)
Temperatures:
    1. Column 220°C, detector 300°C, injector 290°C
    2. Column 220°C, detector 300°C, injector 290°C
    3. Column 260°C, detector 300°C, injector 290°C
Flow: Nitrogen, 30 mℓ/min

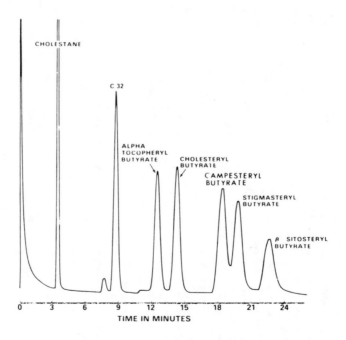

FIGURE GC 2.   Separation of sterols. Instrument: Barber Colman Model 5000 with flame ionization detector; column: 1% SE-30 on Gas Chrom Q (100—120 mesh) (2 m × 4 mm); temperatures: column 250—265°C, detector 300—315°C, injector 300—315°C; flow: 30 mℓ/min nitrogen. (From Sheppard, A. J. et al., *Assoc. Off. Anal. Chem.*, 60, 1302, 1977. With permission.)

Table GC 6
RETENTION TIMES FOR STEROL
TRIMETHYLSILYL ETHERS[8]

Steroid	Retention time (min)
5α-Cholestane (internal standard)	10.0
5β-Cholestan-3α-ol	16.5
5α-Cholestan-3β-ol	20.4
5α-Cholestan-3α-ol	16.2
Cholesterol	19.8
5β-Cholestan-3-one	17.3
5α-Cholestan-3-one	18.9
4-Cholesten-3-one	21.6

Instrument:	Shimadzu-LKB 9000 with multiple ion detector
Column:	1.5% SE-30 on Chromosorb W (60—80 mesh; 2 m × 3 mm)
Temperatures:	Column 255°C, injector 290°C, separator 270°C
Flow:	Helium, 30 mℓ/min

Table GC 7
SEPARATION OF STEROL ESTERS[9]

Acid ester	Sterol as ester			
	Cholesterol	β-Sitosterol	Campesterol	Stigmasterol
12:0	0.44	0.58	0.53	0.51
14:0	0.57	0.78	0.70	0.68
16:0	0.75	1.01	0.94	0.88
18:0	1.00	1.32	1.20	1.14
20:0	1.30	—	—	—
22:0	—	2.33	2.09	—
16:1	0.89	—	—	1.04
18:1	1.16	1.55	1.40	1.35
20:1	1.53	2.03	1.83	1.79
22:1	1.96	2.64	2.39	2.31
18:2	1.40	1.87	1.68	1.61
18:3	1.71	2.29	2.07	1.97
20:4	2.10	—	—	—
20:5	2.52	3.43	3.10	3.03
22:6	3.64	5.00	4.59	4.52

Instrument:	Shimadzu GC 6 AM with flame ionization detector
Column:	3% Silar 10°C on Gas Chrom Q (100—120 mesh) (1 m × 3 mm)
Temperatures:	Column 260°C, detector and injector 320°C
Flow:	Nitrogen, 40 mℓ/min

Table GC 8
RETENTION TIMES OF BORONATE DERIVATIVES OF BRASSINOLIDE AND ITS ANALOGUES[10]

	RRT	
Compound	**1**	**2**
Bismethaneboronate		
Brassinolide	24.5	24.67
24-Epibrassinolide	27.0	26.50
Homobrassinolide	29.0	28.08
(22R,23R)-Norbrassinolide	20.5	20.90
(22R,23S)-Norbrassinolide	24.0	23.84
(22S,23R)-Norbrassinolide	22.5	22.48
(22S,23S)-Norbrassinolide	20.5	20.90
Methaneboromate trimethylsilyl ether		
(22R,23R)-Dihydroxycholesterol		14.41
(22R,23S)-Dihydroxycholesterol		16.46
(22S,23R)-Dihydroxycholesterol		15.17
(22S,23S)-Dihydroxycholesterol		14.41
(20R,22R)-Dihydroxycholesterol		15.94
(20R,22S)-Dihydroxycholesterol		15.49
(20S,22R)-Dihydroxycholesterol		15.07
(20S,22S)-Dihydroxycholesterol		15.94

Instrument: Shimadzu GC-7A with flame ionization detector
Column: 1. Packed, 2% OV-17 on Chromosorb W (80—100 mesh) (150 cm × 4 mm)
                2. OV-17 capillary column (40 m × 0.25 mm)
Temperature: 1. Column 290°C
                2. Column 270°C
Flow: Nitrogen
                1. 30 mℓ/min
                2. 0.4 mℓ/min

Table GC 9
SEPARATION OF STEROLS AS TRIMETHYL SILYL ETHERS OR OXIMES (KETONES AS OXIMES AND HYDROXYLS AS SILYL ETHER)[11]

Steroid	1	2	3
Coprostanone	2.00	1.72	5.38
Methyl coprostanone	2.59	—	7.22
Ethyl coprostanone	2.84	—	7.63
Ethyl coprostanone	3.27	2.81	9.00
Cholestanone	2.28	—	—
Δ^7-Cholestanone	2.57	—	—
Campestanone	2.95	—	—
Stigmastenone	3.29	—	—
β-Sitostanone	3.78	—	—
Epicoprostanol	1.82	1.69	—
Dihydrolanosterol	3.19	2.81	2.57
Δ^7-Lanosterol	3.47	—	3.08
Lanosterol	3.54	3.06	3.32

Table GC 9 (continued)
SEPARATION OF STEROLS AS
TRIMETHYL SILYL ETHERS
OR OXIMES (KETONES AS
OXIMES AND HYDROXYLS AS
SILYL ETHER)[11]

Steroid	1	2	3
Coprostanol	1.75	1.65	1.38
Methylcoprostanol	2.27	2.06	1.85
Ethylcoprostanol	2.48	2.25	1.95
Ethylcoprostanol	2.88	2.55	2.31
Δ^7-Coprostanol	1.80	1.71	1.41
Methostenol	3.16	2.79	3.16
Δ^8-Methostenol	2.95	2.56	2.66
$\Delta^{7,24}$-Methostenol	3.25	—	—
$\Delta^{8,24}$-Methostenol	3.25	—	—
Cholestanol	2.32	2.07	—
Campestanol	3.03	2.60	—
Stigmastenol	3.27	2.79	—
β-Sitostanol	3.81	3.23	—
Δ^7-Cholestenol	2.60	2.29	2.62
Cholesterol	2.25	2.03	2.17
Campesterol	2.94	2.55	2.90
Stigmasterol	3.18	2.74	3.05
β-Sitosterol	3.70	3.17	3.60
Desmosterol	2.52	2.18	—
$\Delta^{5,7}$-cholesterol	2.58	—	—

Instrument:	Packard 873 with flame ionization detector
Columns:	1. 1% DC 560
	2. 3% SE 30
	3. O.83% HiEFF 8 BC all on Gas Chrom P (100—120 mesh) (all 360 × 4 mm)
Temperatures:	1. 245°C
	2. 265°C
	3. 224°C
	Flash 50°C, detector 40°C with above column temperatures

Table GC 10
SEPARATION OF STEROLS AS ACETATES
(RELATIVE TO CHOLESTEROL ACETATE)[12]

Steryl acetate	Column			
	1	2	3	4
5β-Cholest-7-en-3β-ol	0.86	0.88	0.76	0.77
5β-Cholestanol	0.86	0.90	0.76	0.74
Z-22-Dehydrocholesterol	0.87	0.88	0.88	0.87
5α-Cholestan-3α-ol	0.91	0.93	0.80	0.82
E-22-Dehydrocholesterol	0.91	0.89	0.91	0.92
5,22,24-Cholesta-trien-3β-ol	0.94	0.91	1.35	1.37
5,7,22-Cholesta-trien-3β-ol	0.99	1.00	1.21	1.14

Table GC 10 (continued)
SEPARATION OF STEROLS AS ACETATES
(RELATIVE TO CHOLESTEROL ACETATE)[12]

	Column			
Steryl acetate	**1**	**2**	**3**	**4**
Cholesterol acetate	1.00	1.00	1.00	1.00
5α-Cholestanol	1.03	1.05	1.00	1.02
14α-Methyl-5α-cholest-8(14)-en-3β-ol	1.04	1.08	0.94	0.94
5,25-Cholesta-dien-3β-ol	1.07	1.12	1.31	1.27
Desmosterol	1.09	1.12	1.29	1.29
5,7-Cholesta-dien-3β-ol	1.09	1.12	1.29	1.09
Brassicasterol	1.12	1.09	1.10	1.09
5α-Cholest-7-en-3β-ol	1.12	1.11	1.21	1.26
Zymosterol	1.13	1.09	1.30	1.39
Pollinastanol	1.16	1.23	1.16	1.17
14α-Methyl-5α-cholest-7-en-3β-ol	1.17	1.21	1.17	1.16
Ergosterol	1.22	1.22	1.44	1.33
24-Methylenecholesterol	1.26	1.28	1.43	1.39
Lophenol	1.27	1.25	1.32	1.32
14α-Methyl-5α-cholesta-7,22-dien-3β-ol	1.29	1.30	1.26	1.25
Campesterol	1.30	1.29	1.32	1.29
Campestanol	1.34	1.35	1.30	1.32
5,7-Ergosta-dien-3β-ol	1.42	1.45	1.69	1.60
Stigmasterol	1.42	1.32	1.34	1.32
24-Methylenepollinastanol	1.46	1.58	1.68	1.63
7-Ergost-en-3β-ol	1.46	1.42	1.59	1.63
24-Methylpollinastanol	1.51	1.61	1.53	1.52
14α-Methyl-5α-ergost-8-en-3β-ol	1.52	1.53	1.52	1.49
Lanosterol	1.54	1.62	1.47	1.41
Spinasterol	1.58	1.46	1.61	1.65
Cycloartanol	1.61	1.73	1.45	1.40
Sitosterol	1.63	1.56	1.60	1.54
Fucosterol	1.63	1.50	1.76	1.68
Stigmasterol	1.67	1.62	1.58	1.57
28-Isofucosterol	1.69	1.55	1.85	1.79
Cycloartenol	1.75	1.85	1.88	1.81
5α-Stigmast-7-en-3β-ol	1.83	1.71	1.93	1.93
24-Methylcycloartanol	2.06	2.24	1.89	1.79

Instrument:	Glowall; Chromalab 810 with argon ionizing detector
Column:	1. 3% SE-30 on Gas Chrom Q, 244°C, 20 psi, argon flow rate 150 mℓ/min
	2. 1% QF-1 on Gas Chrom P, 231°C, 20 psi, argon flow rate 50 mℓ/min
	3. 3% Hi-Eff 8BP on Gas Chrom Q, 238°C, 25 psi, argon flow rate 95 mℓ/min
	4. 2% PMPE on Gas Chrom Q, 250°C, 20 psi, argon flow rate 95 mℓ/min
Temperatures (°C):	1. 244
	2. 231
	3. 238
	4. 250
Flow: argon	1. 50 mℓ/min (20 psi)
	2. 50 mℓ/min (20 psi)
	3. 95 mℓ/min (20 psi)
	4. 95 mℓ/min (20 psi)

REFERENCES (GC)

1. **Nordby, H. E. and Nagy, S.,** *J. Chromatogr.,* 75, 187, 1973.
2. **Teng, J. I., Kulig, M. J., and Smith, L. L.,** *J. Chromatogr.,* 75, 108, 1973.
3. **Schwartz, R. D., Mathews, R. G., Ramachandran, S., Henly, R. S., and Doyle, J. E.,** *J. Chromatogr.,* 112, 111, 1975.
4. **Ishikawa, T. T., MacGee, J., Morrison, J. A., and Glueck, C. J.,** *J. Lipid Res.,* 15, 286, 1974.
5. **Ballantine, J. A., Roberts, H. C., and Morris, R. J.,** *J. Chromatogr.,* 103, 289, 1975.
6. **Ishikawa, T. T., Brazier, J. B., Stewart, L. E., Fallat, R. W., and Glueck, C. J.,** *J. Lab. Clin. Med.,* 87, 1345, 1976.
7. **Sheppard, A. J., Newkirk, D. R., Hubbard, W. D., and Osgood, T.,** *J. Assoc. Off. Anal. Chem.,* 60, 1302, 1977.
8. **Tohma, M., Nakaty, Y., and Kurosawa, T.,** *J. Chromatogr.,* 171, 469, 1979.
9. **Takagi, T., Sakai, A., Hayachi, A., and Itabashi, Y.,** *J. Chromatogr. Sci.,* 17, 212, 1979.
10. **Takastutoso, S., Ying, B., Marisaki, M., and Ikekawa, N.,** *J. Chromatogr.,* 239, 233, 1982.
11. **McNamara, D. J., Proia, A., and Miettinen, T. A.,** *J. Lipid Res.,* 22, 474, 1981.
12. **Thompson, M. J., Patterson, G. W., Dutky, S. R., Svoboda, J. A., and Kaplanis, J. N.,** *Lipids,* 15, 719, 1980.

SECTION III. HIGH PERFORMANCE LIQUID CHROMATOGRAPHY

The use of HPLC in separation of sterols was relatively unsuccessful until recently when better columns became available.

Table HPLC 1
RETENTION TIMES FOR STEROLS[1]

Sterols	Retention time
Ergosta-5,7,22-trienol (ergosterol)	1.90
Squalane	2.19
Ergosta-7,22-dienol	2.20
Ergosta-7,24(28)-dienol (episterol)	2.02
Ergost-7-enol	2.35
Ergosta-8,22-dienol	2.19
Ergost-8-enol	2.32
Ergosta-8,24(28)-dienol (fecosterol)	2.00
Cholestanol	
Cholesterol	2.20
14-Methylergosta-8,24(28)-dienol (14-methylfecosterol)	1.94
Ergosta-8,22,24(28)-trienol	1.83
Ergosta-7,22,24(28)-trienol	1.83
4,14-Dimethylergosta-8,24(28)-dienol (obtusifoliol)	1.96
4,14-Dimethylcholesta-8,24-dienol (4,14-dimethylzymosterol)	2.07
4,4-Dimethylcholesta-8,24-dienol	2.09
Lanosterol	2.12
24,25-Dihydrolanosterol	2.45
Lanosta-7,9,24-trienol	1.94
Lanosta-7,9-dienol	2.21

Instrument: Waters Model ALC/202 with UV detector
Column: μBondapak C_{18}
Mobile phase: Tetrahydrofuran-acetonitrile-water (5:5:2)
Flow rate: 0.5 mℓ/min
Detection: 254 nm

Table HPLC 2
SEPARATION OF STEROLS BY REVERSED-PHASE HPLC[2]

Sterols with one double bond	Retention time (min)	Sterols with two double bonds	Retention time (min)
Cholesterol	7.48	Desmosterol	5.86
4,4-Dimethyl-Δ^8-cholestenol	7.60	4,4-Dimethyl-$\Delta^{8,24}$-cholestadienol	5.96
Δ^7-Cholestenol	7.68	$\Delta^{3,7}$-Cholestadienol	5.94
Dihydrolanosterol	7.86	Lanosterol	6.12

Instrument: Waters Model ALC/GPC
Column: μBondapak C_{18} (30 cm × 3.9 mm)
Mobile phase: Acetonitrile
Flow: 2.5 mℓ/min
Detection: Refractive index

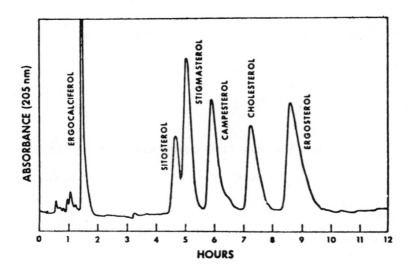

FIGURE HPLC 1.   Separation of C_{27}, C_{28}, and C_{29} sterols. Instrument: modular, Tracor Model 990 pump, variable wavelength Model 970 detector; column: μBondapak C_{18} (16 ft × 1.8 in.) with Porasil B (37—75 μm) guard column; mobile phase: hexane-isopropanol (99.5:0.5); flow: 0.4 mℓ/min; detection: 205 nm. (From Hunter, K. R. et al., *J. Chromatogr.*, 153, 57, 1978. With permission.)

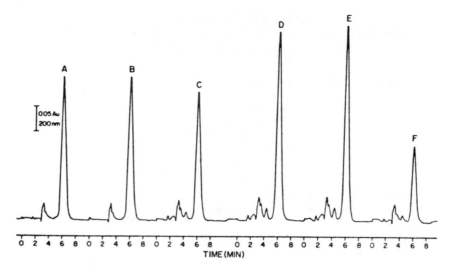

FIGURE HPLC 2.   Separation of total serum cholesterol. Instrument: Varian Model 4200 with Vari-chrom detector; column: μBondapak C$_{18}$ (10μm) (30 cm × 4 mm), Porasil B (37—75 μm) precolumn (5 cm × 2 mm); mobile phase: isopropanol-acetonitrile (50:50); flow: 1 mℓ/min (500 psi); detection: 200 nm; chromatograms: (A) (B) represent 3.0 g/ℓ (7.76 mmol/ℓ) cholesterol standards, (C, D, E, F) represent samples of serum that contain 2.65, 3.91, 4.00, and 1.53 g/ℓ (6.8, 10.12, 10.35, and 3.96 mmoll/ℓ) of total cholesterol, respectively. (From Duncan, I. W. et al., *J. Chromatogr.*, 162, 281, 1979. With permission.)

Table HPLC 3
SEPARATION OF CHOLESTEROL DERIVATIVES AND THEIR OXIDATION PRODUCTS[5]

Sterol	RRT[a] 1	2	3	4	5
Cholesterol	1.00	1.00	1.00	—	—
Cholesta-3,5-dien-7-one	0.13	0.06	0.77	—	—
Cholesta-4,6-dien-3-one	0.44	—	0.75	—	—
Cholest-5-en-3-one	0.12	0.11	1.02	—	—
Cholest-4-en-3-one	0.44	0.39	0.91	—	—
Cholest-4-ene-3,6-dione	0.54	0.47	0.34	—	—
Cholest-5-ene-3β,25-diol	2.01	—	—	—	—
3β-Hydroxycholest-5-en-7-one	5.93	2.80	0.28	1.21	1.21
3β-Hydroxy-5α-cholest-6-ene-5-hydroperoxide	5.35	—	0.24	1.12	1.13
3β-Hydroxycholest-5-ene-7α-hydroperoxide	7.82	3.05	0.21	1.00	1.00
3β-Hydroxycholest-5-ene-7β-hydroperoxide	7.28	2.86	0.23	1.04	1.04
5α-Cholest-6-ene-3β,5-diol	8.51	3.21	0.30	—	—
Cholest-5-ene-3β,7α-diol	19.4	4.27	0.24	—	—
Cholest-5-ene-3β,7β-diol	16.3	3.95	0.25	—	—

Instrument: Waters 6000 system, Perkin-Elmer LC 55 detector
Columns: 1. μPorasil
               2. μPorasil
               3. μBondapak C$_{18}$
               4. μBondapak C$_{18}$
               5. μBondapak C$_{18}$

Table HPLC 3 (continued)
SEPARATION OF CHOLESTEROL DERIVATIVES AND
THEIR OXIDATION PRODUCTS[5]

Mobile Phase: 1. Hexane-isopropanol (24:1) 2 mℓ/min
2. 90% hexane-isopropanol (99:1), 10% hexane-isopropanol (9:1) to
   80% hexane-isopropanol (99:1) 20% hexane-isopropanol (9:1) over
   30 min at 2 mℓ/min
3. Acetonitrile-water (9:1) 1 mℓ/min
4. Acetonitrile-water (17:3) 1 mℓ/min
5. Acetonitrile-water (4:1) 1 mℓ/min

Detection: 212 nm, 230 nm for epoxides, 284 nm for dienes (416)

[a] Retention data relative to cholesterol as unity for systems I-III, relative to the 7α-hydroperoxide[6] as unity for systems 4 and 5.

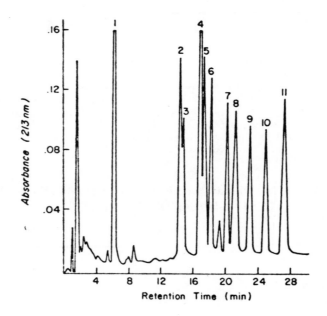

FIGURE HPLC 3.   Separation of cholesterol and esters. Instrument: Dupont 850 with 837 detector; column: Zorbax ODS (15 cm × 4.6 mm) plus Zorbax ODS (25 cm × 4.6 mm) with a Pell ODS guard column (5 cm × 2.1 mm); mobile phase: acetonitrile-tetrahydrofuran (65:35) with water reduced from 3% to 0% over a period of 20 min; temperature: 57°C; flow: 2 mℓ/min; cholesterol ester: (1) unesterified cholesterol, (2) linolenate, (3) arachidonate, (4) linoleate, (5) palmitoleate, (6) myristate, (7) oleate, (8) palmitate, (9) heptadecanoate, (10) stearate, (11) erucate. (From Carroll, R. W. and Rudel, L. L., *J. Lipid Res.*, 22, 359, 1981. With permission.)

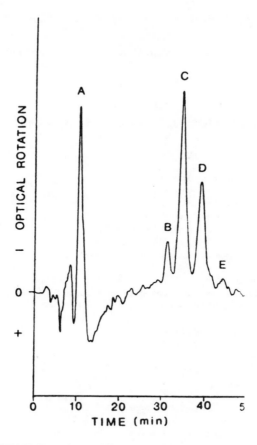

FIGURE HPLC 4.   Separation of cholesterols in serum. Instruments: Milton Roy 196 pump, Rheodyne 1070 injector, and laser detection; column: Alltech 10 μm C_{18} reversed-phase (25 cm × 4.6 mm); mobile phase: tetrahydrofuran-water (76:24); flow: 0.5 mℓ/min; detection: 514 nm. (A) cholesterol and cholestanol, (B) cholesteryl/linolenate and arachidonate, (C) cholesteryl palmitoleate and linoleate, (D) cholesteryl/palmitate and oleate, (E) cholesteryl/stearate. (From Kuo, J. C. and Yeung, E. S., *J. Chromatogr.*, 229, 293, 1982. With permission.)

REFERENCES (HPLC)

1. **Tocha, P. O., Janse, S. J., and Srinson, D. B.,** *Biochemistry,* 16, 4721, 1977.
2. **Hansbury, E. and Scallen, T.,** *J. Lipid Res.,* 19, 742, 1978.
3. **Hunter, K. R., Walden, M. K., and Heftmann, E.,** *J. Chromatogr.,* 153, 57, 1978.
4. **Duncan, I. W., Culbreth, P. H., and Burtis, C. A.,** *J. Chromatogr.,* 62, 281, 1979.
5. **Ansari, G. A. S. and Smith, L. L.,** *J. Chromatogr.,* 175, 307, 1979.
6. **Carroll, R. M. and Rudel, L. L.,** *J. Lipid Res.,* 22, 359, 1981.
7. **Kuo, J. C. and Yeung, E. S.,** *J. Chromatogr.,* 229, 293, 1982.

SECTION IV. THIN LAYER CHROMATOGRAPHY

TLC of cholesterol has been widely used. Detection at the low nanogram level is possible with high performance layers.

Table TLC 1
SEPARATION OF STEROLS AS ACETATES[1]

Sterol acetate		Relative R_f
5α-Cholestan-3β-01 (cholestanol)		1.20
5β-Cholestan-3β-01 (coprostanol)		1.20
5β-Cholestan-7β-01		1.18
5β-Cholestan-3α-01 (epicoprostanol)		1.15
(24R)-24-Ethylcholest-5-en-3β-01 (β-sitosterol)		1.09
(24R)-24-Methylcholest-5-en-3β-01 (campesterol)		1.02
(24S)-24-Ethylcholestra-5,22-dien-3β-01 (stigmasterol)		1.02
Cholesterol	(R_f 0.65)	1.00
Cholest-4-en-3β-01		1.00
(24R)-24-Methylcholest-5,22-dien-3β-01 (brassicasterol)		0.82
Cholesta-5,22-dien-3β-01 (*cis* and *trans*)		0.76
22-*trans*-24-Norcholesta-5,22-dien-3β-01 (C_{26} sterol)		0.69
Cholesta-5,24-dien-3β-01 (demosterol)		0.67
24-Ethylidenecholest-5-en-3β-01 (fucosterol)		0.61
24-Methylenecholest-5-en-3β-01		0.44
Cholesta-5,25-dien-3β-01		0.33
Cholesta-5,7-dien-3β-01		0.32

Layer: Silica Gel HF_{254} impregnated with silver nitrate (50 g gel and 10 g silver nitrate in 110 mℓ water for slurry)

Mobile phase: Hexane-benzene (50:20) 2 developments

Table TLC 2
SEPARATION OF STEROL ACETATES[3]

	R_f	
	1	2
4,4-Dimethylcholest-7-enyl acetate	0.30	0.20
4,4-Dimethylcholest-8-enyl acetate	0.30	0.21
4,4-Dimethylcholest-8(14)-enyl acetate	0.30	0.27
24,25-Dihydrolanosteryl acetate	0.30	0.34
Lanost-7-enyl acetate	0.30	0.29
Lanosteryl acetate	0.23	0.08
Lanosta-7,24-dienyl acetate	0.23	0.06
4,4-Dimethylcholesta-8,14-dienyl acetate	0.14	

Layer: Silica Gel H or alumina G (slurries below)
                1. 4 g silver nitrate in 80 mℓ water was mixed with 40 g Silica Gel H
                2. 14 g silver nitrate and 40 g alumina G in 57 mℓ water

Mobile phase: 1. Benzene-hexane (50:50)
                2. Benzene-hexane (15:85) at 5°C

Table TLC 3
R_f VALUES FOR STEROLS[3]

Compound	Mobile phases					
	1	2	3	4	5	6
Geraniol	0.14	0.24	0.48	0.63	0.72	0.60
Nerol	0.15	—	0.56	0.65	0.71	—
Linalool	—	—	0.81	0.46	0.71	—
Myrcene	—	—	1.00	0.64	0.42	—
α-Terpineol	—	—	0.64	0.65	0.76	—
Terpinolene	—	—	1.00	0.65	0.78	—
Eugenol	0.15	—	0.78	0.72	0.77	0.73
trans,trans-Farnesol	—	0.26	0.51	0.46	0.60	0.52
cis,trans-Farnesol	—	0.33	0.60	0.46	0.60	0.52
Pristane	1.00	1.00	1.00	—	—	—
Phytol	—	—	0.72	—	0.46	0.27
Cholesterol	0.11	0.22	0.59	0	0.45	0.22
5α-Cholestanol	0.11	—	0.57	0	0.40	0.17
Lanosterol	0.20	0.35	0.79	0	0.45	0.22
24,25-Dihydrolanosterol	0.20	—	0.79	0	0.40	0.17
Stigmasterol	0.10	—	0.51	0	0.39	—
β-Sitosterol	0.10	—	0.51	0	0.39	0.20
Cycloartenol	0.21	—	0.82	0	0.39	0.19
24-Methylenecycloartanol	0.21	—	0.82	0	0.39	0.20
Cholesteryl acetate	0.74	—	0.98	0	—	—
Lanosteryl acetate	0.77	—	0.99	0	—	—
Squalene	0.98	0.99	1.00	0	0.21	0.07
Squalene	1.00	1.00	1.00	0	—	—
2,3-Oxidosqualene	0.71	0.92	—	0	0.32	—

Layers: Silica Gel G: I, II, and III; silanized silica gel: IV, V, and VI (reversed-phase)

Mobile phase:
1. Hexane-ethyl acetate (95:5)
2. Hexane-ethyl acetate (90:10)
3. Hexane-ethyl acetate (80:20)
4. Methanol-water (70:30)
5. *p*-Dioxane-water (75:25)
6. *p*-Dioxane-acetic acid-water (50:30:20)

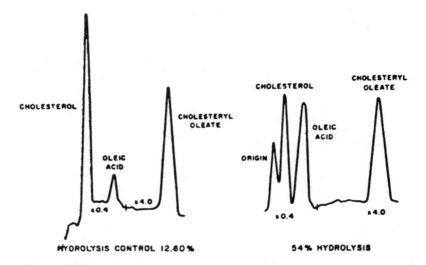

FIGURE TLC 1. Separation of cholesterol and acetates from rat aorta. Layer: Whatman HP; mobile phase: hexane-diethyl ether-acetic acid (83:16:1); detection: spray with 3% copper acetate in 8% phosphoric acid, heat at 130°C for 30 min. Nanogram quantities detected. (From Touchstone, J. C. and Dobbins, M. F., *Clin. Chem.*, 24, 1496, 1978. With permission.)

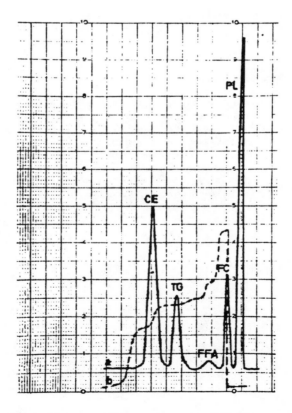

FIGURE TLC 2. Separation of sterols on Chromarods. Layer: Silica Gel G on chromarods; mobile phase: 1,2-dichloroethane-chloroform-acetic acid (92:8:0.1); detection: Flame ionization detector in Iatroscan Instrument. (PC) Phosphatidyl choline, (1,2-D) diglyceride, (1,3) diglyceride, (C) cholesterol, (FFA) free fatty acid, (TG) triglyceride, (CE) cholesterol esters. (From Van Damme, D. et al., *Clin. Chim. Acta*, 89, 231, 1978. With permission.)

Table TLC 4
SEPARATION OF STEROLS[6]

Compound	R_f Values						
	1	2	3	4	5	6	7
Cholesterol	57	66	28	5	8	27	21
Δ^2-Cholestenol	79	95	94	73	73	72	91
3-Ethyl-1^2-cholestenol	80	96	95	73	75	74	91
Ergosterol	63	72	35	6	9	30	23
$2\alpha,3\beta$-Oxidocholestane	82	98	92	66	67	74	87
$2\beta,3\beta$-Oxidocholestane	85	98	94	71	72	76	88
Cholestanol	69	70	28	7	12	37	27
5α-Hydroxycholestanol	29	24	5	0	1	3	4
6β-Hydroxycholestanol	70	14	0	0	0	1	6
β-Sitosterol	75	73	29	9	13	38	20
Stigmasterol	70	77	33	10	14	37	27
Δ^7-Stigmasterol	89	79	36	9	12	36	26
$\Delta^8(14)$-Stigmastenol	88	76	34	8	10	34	25
Δ^{22}-Stigmastenol	70	77	35	11	14	34	30
Stigmastanol	69	78	36	9	14	35	29
Lanosterol	76	89	58	22	27	53	54

Layer: Alumina sintered glass
Mobile phase: 1. Benzene-acetone (4:1)
                  2. Chloroform-ethanol (50:1)
                  3. Chloroform
                  4. Hexane-diethyl ether (5:1)
                  5. Hexane-diethyl ether (3:1)
                  6. Hexane-diethyl ether (1:1)
                  7. Hexane-ethyl acetate-acetic acid (180:30:1)
Detection: Spray with concentrated sulfuric acid and heat at 130°C

Table TLC 5
SEPARATION OF CHOLESTEROL ESTERS[7]

No.	Substance	Mobile phase and R_f			
		1	2	3	4
1	Cholesterol (chol.)	0.03	0.032	start	0.035
2	Chol. formate	0.625	0.447	0.215	0.675
3	Chol. acetate	0.57	0.345	0.10	0.41
4	Chol. propionate	0.68	0.445	0.16	0.53
5	Chol. butyrate	0.64	0.495	0.23	0.71
6	Chol. valerate	0.647	0.662	0.245	0.918
7	Chol. caproate	0.645	0.54	0.252	0.809
8	Chol. caprylate	0.80	0.59	0.262	0.775
9	Chol. caprinate	0.83	0.637	0.28	0.81
10	Chol. undecanoate	0.737	0.657	0.325	0.90
11	Chol. laurate	0.717	0.647	0.35	0.905
12	Chol. tridecanoate	0.767	0.535	0.335	0.75
13	Chol. myristate	0.747	0.675	0.33	0.92
14	Chol. palmitate	0.86	0.717	0.38	0.87
15	Chol. stearate	0.84	0.712	0.395	0.85
16	Chol. undecylen-10-ate	0.712	0.587	0.252	0.862

Table TLC 5 (continued)
SEPARATION OF CHOLESTEROL ESTERS[7]

		Mobile phase and R_f			
No.	Substance	1	2	3	4
17	Chol. oleate	0.84	0.677	0.335	0.845
18	Chol. benzoate	0.657	0.537	0.27	0.83
19	Chol. cinnamate	0.50	0.435	0.155	0.707

Layer: Silica Gel G
Mobile phase: 1. Cyclohexane-methanol (99:1)
                      2. Cyclohexane-chloroform (6:4)
                      3. Cyclohexane-chloroform (8:2)
                      4. Benzene-petroleum ether (1:1)
Detection: Heat after spray with orthophosphoric acid-water (1:1)

REFERENCES (TLC)

1. **Idler, D. R. and Safe, L. M.,** *Steroids,* 19, 315, 1972.
2. **Gibbons, G. F., Mitropoulis, K. A., and Ramanand, K.,** *J. Lipid Res.,* 14, 589, 1973.
3. **Vidrine, D. W. and Nicholas, H. J.,** *J. Chromatogr.,* 89, 92, 1974.
4. **Touchstone, J. C. and Dobbins, M. F.,** *Clin. Chem.,* 24, 1496, 1978.
5. **Van Damme, D., Vankerckhoven, G., Vercaemst, R., Soctewey, F. L., Blaton, V., Peeters, H., and Rosseneu, M.,** *Clin. Chim. Acta,* 89, 231, 1978.
6. **Okamura, T. and Kadono, T.,** *J. Chromatogr.,* 86, 57, 1973.
7. **Szoni, G. D. and Matkovics, B.,** *Biochem. J.,* 20, 269, 1975.

Chapter 13

VITAMIN D

Synthetic ergocalciferol or cholecalciferol, ergosterol or 7-dehydrocholesterol, respectively, is irradiated at low temperatures, and the resultant pre-vitamin is isolated and thermally converted to the vitamin D during their preparation. This irradiation process results in a mixture of pre-vitamin, lumisterol, and tachysterol. The structures for the various forms are shown in Figure 1.

The importance of determination of the D vitamins in pharmaceutical preparations, enriched foods, and biological materials dictates the need for methods to separate the various analogues. Any method for this analysis must be capable of distinguishing the vitamin from the various isomers as well as the pre-vitamin.

SECTION I: PREPARATIONS

Vitamin D: Extraction Methods for Tablets

Extraction of vitamin D_2, E-acetate, and A-acetate (method of Barnett and Frick[1]) — An accurately weighed sample, estimated to contain 10,000 to 20,000 units of vitamin A, 500-10,000 units of vitamin D_2, and 35 to 65 units of vitamin E acetate, was transferred to a 250-mℓ separatory funnel containing 25 mℓ water, 23 mℓ absolute ethanol, and 2 mℓ pyridine. The funnel was shaken vigorously for 2 min and placed in an oven at 50°C for 30 min. The sample was removed from the oven, shaken vigorously for 2 min, and extracted 3 times by gentle agitation with 50 mℓ portions of warm hexane (45 to 50°C). The hexane extracts were collected and water-washed 5 times with 50 mℓ water, filtered through anhydrous sodium sulfate, and evaporated to dryness under reduced pressure with the aid of a warm (40°C) water bath. A solution of methanol-tetrahydrofuran (62.5:37.5) was used to transfer the residue to a 10-mℓ volumetric flask containing 2.0 mℓ cholesterol benzoate internal standard solution (5 mg/mℓ) dissolved in tetrahydrofuran.

Fat soluble vitamins in tablets (method of Dolan et al.[2]) — 1 tablet equivalent of the granular pool or liquid standard and 5 mℓ hexane were added to 25 mℓ 65% ethanol and heated for 5 min on a steam bath in a sealed container with intermittent shaking. After cooling to room temperature, 25 mℓ 5% NaCl were added, and the phases were allowed to separate after further gentle shaking. An aliquot (10 or 25 $\mu\ell$) of the hexane layer was injected onto an HPLC column.

Extraction of vitamin D from tablets in presence of fat soluble vitamins (method of Osadca and Araujo[3]) — Vitamin D (D_2 or D_3) or vitamin AD beadlets and powders in gelatin carbohydrate base — dissolve the sample (50,000 IU vitamin D) in 25 mℓ warm 0.3% aqueous ammonia. Pipette 4 mℓ into a 50 mℓ glass-stoppered centrifuge tube, add 4.0 mℓ ethanol, swirl, and add 20 mℓ hexane. Shake mechanically for 5 min. Centrifuge for 5 min at 2000 rpm. Pipette 1.0 mℓ clear hexane extract into a 25 mℓ conical, glass stoppered centrifuge tube. Evaporate the hexane under N_2 in a water bath (55°C). Dissolve the residue in 4 mℓ methanol and inject 20 $\mu\ell$ into liquid chromatograph.

Vitamin AD or ADE, injectable (with antioxidants in emulsifiable solution) or liquid dispersible concentrates — into a 50 mℓ glass-stoppered centrifuge tube place the sample (8000 IU vitamin D) and 3 mℓ ethanol. Swirl and add 3 mℓ water, then add 20 mℓ hexane. Stopper and shake mechanically 5 min and continue as described above (spin, evaporate, dissolve).

Multivitamin chewable or decavitamin (USP-style) tablets — transfer 2 ground tablets to a 50 mℓ glass-stoppered centrifuge tube. Rinse the mortar with 5 mℓ warm 0.3% aqueous ammonia and add the rinsings to the centrifuge tube. Disperse by swirling in a hot water

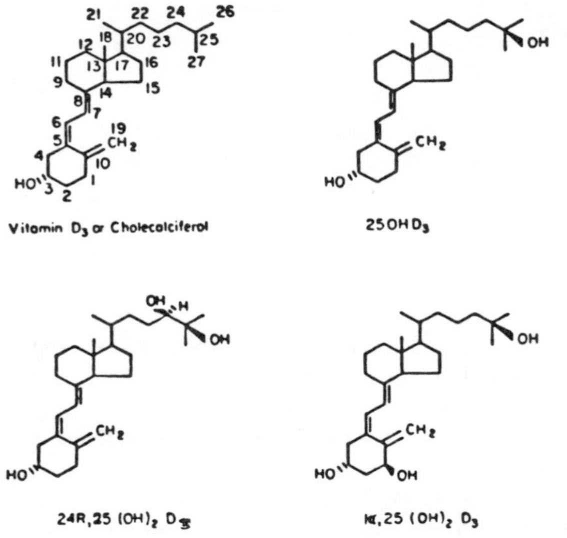

FIGURE 1. Structure of vitamin D_3 and its metabolites. Vitamin D_2 or ergocalciferol and its metabolites have an additional double bond between C-22 and C-23. Systematic and trivial names of vitamin D and its derivatives are as follows: vitamin D_2 (9,10-*seco*-5,7,10(19), 22-ergostatetraen-3β-ol): D_2; vitamin D_3 (9,10-*seco*-5,7,10 (19)-cholestatrien-3β-ol): D_3; 25-hydroxy-vitamin D_2: 25-OHD$_2$; 25-Hydroxyvitamin D_3: 25-OHD$_3$; isotachysterol isomer formed from 25-OHD$_2$: 25-OHITS$_2$ (Isotachysterol isomer formed from 25-OHD$_3$ 25-OHITS$_3$).

bath (55°C). Add 5 mℓ ethanol, swirl, and add 10 mℓ hexane. Shake mechanically for 5 min. Centrifuge for 5 min (2000 rpm), pipette 2 mℓ clear hexane extract into a 25 mℓ conical, glass-stoppered centrifuge tube, and evaporate the hexane under N_2 in water bath. Dissolve the residue in 2 mℓ methanol and inject 20 µℓ into a liquid chromatograph.

Vitamin D resins and Vitamin AD or ADE oil solutions — disperse or dissolve a sample portion in methanol and dilute further with methanol to obtain about 100 IU vitamin D per milliliter. Spin, if necessary, to obtain a clear methanol layer. Use 20 µℓ for injection.

Extraction of vitamin D_3 from milk (method of Koshy and Van Slik[4]) — A 100 g sample was used for analysis. The thawed sample was transferred to a 1-ℓ mixing cylinder to which 0.5 g NaHCO$_3$, 0.25 g sodium ascorbate, and 200 mℓ alcohol were added. The mixture was shaken to precipitae proteins and then further shaken with 300 mℓ anhydrous ethyl ether. The precipitates were allowed to settle. The supernatant extract was poured into a 1-ℓ separatory funnel and gently shaken with 100 mℓ ethanol and 200 mℓ ethyl ether and allowed to settle. The supernatant was added to the separatory funnel, and the combined extract was separated from residual water. The extract was filtered through cotton into a 1-ℓ round-bottom flask and rotary evaporated under vacuum to about 25 mℓ milky residue in a water bath (40°C).

Solvent partition — 100 mℓ 5% NaHCO$_3$ was added to the concentrated extract; the mixture was transferred to a 250-mℓ separatory funnel. It was extracted with 4 × 50 mℓ methylene chloride using each portion to first rinse the flask. The extracts were combined in a 500 mℓ round-bottom flask and evaporated just to dryness under vacuum. The residue was immediately dissolved in 100 mℓ hexane, transferred to a 250 mℓ separatory funnel, and extracted with 4 × 50 mℓ acetonitrile using each portion to rinse. (The hexane and acetonitrile were mutually saturated with each other prior to use.) The acetonitrile extracts were combined in a 500-mℓ round-bottom flask and evaporated just to dryness under vacuum in a water bath (40°C).

Vitamin D in fortified milk (method of Thompson et al.[5]) — Saponification — Mix 50 mℓ milk with an equal volume of 1% ethanolic pyrogallol in a 500-mℓ boiling flask. To this is added 17.5 g KOH and swirl flask in cold water until all KOH has dissolved. The mixture is allowed to stand overnight in dark, then transfer the saponified milk to 250 mℓ separatory funnel, using 50 mℓ water, 10 mℓ ethanol, and finally 20 mℓ ether to rinse boiling flask. Then extract the mixture with 5 × 50 mℓ ether and 50 mℓ hexane. Wash the pooled extracts with water (100 mℓ total); back-extract first wash with ether. The combined extracts are evaporated under reduced pressure and dissolved in 0.3 to 0.5 mℓ hexane for chromatography.

Vitamin D in fortified milk (method of Henderson and Wickroski[6]) — Perform saponification according to procedure described above with minor modifications. Add 120 mℓ 1% ethanolic pyrogallol to 100 mℓ milk in a 500 mℓ Phillips flask. Add 40 g KOH to 20 mℓ water, cool, and add to milk-pyrogallol mixture. Let mixture saponify overnight at room temperature with constant stirring.

Transfer saponified mixture to a 1-ℓ separatory funnel, using 100 mℓ water, 20 mℓ ethanol, and 100 mℓ ethyl ether-petroleum ether (1:1) to rinse boiling flask. Extract mixture with 2 additional 100 mℓ portions of ethyl ether-petroleum ether, and wash pooled ether extracts with 50 mℓ water. Re-extract water by washing with 100 mℓ ethyl ether-petroleum ether, and add to other ether extracts. Continue washing ether extracts with 50 mℓ portions of water until there is no further reaction to phenophthalein. (If emulsion forms disperse with small spray of ethanol.)

Dry the stem of the separatory funnel with filter paper and filter extract through phase separating silicone-treated paper (Whatman 1PS) to eliminate all moisture from ether extract. Evaporate filtered ether extract to dryness under reduced pressure at 50°C using rotary evaporator, add 5 to 10 mℓ ethanol, and evaporate just to dryness. Immediately add 5 mℓ CHCl$_3$ in preparation for chromatography.

To the flask containing the unsaponifiables and vitamin D in CHCl$_3$, add 2 g deactivated neutral alumina and 0.2 mℓ aminoazobenzene dye. Dye will act as a marker for vitamin D. Evaporate CHCl$_3$ from alumina using rotary evaporator. Immediately pour dried alumina containing vitamin D on top of previously prepared neutral dry alumina column, using vibrator to settle it, and then add a few glass beads or very thin pad of glass wool. Rinse flask once with 5 mℓ CHCl$_3$, pouring rinsings onto column, and continue to elute vitamin D by dripping chloroform from separatory funnel, while maintaining the 3 to 4 cm head of solvent. First yellow eluate is carotenes from milk. This is followed by yellow of aminoazobenzene dye. After all dye has eluated, discard additional 10 mℓ eluate and collect following 15 mℓ eluate, which contains vitamin D. Evaporate chloroform under reduced pressure at 50°C to about 1 mℓ. This is transferred quantitatively to a small, tapered vial and evaporated to dryness under N$_2$. 200 $\mu\ell$ methanol are added for HPLC analysis.

Vitamin D$_3$ concentrates in resin (method of Hofsass et al.[7]) — Vitamin D in resin — use either method A or B. (Method A) Let the resin come to room temperature, pull the filaments off the main lump with narrow stainless steel spatula, and introduce amount equivalent to 180 to 220 mℓ, accurately weighed, into tared 10 mℓ beaker. (Method B) Freeze resin, and break off fragments by striking. Select single fragments of resin (none less than 30 mg) and transfer amount equivalent to 180 to 220 mg, measured to 0.1 mg, to bottom of tared beaker with spatula. Discard fines produced during fracture.

Heating resin on steam bath is not recommended because this may produce changes in the material. However, if Method A or B proves to be impossible, resin may be heated slightly and transferred to a tared 10 mℓ beaker with glass rod. It is helpful throughout the weighing procedure to keep the beaker of acetone and acetone wash bottle handy, so that spatulas and rods may be washed free of resin, if necessary.

Add about 1 mℓ ethanol followed by 6 mℓ washed isooctane and let resin dissolve with agitation (5 min). Quantitatively transfer to 100 mℓ volumetric flask and dilute to volume with washed isooctane. Use this solution directly for injection. The concentration of vitamin D_3 will be about 1.25 mg/mℓ for resin at 25 × 10^6 IU/g label potency.

Vitamin D_3 in oil — accurately weigh 1 g oil containing 1 × 10^6 IU/g or 0.5 g oil containing 2 × 10^6 IU/g into a tared 10 mℓ beaker. Add about 7 mℓ washed isooctane and stir slightly until completely dissolved. Quantitatively transfer to 25 mℓ flask and dilute to volume with washed isooctane. Use this solution for injection (concentrated vitamin D_3 is about 1.0 mg/mℓ). Dry concentrates — weigh 1.0 g sample of 200,000 IU/g concentrate into a 100 mℓ conical flask. Add 30 mℓ reagent grade isopropanol, followed by 5 mℓ 50% aqueous KOH. Reflux on steam bath 0.5 hr. Cool in ice bath. Transfer to 500 mℓ separatory funnel. Rinse saponification flask with 50 mℓ water and transfer rinse to separatory funnel. Rinse saponification flask with 100 mℓ petroleum ether, add rinse to separatory funnel, and shake gently for 1 min. Let layers separate, and transfer petroleum ether extract to another separatory funnel. Repeat extraction with a second 100 mℓ portion of petroleum ether. Wash pooled petroleum ether extracts with 50 mℓ portions of water until they are neutral to phenolphthalein (3 to 4 portion). Dry ether extract by filtering through 1 g anhydrous Na_2SO_4 into a 250 mℓ conical flask. Evaporate to about 50 mℓ under N_2 and transfer to a 100 mℓ round-bottom flask. Continue evaporation under N_2 using a minimum of steam. Transfer evaporated extract to 5 mℓ amber volumetric flask, using several 1 mℓ washings of isooctane, and dilute to volume. Use this solution for injection. The concentration of vitamin D_3 will be about 1.0 mg/mℓ.

Vitamin D in vitamin preparations (method of Vanhaelen-Fastre and Vanhaelen[8]) — Resins and oils — dissolve in light petroleum to give a solution containing approximately 2000 IU/mℓ. Dry concentrates — dry concentrate equivalent to about 20,000 IU vitamin D was accurately weighed into a centrifuge tube (40 mℓ), and 12 mℓ *N,N*-dimethylformamide or DMSO was added. After vigorous shaking, the tube was placed in an ultrasonic bath for 15 min and shaken every 5 min. Then 10 mℓ isooctane and 8 mℓ ice-cooled water were added to the dispersion (the tube being kept in an ice bath). The tube was then vigorously shaken for 3 min and spun at 4000 rpm for 10 min. A few milliliters of the upper phase were pipetted and immediately used for HPLC analysis. Multivitamin formulations — oily solutions were diluted with light petroleum to about 200 IU/mℓ.

With aqueous solutions, a volume equivalent to about 2000 IU vitamin D was pipetted into a centrifuge tube (40 mℓ capacity) and diluted to 8 mℓ with ice-cooled water; 10 mℓ isooctane and 12 mℓ DMF or DMSO were added. Further extraction steps were similar to those described for dry concentrates.

If multivitamin tablets and coated tablets did not disintegrate in DMF or DMSO after shaking and ultrasonic treatment, they were ground and the resulting powder immediately used for the determination. A number of tablets or powder weight equivalent to 2000 IU vitamin D was treated as described for dry concentrates. Disintegration time in the ultrasonic bath was 15 to 30 min.

A number of capsules (2000 IU of vitamin D) were extracted as described for dry concentrates, and again, disintegration time in the ultrasonic bath was 15 to 30 min.

Vitamin D_3 in cod liver oil (method of Ali[9]) — Saponification — about 10 g cod liver oil were weighed in a round-bottomed 200 mℓ flask, and 50 mℓ 2 *N* ethanolic KOH solution were added and refluxed at 70°C for 30 min. A stream of nitrogen was passed through the solution during saponification. After cooling, the solution was extracted 1 time with 100 mℓ and 3 times with 30 mℓ portions of petroleum (40 to 60°C). The combined petroleum-ether extracts were washed with water to remove traces of alkali, dried over anhydrous sodium sulfate, and the solvent was evaporated under vacuum at about 40°C.

Precipitation of cholesterins — the residue from the sponification procedure was taken in

20 mℓ of 90% ethanol and heated on a water bath for 5 min at 70°C; 50 mℓ of 1% digitonin solution were added, and the solution was kept at this temperature for another 15 min and then set aside for another 3 hr. The precipitation was removed by filtration through a G3 sintered glass crucible, and the filtrate mixed with 50 mℓ water and extracted 3 or 4 times with 50 mℓ portions of petroleum-ether. Extracts were washed with water twice, dried over anhydrous sodium sulfate, and distilled under vacuum to a volume of about 2 mℓ.

Vitamin D in resin (method of Tartivita et al.[10]) — Accurately weigh enough resin as to be equivalent with 30 mg of vitamin D_3 into a 25 mℓ amber volumetric flask and dissolve in and dilute to volume with mobile phase ($CHCl_3$-n-hexane-tetrahydrofuran (70:30:1). Pipette 5 mℓ solution into a 25 mℓ amber volumetric flask and dilute to volume with mobile phase. Prepare a 1:1 mixture of the final sample solution and p-dimethylaminbenzaldehyde (30 mg/dℓ) in a suitable amber vessel and mix well. The DMABA solution is made up in mobile phase.

Vitamin D in multivitamin preparations (method of Strong[11]) — Carry out the assays while avoiding bright daylight and allow only the minimum necessary exposure to artificial light, particularly with solutions containing Vitamin A.

Alkaline reflux — place sample without pretreatment (intact capsules or tablets) equivalent to 2000 IU (50 μg) vitamin D in a 250-mℓ low actinic conical flask containing a 5.1 cm stirring bar. Add 48 mℓ 95% ethanol, 2 mℓ of 5% pyrogallol in 95% ethanol, and 10 mℓ ether. Mount under a condenser and heat to boiling using a hot plate.

Add 20 mℓ 20% (w/v) aqueous KOH through the condenser and reflux with stirring for 45 min for oil-based formulations, 30 min for tablets or powder-filled capsules, or 20 min for powder samples or water-based formulations. Cool and, without delay, rinse the condenser with 20 mℓ water. Collect in the flask and add 40 mℓ petroleum ether.

Extraction — transfer the contents of the flask to a 250-mℓ separatory funnel. Add 40 mℓ water, 15 mℓ ether, and 35 mℓ petroleum ether and shake vigorously. Drain the lower aqueous-ethanol layer into a second separator and shake with 80 mℓ ethyl ether-petroleum-ether (1:3). Discard the lower layer. Combine the extracts in the first separator and wash with 50 mℓ 40% ethanol, followed by 50 mℓ water. Discard the wash solutions.

If necessary, add a few drops of 15% (w/v) aqueous sodium sulfate solution to break emulsions. Filter the ether-petroleum-ether extracts through cotton into a boiling flask containing 1.4 mℓ light mineral oil. (*Note:* Pour through neck of separator after draining bottom layer to avoid contamination with aqueous phase.)

Vitamin D Extraction Methods for Tissues

Cow plasma sample preparation after extraction (method of Koshy and Van Slik[12]) — Extraction: 25 frozen sample was thawed and transferred to a 250-mℓ mixing cylinder to which 250 mg sodium bicarbonate, 150 mg sodium ascorbate, and 50 mℓ alcohol were added and mixed. To this, 75 mℓ anhydrous ethyl ether was added, mixed thoroughly, and allowed to settle (2 to 3 min). The supernatant layer was decanted into a 250 mℓ separatory funnel to which 25 mℓ water were added, mixed, and allowed to separate (2 min). The lower aqueous phase was discarded. The solid residue in the mixing cylinder was extracted with 50 mℓ ether and 25 mℓ alcohol. The supernatant was transferred to the original extract in the separatory funnel. Solvent partitions — saturated sodium chloride solution, pH 8 (25 mℓ) was added to the combined extracts, mixed gently, and allowed to separate (15 min). The lower aqueous phase was discarded, and the extract was transferred to a 500-mℓ round-bottom flask. The separatory funnel was rinsed with 30 mℓ ether and rinsings were added to the extract in the flask. The extract was evaporated until a milky liquid was left. This phase was transferred to a 250-mℓ separatory funnel, and 50 mℓ of 5% NaHCO$_3$ were added. The solution was extracted 4 times with 25 mℓ methylene chloride each, first using the methylene chloride to rinse the flask. The methylene chloride was evaporated using a rotary evaporator and water at 45 to 50°C. The residue was transferred to a 250-mℓ separatory funnel with 50 mℓ *n*-hexane and extracted 4 times with 25 mℓ of acetonitrile each (solvents mutually saturated with each other). The acetonitrile extracts were collected in a 200-mℓ round-bottom flask and evaporated just to dryness in a rotary evaporator.

Partition column chromatography — 2 g Celite 545 was transferred to a 20-mℓ test tube and mixed well with 1.6 mℓ methanol-water (80:20). A small wad of cotton was placed in the outlet of a disposable serological pipette. The wide end was inserted into the test tube containing the wet Celite, and a portion of the absorbent was collected in the column and tamped down gently with a glass rod. In this manner, all the Celite was packed in the column. The effluent from the column separation of the cow plasma or serum was evaporated to dryness and immediately transferred to a small, tapered tube using *n*-pentane and evaporated under N_2 to a volume of 50 to 100 μℓ. It was transferred to the top of the Celite column with a Pasteur pipette using two or more 100 μℓ washings of the mobile phase. The reservoir tube was attached to the column and filled with the mobile phase, and the column was eluted under a small N_2 pressure (flow rate 0.75 mℓ/min). The first 6.5 mℓ were discarded (contains less polar impurities and any 25-OH-D_2). The next 8 mℓ (containing 25-OH-D_3) were collected in a tapered glass tube. It was evaporated to dryness under N_2 while the tube was immersed in water at 30 to 35°C. The residue was reconstituted in 100 μℓ ethyl acetate or acetonitrile for the HPLC analysis.

Plasma vitamin D_3 (method of DeLeenhur and Cruyl[13]) — Plasma can be extracted by the scheme below:

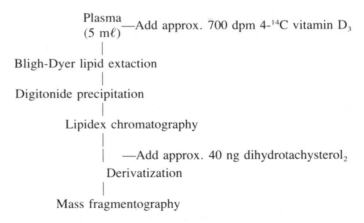

Plasma (5 mℓ) —Add approx. 700 dpm 4-^{14}C vitamin D_3
|
Bligh-Dyer lipid extaction
|
Digitonide precipitation
|
Lipidex chromatography
|
| —Add approx. 40 ng dihydrotachysterol$_2$
Derivatization
|
Mass fragmentography

Extraction — plasma carrier protein-vitamin D_3 complex was broken by denaturation of the protein using alcoholic solvents such as ethanol, followed by chloroform or ether extraction.

Human serum vitamin D_3 (method of Schaeffer and Goldsmith[14]) — Serum from fasting subjects (0.8 mℓ) was labeled by addition of 1.02 mℓ ethanol containing 5000 dpm (100 pg) [3]H-25-OH-D_3 (9.3 Ci/mmol, New England Nuclear, Boston). Addition of 3 mℓ methanol-dichloromethane (2:1) to the labeled serum caused precipitation of serum proteins, which were removed by centrifugation for 10 min at 1000 g. The supernatant was decanted into a 10 mℓ separatory funnel, and 2 mℓ dichloromethane were added. The lower layer was collected and evaporated in a stream of N_2.

Plasma vitamin D (method of Jones[15]) — To 2 mℓ plasma were added 5000 cpm ^{3}H-D_2 and 5000 cpm [3]H-25-OH-D_3, each in 10 μℓ ethanol, and the mixture was allowed to equilibrate at 4°C for 30 min. The lipids were extracted with 7.5 mℓ methanol-chloroform (2:1) according to the method of Bligh and Dyer. Phase separation was accomplished by addition of 2.5 mℓ saturated KCl and 2.5 mℓ chloroform, followed by spinning at 100 g for 10 min. The aqueous (upper) layer was re-extracted with an additional 6 mℓ chloroform.

The combined chloroform layers were evaporated under reduced pressure to yield a yellow lipid extract. When this appeared cloudy, it was redissolved in 1 mℓ isopropanol-hexane (5.5:94.5) and filtered through a syringe fitted with a Swinney filter and Millipore® 0.4 μm Teflon® filter. The clear lipid extract was finally transferred to a 5 mℓ conical screw-capped vial and evaporated in a stream of N_2. The residue was dissolved in 200 $\mu\ell$ isopropanol-hexane mixture. Flocculent precipitates were sometimes observed after refrigeration of the extracts; these were spun at 2000 g for 2 min to clear.

Human serum vitamin D (method of Styrd and Gilbertson[16]) — 1 mℓ serum was fortified with 100 $\mu\ell$ (10^4cpm) of the purified ^{3}H-25-OH-D_3 solution; 2 mℓ absolute ethanol were added, vortex mixed, and allowed to stand for 10 min. Hexane (6 mℓ) and 1.5 mℓ sodium carbonate solution (20 g/ℓ) were then added and the sample vigorously mixed.

The mixture was spun (10 min at 3000 g) and the hexane layer removed. The aqueous layer was re-extracted with 5 mℓ hexane. The combined hexane layers were evaporated to 1 mℓ in a 40°C water bath under N_2. Toluene (2 mℓ) was added and the sample evaporated to 1 mℓ. Another 2 mℓ toluene were added and evaporated to 0.5 mℓ.

Purification column — a 0.5 × 2 cm column of washed silica gel was prepared by adding the hexane slurry to a Pasteur pipette containing a small plug of glass wool. The toluene solution from the extraction was applied to the column, followed by a 0.5 mℓ hexane rinse. The column was eluted with 2 mℓ hexane followed by 7 mℓ ethanol-hexane (1.5:98.5) which was discarded. The following eluent, 7 mℓ absolute ethanol-hexane was collected and may be stored at 4°C overnight.

This fraction was evaporated under N_2 in a 40°C water bath to a volume of 0.5 mℓ, which was transferred to a 1 mℓ Reacti-vial. The sample was evaporated to dryness under N_2 and redissolved in 50 $\mu\ell$ ethanol-hexane (2.5:97.5) before injection into the HPLC instrument.

Human plasma vitamin D (method of Lambert et al.[17]) — Extraction — all glassware involved in extraction and chromatography was siliconized. Plasma (5 mℓ) was placed in a 30 mℓ polypropylene centrifuge tube in an ice bath, and a radioactive internal standard of each metabolite (10,500 dpm) was added to monitor recovery. The sample was extracted with 3 vol cold methanol-methylene chloride (2:1) per vol plasma and vortexed for 2 min; 1 vol cold methylene chloride was then added, vortexed for 2 min, and spun at 23,300 g for 10 min at 4°C. The lower organic phase was siphoned off via a Teflon® tube into a polypropylene tube, and the aqueous phase was re-extracted with 2 vol cold methylene chloride. All organic phases were combined and evaporated to dryness under N_2.

Gel chromatography — the residue was dissolved in 100 $\mu\ell$ n-hexane-chloroform-methanol (9:1:1) and applied to a Sephadex® LH-20 column (0.9 × 15 cm) equilibrated with the same solvent. The polypropylene tubes were washed twice with 100 $\mu\ell$ solvent and these washes were used to wash the sample into the column bed. Vitamin D and its metabolites were eluted with a nonlinear ''flow gradient'' consisting of the following incremented changes in flow rate: 0.5 mℓ/min for the first 10 min of eluant (vitamin D); 1.3 mℓ/min for the next 15 mℓ (25-OH-D); and 1.5 mℓ/min for the remainder of the gel chromatography elution (24,25-$(OH)_2$D and 1α-25$(OH)_2$D).

Human serum vitamin D (method of Gilbertson and Styrd[18]) — 1 mℓ serum was spiked with ^{3}H-25-OH-D_3 solution (100 $\mu\ell$, 10^4 cpm) and mixed with 7 mℓ chloroform-methanol (2:1) with a vortex mixer. This solution was vigorously mixed with 1.5 mℓ sodium carbonate solution (20 g/ℓ) and centrifuged for 10 min at 3000 g. The chloroform layer was removed and the aqueous layer was re-extracted with 7 mℓ chloroform-methanol (2:1). The two chloroform layers were combined and evaporated to 1 mℓ in a 40°C water bath under N_2. The residue was mixed with 2 mℓ benzene and evaporated to about 0.5 mℓ. This procedure was repeated once, and the final 0.5 mℓ benzene solution used for further purification by gel chromatography. The mobile phase was chloroform-hexane (3:2). The silica gel was purified by washing with a generous amount of methylene chloride-ethanol (92:8) followed by hexane.

Bovine liver, kidney, and muscle vitamin D (method of Koshy and Van Der Slik[19]) — Extraction — 50 thawed tissue was transferred into a 250-mℓ stainless steel blender cup to which about 0.25 g sodium ascorbate, 0.5 g sodium bicarbonate, and 100 mℓ alcohol (95% alcohol) were added. The contents were blended at high speed for 2 to 3 min. The slurry was transferred to a 500-mℓ mixing cylinder with 150 mℓ anhydrous ethyl ether AR, mixed thoroughly, and poured into two 250-mℓ centrifuge cups. After centrifugation at about 2000 rpm for 5 min, the clear extract was poured into a 500-mℓ separatory funnel to which 50 mℓ water was added. The residue in the centrifuge cups was re-extracted with 75 mℓ alcohol and 130 mℓ ethyl ether in a 500-mℓ mixing cylinder (liver and kidney) or in a blender (muscle), and the extract was spun as above. The supernatant was added to the 500-mℓ separatory funnel, mixed, and allowed to separate. The aqueous phase and any solids that separated at the bottom were discarded. Solvent partition — 60 mℓ saturated NaCl (pH 8) were added to the extract, mixed, allowed to separate, and the aqueous phase discarded. The extract was evaporated to about 40 to 50 mℓ in a 1-ℓ round-bottom flask under vacuum in a water bath at about 40°C. To this, 100 mℓ 5% NaHCO$_3$ were added and transferred to a 250-mℓ separatory funnel and extracted with 4 × 50 mℓ methylene chloride using each methylene chloride portion to first rinse the flask. The combined methylene chloride extract was evaporated from a 500-mℓ round-bottom flask just to dryness under vacuum. The residue was immediately dissolved in 100 mℓ *n*-hexane and extracted with 4 × 50 mℓ acetonitrile using each acetonitrile portion to rinse the round-bottom flask (the hexane and acetonitrile were mutually saturated with each other prior to use). The acetonitrile extract was evaporated just to dryness under vacuum in a water bath at about 40°C from a 500-mℓ round-bottom flask. The test tube using hexane-ether (1:1) was concentrated to about 0.5 mℓ under a stream of nitrogen with tube immersed in water at about 40°C.

Silica gel column chromatography — neutral Silic Ar (1.8 g), 200 to 325 mesh or equivalent, was slurry packed with *n*-hexane-ether (1:1) in a 5 mℓ disposable serological pipette with a wad of absorbent cotton at the bottom. The extract from the solvent partition step was transferred to the top of the column using a 9-in. disposable Pasteur pipette and allowed to drain to the top of the packing. The tube was rinsed with small portions of hexane-ether and the rinsing transferred to the column, allowing each rinsing to drain to the top of the column before the addition of the solvent. The column was expanded using a Teflon® adaptor so that the head of the column had a capacity of about 7 to 8 mℓ. The tube was filled with 3.5 mℓ solvent, eluted under N$_2$ pressure at a flow rate of about 0.75 mℓ/min, and the effluent discarded. The mobile phase was changed to ethyl ether-ethyl acetate (9:1). The first 2 mℓ of the effluent were rejected and the next 4 to 5 mℓ collected in a tapered glass tube. The solvent was removed under N$_2$ while the tube was immersed in a 40°C water bath. The residue was reconstituted in a small volume of *n*-pentane.

Celite partition column chromatography — the column was made from 6 mm Pyrex® glass tubing. It was about 12-in. long with a 10/18 standard taper female joint at the top to fit a small reservoir to hold about 12 to 13 mℓ solvent. The lower end of the column was drawn to a bore of about 1 mm diameter. The reservoir was made from a 15-mℓ volumetric pipette whose stem was cut off below the bulb and fitted with a 10/18 standard taper male joint. Alternately, a 5-mℓ disposable serological pipette has been used as a column. A glass tube of the same dimension was attached by a Teflon® adapter to hold about 8 to 10 mℓ of the mobile phase.

Sodium ascorbate (250 mg) was dissolved in 20 mℓ water and mixed with 80 mℓ methanol in a 500-mℓ separatory funnel and shaken with 400 mℓ *n*-pentane to effect mutual saturation. The former was the stationary phase and the latter, the mobile phase. The methanol water will discolor with age, but is usable for at least a week if protected from light.

REFERENCES

1. **Barnett, S. A. and Frick, L. W.,** *Anal. Chem.,* 51, 643, 1979.
2. **Dolan, J. W., Gant, I. R., Giese, R. W., and Karger, B. L.,** *J. Chromatogr. Sci.,* 16, 616, 1978.
3. **Osadca, M. and Araujo, M.,** *J. Assoc. Off. Anal. Chem.,* 60, 993, 1977.
4. **Koshy, K. T. and Van Der Slik, A. C.,** *J. Agric. Food Chem.,* 27, 560, 1979.
5. **Thompson, J. N., Maxwell, W. B., and L'Abbe, M.,** *J. Assoc. Off. Anal. Chem.,* 60, 998, 1977.
6. **Henderson, S. K. and Wickroski,** *J. Assoc. Off. Anal. Chem.,* 61, 1130, 1978.
7. **Hofsass, H., Alicino, N. J., Hirsch, A. C., Ameika, L., and Smith, L. D.,** *J. Assoc. Off. Anal. Chem.,* 61, 735, 1978.
8. **Vanhaelen-Fastre, R. and Vanhaelen, M.,** *J. Chromatogr.,* 153, 219, 1978.
9. **Ali, S. L.,** *Fresenius' Z. Anal. Chem.,* 293, 131, 1978.
10. **Tartivita, K. A., Sciarello, J. P., and Rudy, B. C.,** *J. Pharm. Sci.,* 65, 1024, 1976.
11. **Strong, J. W.,** *J. Pharm. Sci.,* 65, 968, 1976.
12. **Koshy, K. T. and Van Der Slik, A. C.** *J. Agric. Food Chem.,* 25, 1246, 1977.
13. **De Leenhur, L. P. and Cruyl, A. A. M.,** *J. Chromatogr. Sci.,* 14, 434, 1976.
14. **Schaeffer, P. C. and Goldsmith, R. S.,** *J. Lab. Clin. Med.,* 91, 104, 1978.
15. **Jones, G.,** *Clin. Chem.,* 24, 287, 1978.
16. **Styrd, R. P. and Gilbertson, R. P.,** *Clin. Chem.,* 24, 937, 1978.
17. **Lambert, P. W., Syverson, B. J., Arnaud, C. D., and Spelsberg, T. C.,** *J. Steroid Biochem.,* 8, 929, 1977.
18. **Gilbertson, T. J. and Styrd, R. P.,** *Clin. Chem.,* 23, 1700, 1977.
19. **Koshy, K. T. and Van Der Slik, A. C.,** *Anal. Biochem.,* 74, 282, 1976.

SECTION II. GAS CHROMATOGRAPHY

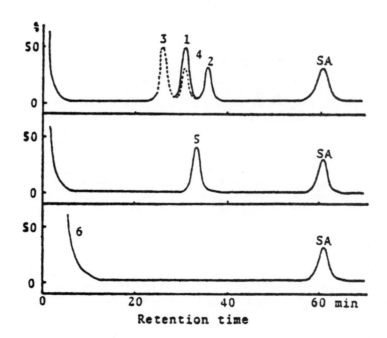

FIGURE GC 1.   Separation of D vitamins. Instrument: Shimadzu GC-4B PFE with flame ionization detector; column: 15% OV-17 on Shimalite (80—100 mesh) (150 cm × 0.4 mm); temperatures: column 225°C, detector 300°C, injector 250°C; flow: 120 mℓ/min nitrogen. (1) Pyro-D_2, (2) isopyro-D_2, (3) pyro-D_3, (4) isopyro-D_3, (5) DL-α-tocopherol, (6) retinol. SA = Stimasteryl acetate. (From Kobayashi, T. and Adachi, A., *J. Nutr. Sci. Vitaminol.,* 24, 41, 1976. With permission.)

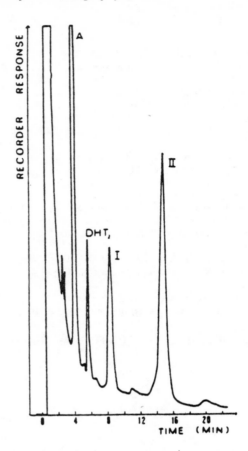

FIGURE GC 2.   Separation of some D vitamins. Instrument: Hewlett-Packard 5750 G with flame ionization detector; column: 2% QF-1 on Gas Chrom Q (80—100 mesh); temperatures: all 225°C; flow: nitrogen 6 mℓ/sec. (A) Artifact (reagent blank), (DHT_2) 3-methyl-dihydrotachysterol$_2$, (I) (minor 3-methyl 9,10-seco-all trans-cholesta-5(10),6,8-(14), 24-tetraene, (II) (major) 3,25-bismethoxy-isotachysterol$_3$. (From DeLeenhur, A. P. and Cruyl, A. A. M., *J. Chromatogr. Sci.*, 14, 434, 1976. With permission.)

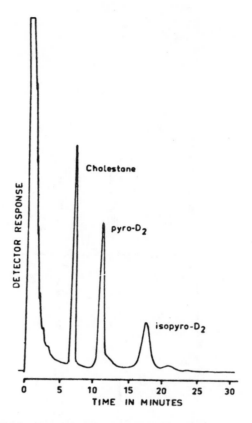

FIGURE GC 3.   Gas chromatography of vitamin D trimethylsilyl ethers. Instrument: Varian Aerograph 2100 with flame ionization detector; column: 4% OV-225 on Chromosorb W acid washed and DMCS treated; temperature: column 230°C, detector and injector 300°C; flow: 55 mℓ/min. (From Fisher, A. L. et al., *J. Chromatogr.*, 65, 493, 1972. With permission.)

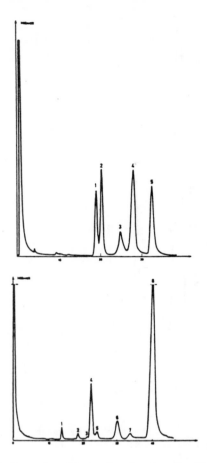

FIGURE GC 4.   Separation of D vitamins after photochemical reaction. Instrument: Varian Aerograph-2100 with flame ionization detector; column: (A) 3% JXR on Gas Chrom P-100/120: 6 ft × 0.08 in., i.d. metal, (B) 3% XE 60 on Gas Chrom Q-100/120: 10 ft × 0.08 in., i.d. Pyrex; temperature: (A) 250°C, (B) 230°C, detector and injector equivalent; flow: helium 30 mℓ/min. A: (1) Pyrocalciferol, (2) lumisterol 2, (3) isopyrocalciferol, (4) ergosterol, (5) tachysterol 2. B: (1, 2, 3) Toxisterols, (4) pyrocalciferol, (5) lumisterol, (6) isopyrocalciferol, (7) ergosterol, (8) tachysterol. (From Mermet-Bouvier, R., *J. Chromatogr. Sci.*, 10, 733, 1972. With permission.)

REFERENCES (GC)

1. **Kobayashi, T. and Adachi, A.,** *J. Nutr. Sci. Vitaminol.,* 24, 41, 1976.
2. **De Leenhur, A. P. and Cruyl, A. A. M.,** *J. Chromatogr. Sci.,* 14, 434, 1976.
3. **Fisher, A. L., Parfitt, A. M., and Lloyd, H. M.,** *J. Chromatogr.,* 65, 493, 1972.
4. **Mermet-Bouvier, R.,** *J. Chromatogr. Sci.,* 10, 733, 1972.

SECTION III. HIGH PERFORMANCE LIQUID CHROMATOGRAPHY

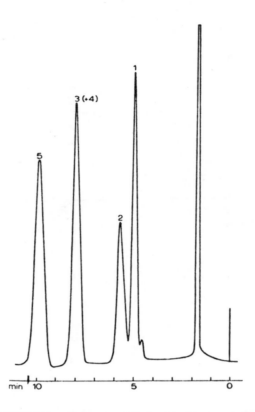

FIGURE HPLC 1. Separation of D vitamins. Instrument: Waters Model 6000 pump with Model 440 UV detector; column: μPorasil (30 cm × 0.25 mm) (steel); mobile phase: light petroleum — 1,2-dichloroethane-tetra-hydrofuran (85:8:7); flow: 2 mℓ/min; detection: 254 nm. (1) Pre-vitamin D_2/D_3, (2) lumisterol, (3) (+4) tachysterol$_3$, (5) pro-vitamin D_2/D_3. (From Vanhaelen-Fastre, R. and Vanhaelen, M., *J. Chromatogr.*, 153, 219, 1978. With permission.)

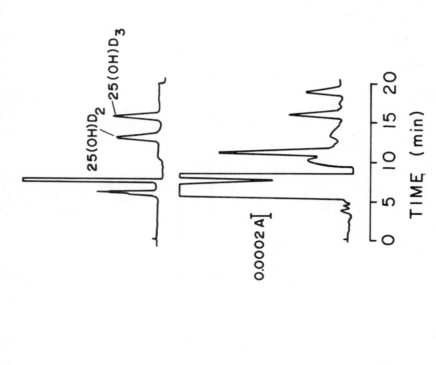

FIGURE HPLC 2.   Separation of D vitamins and derivatives. Instrument: Perkin-Elmer 1220 with LC-55 detector; column: Partisil 10 PXS (50 cm × 4.6 mm); mobile phase: chloroform-hexane-acetic acid (70:30:1); flow: 1 mℓ/min; detection: 268 nm. (1) Vitamin A-alcohol, (2) vitamin A-ester, (3) vitamin E-acetate, (4) provitamin D₂, (5) provitamin D₃, (6) lumisterol₃, (7) tachysterol, (8) vitamin D₃, (9) 7-dehydrocholesterol, (10) ergosterol. (From Ali, S. L., *Fresenius' Z. Anal. Chem.*, 293, 131, 1978. With permission.)

FIGURE HPLC 3.   Separation of 25-hydroxy vitamin D from serum. Instrument: Waters Model 204, Model 440 detector; column: μPorasil (25 cm × 0.4 mm), two columns in series; mobile phase: isopropanol-methylene chloride (1:49); flow: 1 mℓ/min; detection: 254 nm. (A) Standard mixture of 25(OH)D₂ and 25(OH)D₃, (B) human serum extract 25(OH)D₃ peak corresponds to 6.80 ng. (From Schaeffer, P. C. and Goldsmith, R. S., *J. Lab. Clin. Med.*, 91, 104, 1978. With permission.)

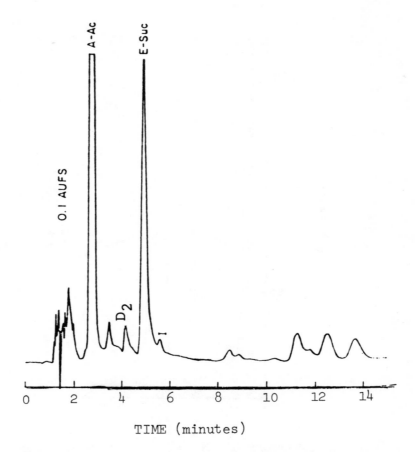

FIGURE HPLC 4. Separation of vitamins A, E, and D. Instrument: Waters 6000 pump with Valco loop injector; column: LDC-C$_8$RP (25 cm × 4.6 mm); mobile phase: methanol-water (92:8) containing 0.1% phosphoric acid; flow: 2.8 mℓ/min; detection: 254 nm. Vitamin A acetate, D$_2$: vitamin D$_2$; vitamin E succinate. (From Dolan, J. W. et al., *J. Chromatogr. Sci.,* 16, 616, 1978. With permission.)

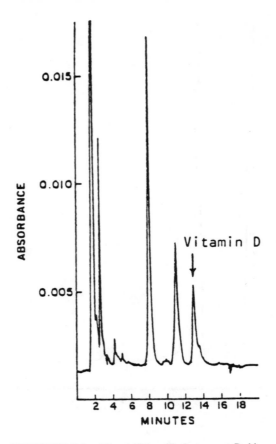

FIGURE HPLC 5.   Vitamin D in milk. Instrument: Perkin-Elmer 601 and Model LC55 detector; column: LiChrosorb Si-60 (5 μm) (25 cm × 3.2 mm); mobile phase: isopropanol-hexane (0.6:99.4); flow: 1 mℓ/min; detection: 265 nm. (From Thompson, J. N. et al., *J. Assoc. Off. Anal. Chem.*, 60, 998, 1977. With permission.)

Table HPLC 1
RELATIVE RETENTION OF VITAMIN D ISOMERS[6]

	Mobile phase		
	1 (p-Cresol 1.00)	2 (α-Naphthol 1.00)	3 (p-Cresol 1.00)
Isomers			
5,6-*Trans*-Vitamin D_2	1.59	1.73	0.62
Isovitamin D_2	1.62	1.73	0.58
Pre-vitamin D_2	1.88	1.89	—
Lumisterol$_2$	2.13	2.09	—
Isotachysterol$_2$	2.29	2.42	0.84
Vitamin D_2	2.87	2.56	1.12
Tachysterol$_2$	2.98	3.09	—
Ergosterol	3.64	3.80	—

Instrument:	Shimadzu-Dupont 830 with UV detector
Column:	Zorbax Sil (25 cm × 2.1 mm)
Mobile phase:	1. Methanol-ether-pentane (0.15:2:97.85)
	2. Ether-hexane (10:90)
	3. $CHCl_3$ (distilled)-pentane (55:45)
Flow:	0.6 mℓ/min
Detection:	254 nm

Table HPLC 2
RETENTION TIMES OF VITAMIN D METABOLITES[7]

Steroid	Time (min)	Mobile phase
D_3	3.7	Gradient elution:
24R-(OH)-D_3	10.1	0.02% MeOH-CH_2Cl_2, gra-
24S-(OH)-D_3	10.1	dient rate 0.3% min, to 6%
25-(OH)-D_3	10.7	MeOH-CH_2Cl_2
1α-(OH)-D_3	12.2	
24R,25-(OH)$_2$-D_3	13.4	
24S,25-(OH)$_2$-D_3	13.4	
1α,24R-(OH)$_2$-D_3	15.4	
1α,24S-(OH)$_2$-D_3	15.6	
1α,25-(OH)-D_3	16.9	
1α,24R,25-(OH)$_3$-D_3	21.0	
1α,24S,25-(OH)$_3$-D_3	21.2	
D_3	2.3	2% MeOH-CH_2Cl_2
24R-(OH)-D_3	3.0	
24S-(OH)-D_3	3.0	
25-(OH)-D_3	3.8	
1α-(OH)-D_3	6.2	
24R,25-(OH)$_2$-D_3	7.7	
24S,25-(OH)$_2$-D_3	7.7	
1α,24R-(OH)$_2$-D_3	13.1	
1α,24S-(OH)$_2$-D_3	13.9	
1α,25-(OH)$_2$-D_3	19.2	
24R-(OH)-D_3 di-TMS	13.8	2% CH_2Cl_2-*n*-hexane
24S-(OH)-D_3 di-TMS	12.1	
24R,25-(OH)$_2$-D_3 tri-TMS	21.2	2% CH_2Cl_2-*n*-hexane
24S,25-(OH)$_2$-D_3 tri-TMS	18.4	
1α,24R-(OH)$_2$-D_3	16.9	1.5% MeOH-*n*-hexane
1α,24S-(OH)$_2$-D_3	18.0	
1α,24R.25-(OH)$_3$-D_3	12.5	3.5% MeOH-CH_2Cl_2
1α,24S,25-(OH)$_3$-D_3	13.2	

Instrument: Shimadzu-Dupont 830 with UV detector
Column: Zorbax Sil (25 cm × 2.1 mm)
Mobile phase: As in table
Flow: 0.4 mℓ/min
Detection: 254 nm

Table HPLC 3
SEPARATION OF D
VITAMINS[8]

Secosteroid	Retention time (min)
D_3	2.00
D_2	2.00
25-OHD$_2$	6.80
25-OHD$_3$	8.36
ITS$_3$	1.78
ITS$_2$	1.78
25-OHITS$_2$	8.78
25-OHITS$_3$	11.20

Instrument:	Applied Chromatography Systems 750/03 pump, Rheodyne 7/25 injector, Schoeffel SF 770 detector
Column:	Zorbax Sil (25 cm × 0.46 mm)
Mobile phase:	Hexane-isopropanol (95:5)
Flow:	1—5 mℓ/min
Detection:	264 nm

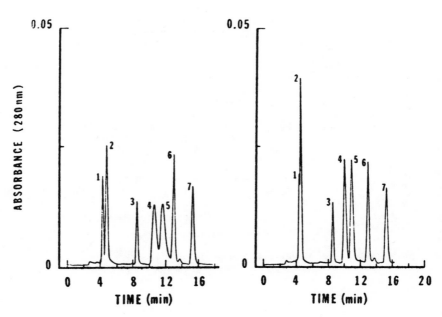

FIGURE HPLC 6.   Separation of D vitamins. Instrument: Waters ALC/GPC 2000 with Model M 440 detector; column: Zorbax ODS (6 μm) (25 cm × 4.6 mm) (steel); mobile phase: (left) acetonitrile-methylene chloride (70:30), (right) same but 0.5 mℓ/ℓ of methanol was added; flow: 1 mℓ/min; detection: 280 nm. (Left) Vitamin mixture: 1 = retinyl acetate (145 ng); 2 = retinol (345 ng); 3 = α-tocopheryl acetate (1280 ng); 4 = vitamin D$_2$ (365 ng); 5 = vitamin D$_3$ (370 ng); 6 = β-carotene (415 ng); 7 = retinyl palmitate (440 ng). (From Landen, M., *J. Chromatogr.*, 211, 155, 1981. With permission.)

REFERENCES (HPLC)

1. **Vanhaelen-Fastre, R. and Vanhaelen, M.,** *J. Chromatogr.,* 153, 219, 1978.
2. **Ali, S. L.,** *Fresenius' Z. Anal. Chem.,* 293, 131, 1978.
3. **Schaeffer, P. C. and Goldsmith, R. S.,** *J. Lab. Clin. Med.,* 91, 104, 1978.
4. **Dolan, J. W., Grant, J. R., Tanaka, N., Giese, R. W., and Kurger, B. L.,** *J. Chromatogr. Sci.,* 16, 616, 1978.
5. **Thompson, J. N., Maxwell, W. B., and L'Abbe, M.,** *J. Assoc. Off. Anal. Chem.,* 60, 998, 1977.
6. **Tsukida, K., Kodama, A., and Saiki, A.,** *J. Nutr. Sci. Vitaminol.,* 22, 5, 1976.
7. **Ikekawa, N. and Koizuma, N.,** *J. Chromatogr.,* 119, 227, 1976.
8. **Trafford, D. J. H., Seamark, D. A., Turnbull, H., and Malcin, H. L.,** *J. Chromatogr.,* 226, 351, 1981.
9. **Landen, M., Jr.,** *J. Chromatogr.,* 211, 155, 1981.

SECTION IV. THIN LAYER CHROMATOGRAPHY

Table TLC 1
R_f VALUES FOR VITAMINS ON TLC[1]

Vitamin	Layer 1	Layer 2	Layer 3	Color*
A-Acetate	0.82	0.85	0.86	Blue
A-Palmitate	0.25	0.33	0.27	Blue
K_1	0.40	0.53	0.45	Yellow
E (α-tocopherol)	0.63	0.72	0.67	Orange
E_{Ac}(α-tocopherol-acetate)	0.52	0.63	0.56	Light yellow
D_2	0.79	0.82	0.83	Reddish brown
D_3	0.79	0.82	0.83	Reddish brown

Layer:
1. Starch — spread from 66% in paraffin oil-ether (2:27)
2. Cellulose — spread of cellulose MN-300 15% in ether (3:10)
3. Talc — slurry of 54.4% adsorbent in ethanol for preparation of layer

Mobile phase: Acetone-acetic acid (30:20)
*Detection: (Color) antimony trichloride-chloroform (1:4) spray

Table TLC 2
R_f VALUES FOR D VITAMINS[2]

	Mobile phase		
Steroid	1	2	3
Cholecalciferol	0.45	0.52	0.45
Ergocalciferol	0.45	0.52	0.40
Cholesterol	0.35	0.41	0.73
Retinol	0.45	ND	0.23
25-Hydroxycholecalciferol	0.18	0.24	0.19

Layer:	1. Silica Gel G
	2. Silica Gel G
	3. Silica Gel G with 5% silver nitrate
Mobile phase:	1. Chloroform
	2. Chloroform-methanol (98:2)
	3. Chloroform-acetone (85:15)
Detection:	Spray with 0.25% rhodamine in alcohol; observe under UV light

Table TLC 3
R_f VALUES OF D VITAMINS[3]

Steroid	R_f
Retinol	0.17
Ergocalciferol	0.35
Cholecalciferol	0.42
Ergosterol	0.64
D,L-α-Tocopherol	0.69
Stigmasterol	0.71
Cholesterol, campesterol, β-sitosterol	0.75
Anhydroretinol	0.97

Layer:	Silica Gel G (Merck) impregnated with silver nitrate
Mobile phase:	Chloroform-acetone (9:1)
Detection:	Spray with 0.05% fluorescein in alcohol; observe under UV light

Table TLC 4
R_f VALUES FOR VITAMIN D METABOLITES ON SILICA GEL[4]

	R_f Values				
Layer	D_3	25-OH-D_3	24,25-$(OH)_2$-D_3	25,26-$(OH)_2$-D_3	1,25-$(OH)_2$-D_3
1	0.69	0.56	0.41	0.30	0.20
2	0.65	0.51	0.38	0.28	0.22
3	0.70	0.56	0.48	0.39	0.32

Layers:	1. Silica Gel G F_{254} (Merck)
	2. Silica Gel G (Merck)
	3. Silica Gel (Whatman)

Mobile phase:	Chloroform-ethanol-water (183:16:1)
Detection:	Fluorescence quench

REFERENCES

1. **Perisic, N., Petrovic, S., and Hadzic, P.,** *Chromatographia,* 9, 130, 1976.
2. **Sklan, D., Budowski, P., and Katz, M.,** *Anal. Biochem.,* 54, 606, 1973.
3. **Sklan D., Budowski, P., and Katz, M.,** *Anal. Biochem.,* 56, 606, 1973.
4. **Justova, V., Wiedtova, Z., and Starka, L.,** *J. Chromatogr.,* 230, 319, 1982.

INDEX

A

D

F

U